François R. Cossec
Igor V. Dolgachev

Enriques Surfaces I

1989

Birkhäuser
Boston · Basel · Berlin

François R. Cossec
MATRA
78182 Saint-Quentin-en-Yven
France

Igor V. Dolgachev
Department of Mathematics
University of Michigan
Ann Arbor, MI 48109-1003
U.S.A.

Library of Congress Cataloging-in-Publication Data
Cossec, François R.
Enriques surfaces/François R. Cossec, Igor V. Dolgachev.
p. cm. — (Progress in mathematics ; v. 76–)
Bibliography: v. 1, p.
Includes index.
ISBN 0-8176-3417-7 (v. 1 : alk. paper)
1. Enriques surfaces. I. Dolgachev, I. (Igor V.) II. Title.
III. Series: Progress in mathematics (Boston, Mass.) ; vol. 76, etc.
QA573.C67 1989
516.3′52—dc19 88-8180

CIP-Titelaufnahme der Deutschen Bibliothek
Cossec, François R.:
Enriques surfaces/François R. Cossec ; Igor V. Dolgachev.—
Boston;Basel:Birkhäuser.
NE: Dolgachev, Igor V.;
1 (1989)
(Progress in mathematics ; Vol. 76)
ISBN 3-7643-3417-7 (Basel) Pb.
ISBN 0-8176-3417-7 (Boston)
NE: GT

Printed on acid-free paper.

ISBN 0-8176-3417-7
ISBN 3-7643-3417-7

Text prepared by the authors in camera-ready form.
Printed and bound by Edwards Brothers, Inc., Ann Arbor, Michigan.
Printed in the U.S.A.

9 8 7 6 5 4 3 2 1

Preface

This is the first of two volumes representing the current state of knowledge about Enriques surfaces which occupy one of the classes in the classification of algebraic surfaces. Recent improvements in our understanding of algebraic surfaces over fields of positive characteristic allowed us to approach the subject from a completely geometric point of view although heavily relying on algebraic methods.

Some of the techniques presented in this book can be applied to the study of algebraic surfaces of other types. We hope that it will make this book of particular interest to a wider range of research mathematicians and graduate students.

Acknowledgements. The undertaking of this project was made possible by the support of several institutions. Our mutual cooperation began at the University of Warwick and the Max Planck Institute of Mathematics in 1982/83. Most of the work in this volume was done during the visit of the first author at the University of Michigan in 1984-1986. The second author was supported during all these years by grants from the National Science Foundation.

During the course of this work, many helpful discussions were held with various mathematicians. We acknowledge our special gratitude to Wolf Barth, Bill Lang, Eduard Looijenga, Chris Peters, and Miles Reid.

Contents

Introduction.....1

Chapter 0. Preliminaries

§1. Double covers.....9

§2. Rational double points.....24

§3. Del Pezzo surfaces.....34

§4. Symmetric quartic Del Pezzo surfaces.....39

§5. Symmetric cubic Del Pezzo surfaces.....57

§6. Prym canonical maps.....62

§7. The Picard scheme.....65

Bibliographical notes.....71

Chapter I. Enriques surfaces: generalities

§1. Classification of algebraic surfaces.....72

§2. The Picard group.....75

§3. The K3-cover.....83

§4. Differential invariants.....88

§5. Riemann-Roch and a vanishing theorem.....95

§6. Examples.....99

Bibliographical notes.....101

Chapter II. Lattices and root bases

§1. Generalities.....103

§2. Root bases and their Weyl groups.....106

§3. Root bases of finite and affine type.....110

§4. Root bases of hyperbolic type.....112

§5. The Enriques lattice.....117

§6. The Reye lattice.....128

§7. The function φ_M 134
§8. 2-congruence subgroups of finite Weyl groups 140
§9. The factor group W/W(2) 145
§10. The structure of W(2) 150
Bibliographical notes 162
Tables 163
Chapter III. The geometry of the Enriques lattice.
§1. Divisors of canonical type 166
§2. The nodal chamber 175
§3. Canonical r-sequences and $U_{[r]}$-markings 178
§4. U-markings 182
§5. $U_{[3]}$-markings 186
§6. Linear systems |C| with $C^2 \le 10$ 221
Bibliographical notes 224
Chapter IV. Projective models.
§1. Preliminaries 226
§2. Linear systems on K3-surfaces 229
§3. Numerical connectedness 232
§4. Base-points 240
§5. Hyperelliptic maps 243
§6. Birational maps 249
§7. Superelliptic maps 253
§8. The branch locus of superelliptic maps 260
§9. Projective models of degree ≤10 268
§10. Applications to linear systems 279
Appendix. A theorem of Igor Reider 281
Bibliographical notes 284
Chapter V. Genus 1 fibrations.
§1. Genus 1 fibrations:generalities 285
§2. The Picard group 292
§3. Jacobian fibrations 302
§4. Ogg-Shafarevich theory 312

§5. Weierstrass models 326
§6. Genus 1 fibrations on rational surfaces 345
§7. Genus 1 fibrations on Enriques surfaces 363
Bibliographical notes 372
Bibliography 376
Index 389
Glossary of notations 394

Introduction

Enriques surfaces bear the name of one of the founders of the theory of algebraic surfaces, the Italian geometer Federigo Enriques. He constructed some of them to give the first examples of nonrational algebraic surfaces on which there are no regular differential forms. At the same time a different construction of such surfaces was given by another Italian geometer, no less famous, Guido Castelnuovo. The original construction of Enriques gives a birational model of an Enriques surfaces represented by a surface of degree 6 in $\mathbb{P}^3$ passing doubly through the edges of the coordinate tetrahedron (see **[En 1,En 4,AS,G-H 1]**). Castelnuovo's model is a surface of degree 7 in $\mathbb{P}^3$ with a triple line and other singularities (see **[Cas,En 4]**. Later on Enriques **[En 2,En 4]** gave a construction of his surfaces as double planes with some branch curve of degree 8. We discuss these models in great detail in Chapter 4. Subsequently G. Fano **[Fa]** constructed a family of Enriques surfaces represented by congruences of lines in $\mathbb{P}^3$ associated to webs of quadrics (see **[Cos 2,[G-H 1]**). We will discuss this construction in Part II of this book. In **[En 3]** Enriques observed that every Enriques surface can be obtained as a quotient of a surface of type K3 (in modern terminology) by a fixed-point-free involution. The first modern treatment of Enriques surfaces over fields of characteristic different from 2 was given by M. Artin **[Art 1]** and B. Averbukh (see **[AS]** Chapter IX, **[Av]**).

It turned out in the works of Enriques and Castelnuovo that Enriques surfaces play a special role in the classification of algebraic surfaces. Recall that, according to a modern interpretation (due to I. Shafarevich) of Enriques's classification of surfaces, all projective nonsingular algebraic surfaces X are divided into four classes with respect to the possible behavior of the function $f(n) = \dim |nK_X|$ defined for $n>0$, where K_X is the canonical divisor class of X. The first class consists of surfaces for which f(n) = 0 for all n>0. This includes rational and ruled surfaces. The second class consists of surfaces for which f(n) is not identically zero but is bounded. This includes surfaces birationally isomorphic to either Enriques surfaces, or surfaces of type K3, or abelian surfaces, or hyperelliptic surfaces. The third class consists of surfaces for which f(n) is asymptotically linear. This class consists of so-called properly elliptic surfaces. Finally, the remaining surfaces are of general type; the function f(n) is asymptotically quadratic in this case. Enriques surfaces can be distinguished from other surfaces by the properties that $2K_X$ is linearly equivalent to 0 and the second Betti number $b_2(X)$ is equal to 10. There are other characterizations if the ground field is of characteristic $p \neq 2$. For example, X is an Enriques surfaces if and only if it belongs to the second class and has no regular differential forms. Or X is an Enriques surface if and only if it is isomorphic to a quotient of a K3-surface by a fixed-point-free involution. The latter property makes the theory of Enriques surfaces a special case of the theory of K3-surfaces. Topologically, over the field of complex numbers an Enriques surface is a smooth 4-manifold whose universal cover is of degree 2 and diffeomorphic to a minimal resolution of the 16 nodes of a Kummer surface.

The classification of algebraic surfaces was extended to the case of ground fields of positive characteristic in the works of E. Bombieri and D. Mumford. They gave a characteristic-free definition of Enriques surfaces and undertook the first study of these surfaces in characteristic 2. As far as we know the first construction of Enriques surfaces in characteristic 2 was given by M. Reid (see **[B-M 2]**). We present these examples in Chapter 1. The case of characteristic 2 turned out to be the most special case for Enriques surfaces. In this case only new pathological phenomena arise, as for example the absence of a K3-cover of an Enriques surfaces. The paper of E. Bombieri and D. Mumford **[B-M 1]** was the first in which these new phenomena were discovered and studied. They showed that the extension of the classification of algebraic surfaces to the case of positive characteristic opens up an absolutely new world, where the familiar objects metamorphose into quite unrecognizable objects with equally rich and beautiful geometrical structure. The inclusion of this case in our treatment of Enriques surfaces is one of the main reasons for the unexpectedly large size of this book

As is mentioned in the Preface we are trying to study Enriques surfaces in all characteristics. We share the opinion that the geometry ends with the application of transcendental methods and the true understanding of the geometrical structure can be obtained only by understanding the case of a ground field of arbitrary characteristic. Unfortunately, in many respects we still depend on transcendental methods and no analogues of the results known for complex algebraic surfaces have been found in the general case. In our case this appears in application of the properties of the period map for K3-surfaces covering Enriques surfaces. This gives important information about automorphisms and moduli of Enriques surfaces, which cannot be obtained thus far by purely algebro-geometrical methods. However, we set as one of the goals for this book the illustration, in the case of Enriques surfaces, of the richness and the beauty which adds to the geometry of algebraic surfaces when the case of fields of arbitrary characteristic is considered. Of course the transcendental methods give us something more, such as the insight into the differential geometry and the topology of algebraic surfaces. There are excellent expositions of the theory of periods of complex Enriques and K3-surfaces (**[B-P-vdV, Nam]**) to which we refer the reader. We will review this theory in the second volume.

Since the theory of K3-surfaces cannot be applied to the study of Enriques surfaces in all characteristics we try to avoid its application whenever possible. The main difference in the study of K3-surfaces and Enriques surfaces is that the latter essentially form one irreducible family of surfaces whereas the structure of K3-surfaces depends very much on the structure of its Picard lattice of algebraic cycles. The Picard lattice of any Enriques surface is isomorphic to an even hyperbolic unimodular lattice of rank 10, which is defined uniquely up to isomorphism. The basic tool of our approach to the study of Enriques surfaces is the use of the rich arithmetic properties of this lattice. This allows us to describe all possible projective models of Enriques surfaces and get a lot of information about the structure of the set of nonsingular rational curves and pencils of elliptic curves on them, and also apply these techniques to the study of automorphisms of the surfaces. The same technique is also applied to the study of K3-surfaces. The complete study of K3-surfaces over fields of arbitrary characteristic similar to the one undertaken here for Enriques surfaces seemed to us an unreasonably difficult and time consuming project. The length of the present work obviously supports this opinion.

Here is a brief description of the content of this volume. In Chapter 0 we collect all the necessary preliminary facts which we use in our study of Enriques surfaces. We emphasise the characteristic 2 case. Thus we include this case in our discussion of such topics as double covers of surfaces, rational double points, Del Pezzo surfaces and others. A new result here is a classification of symmetric Del Pezzo surfaces of degree 3 and 4, where symmetric means the existence of a principal double cover away from the singular points.

In Chapter 1 we recall the Enriques-Bombieri-Mumford classification of surfaces and indicate the place in this classification occupied by Enriques surfaces. The values of the basic numerical invariants of the surfaces are given. Following Bombieri-Mumford we divide Enriques surfaces over fields of characteristic 2 into 3 classes with respect to the three possible structures of their Picard scheme, which is a finite group scheme of order 2. These are classical Enriques surfaces, μ_2-surfaces and α_2-surfaces. The properties of Enriques surfaces over fields of characteristic 2 depend very much on the class to which these surfaces belong. For example, only in the second case is the double cover, defined by the dual of the Picard scheme, unramified and isomorphic to a K3-surface. In other cases the double cover is purely inseparable and can be either a singular model of a K3-surface or a singular rational surface. We study this double cover in §3. The central result of this chapter is a description of the Picard lattice of an Enriques surface which turns out to be a unimodular even lattice of rank 10 and signature (1,9). In the case of complex surfaces the proof of this result is an easy consequence of Hodge theory and Poincaré duality. In the case of an arbitrary characteristic we give a proof using the theory of flat duality of sheaves on surfaces due to J. Milne. The previous proofs of this result used either the existence of a genus 1 fibration (**[B-M 2]**) or the theory of crystalline cohomology (**[I11]**). In **§**4, following the work of W. Lang and others, we study some differential invariants of Enriques surfaces like the Hodge numbers $h^{p,q}$ or the De Rham Betti numbers. Finally we prove a Kodaira-Ramanujam vanishing theorem for Enriques surfaces which is known not to be true for any surface over a field of arbitrary characteristic.

Chapter 2 presents the needed definitions and results about integral quadratic forms (lattices), their root bases and Weyl groups. This techniques was applied for the first time to the study of K3-surfaces by I. Pyatetski-Shapiro and I. Shafarevich. Later it turned out that the same technique applies equally to the study of Enriques surfaces. We study the Picard lattice **E** of an Enriques surface with great details, especially its orthogonal group. The lattice **E** occurs in many other situations, for example in the theory of Kac-Moody algebras **[Kac]**, the theory of surface singularities **[Arn]**) and the theory of Cremona transformations **[D-O]**. We prove here a new result about the generators of the 2-level congruence subgroup of the Weyl group of some lattices, including **E**. This result will be used in Part II for the description of the automorphism group of Enriques or Coble surfaces (cf. **[C-D 2,Do 5]**). We tried to make this chapter as complete as possible to make it available for future references to its material used in questions not considered in this book (see for example **[D-O]**).

In Chapter 3 we give the first applications of the lattice-theoretical approach to the theory of Enriques surfaces. The classical result of Enriques gives the existence of a genus 1 fibration on any Enriques surface. This result was generalized to characteristic 2 by Bombieri and Mumford. From this result they deduced the structure of the Picard lattice of an Enriques surface. Our proof of the latter result is independent of the existence of a genus 1 fibration, and we show, following an idea of

Shafarevich, that the existence of a genus 1 fibration follows simply from the fact that the Picard lattice of an Enriques surface contains isotropic vectors. This approach to the proof of the existence of genus 1 fibrations on Enriques surfaces was first suggested by W. Lang. By similar methods we prove the Reducibility Lemma of Enriques which asserts that every divisor on an Enriques surface is linearly equivalent to an integral combination of nonsingular rational and elliptic curves. Generically one expects that an Enriques surface does not contain rational nonsingular curves (the precise meaning of this will be discussed in Part II of the book). In this case (an unnodal Enriques surface) the set of genus 1 fibration is bijectively equivalent to the infinite set of primitive isotropic vectors in the Picard lattice. This correspondence is based on the observation that the class of the fibre of a genus 1 fibration in the Picard lattice is a twice a primitive isotropic vector. In particular, it follows from the structure of this lattice that one can find sequences of 10 isotropic vectors with mutual scalar product equal to 1. The existence of such sequences plays an important role in the study of nonsingular projective models of the surfaces as surfaces of degree 10 in $\mathbb{P}^5$ (Fano models). This is the minimal possible degree of an Enriques surface embedded into a projective space. The situation becomes much more complicated in the case where an Enriques surface contains nonsingular rational curves (nodal Enriques surface). Then the isotropic vectors corresponding to genus 1 fibrations are restricted by the condition that their intersection with the class of any nonsingular rational curve is non-negative. We define an important non-degeneracy invariant of Enriques surface as the maximal number r such that F contains r isotropic vectors representing fibres of genus 1 fibrations with pairwise scalar product equal to 1. Thus, for unnodal Enriques surfaces d(F) = 10. By a long case-by-case analysis we prove that in general d(F) ≥ 3. This result is applied to the study of projective models in Chapter 4, where we show the existence of non-special double plane constructions (d(F) ≥ 2) and of a sextic surface construction (d(F) ≥ 3). Its proof is greatly simplified if the characteristic of the ground field is different from 2. It is known that d(F) may take the value

4; no examples of Enriques surfaces with d(F) = 3 are known to us. Finally we apply the lattice-theoretical techniques to classifying linear systems |C| on Enriques surfaces with $C^2 \leq 10$.

In Chapter 4 we study projective models of Enriques surfaces. In many respects some of the results here are analogs of the results of B. Saint-Donat on linear systems on K3-surfaces. The difficulties in our case arise mostly in the case of characteristic 2. A recent result of I. Reider shows that in characteristic 0 many results about linear systems on Enriques or K3-surfaces are greatly simplified if one applies the results of F.Bogomolov on stable vector bundles. We give an account of this approach in an Appendix to Chapter 4. The results on linear systems together with the results of Chapter 3 are applied to the study of models of Enriques surfaces of degree at most 10. Thus we give here all the classical constructions of Enriques surfaces together with their analogs in characteristic 2.

Finally, in Chapter 5 of this volume we study elliptic and quasi-elliptic (genus 1) fibrations on Enriques surfaces. To be as self-contained as possible we include in this chapter all the basic facts about genus 1 fibrations on algebraic surfaces with emphasis on the arbitrary characteristic case. This makes our exposition of the theory of elliptic surfaces very different from the previous ones given in **[B-P-vdV, G-H 1]**. It is based on the theory of A. Ogg and I. Shafarevich of prinicipal homogeneous spaces of elliptic curves over function fields. We extend and complete the previous expositions of this theory given in **[AS, Do 3]**. Also, following the work of W. Lang, we extend this theory to the case of quasi-elliptic fibrations on algebraic surfaces. The relationship between the fibres of a genus 1 fibration and of its associated jacobian fibration was studied to the case of characteristic p > 2 by T. Katsura and K. Ueno. We extend their result in characteristic 2. Thus we show that the problem of the description of degenerate fibres of a genus 1 fibration on an Enriques surface is "almost" equivalent to the same problem on a rational surface. This leads us to the study of genus 1 fibrations on rational surfaces which is equivalent to the study of so-called Halphen pencils of plane curves. We classify all possible configurations of reducible degenerate fibres of extremal genus 1 fibrations on rational surfaces and show that it coincides with Dynkin's classification of maximal semi-simple subalgebras of the simple Lie algebra of type E_8. Note that restricting oneself only to the case of elliptic fibrations (or to the case of ground field of characteristic 0) as in **[M-P]** or **[Nar]** leaves two cases of Dynkin's classification unaccounted for. We hope that this chapter together with Chapter 2 will serve as references for needs different from the study of Enriques surfaces. We will return to the study of the set of genus 1 fibrations on Enriques surfaces in Part II.

Each chapter ends with historical and bibliographical notes. We have tried very hard to give credit to all the mathematicians who have contributed to the study of Enriques surfaces. We are fully aware that omissions, incompleteness and inaccuracies in our presentation of these historical notes are possible. We apologize in advance to those of our colleagues whose work has not been adequately presented in our book.

Chapter 0

PRELIMINARIES.

S1. Double covers.

A morphism $f: X \to Y$ of integral schemes over an algebraically closed field κ is called a **double cover** if f is finite and of degree 2. A double cover is said to be **separable** (resp. **inseparable**) if the corresponding extension of the fields of rational functions is separable (resp. inseparable).

From now on we assume that X is Cohen-Macaulay and Y is regular. Let $f:X \to Y$ be a double cover and

$$\mathcal{A} = f_*\mathcal{O}_X .$$

This is a locally free $\mathcal{O}_Y$-Algebra of rank 2 and

$$X \cong \mathrm{Spec}(\mathcal{A}).$$

Let $\{U_i\}_i$ be an open affine covering of Y which trivializes $\mathcal{A}$, $A_i = \mathcal{O}_Y(U_i)$. Then $\mathcal{A}_i = \mathcal{A}|U_i$ is isomorphic to the A_i-algebra:

$$A_i[T]/(T^2+a_iT+b_i),$$

where $a_i, b_i \in A_i$. Let t_i be the image of T in this algebra. Clearly, 1 and t_i generate $\mathcal{A}_i$ as a free A_i-module of rank 2. Replacing i by j, we find that

$$t_i = g_{ij}t_j + c_{ij}$$

for some $g_{ij} \in \mathcal{O}_Y(U_i \cap U_j)^*$ and $c_{ij} \in \mathcal{O}_Y(U_i \cap U_j)$. This shows that $\mathcal{A}$ is isomorphic to an extension

$$0 \to \mathcal{O}_Y \to \mathcal{A} \to \mathcal{L} \to 0,$$

where $\mathcal{L}$ is an invertible sheaf with transition functions $\{g_{ij}\}$.

The functions $\{c_{ij}\}$ define a cocycle $c \in Z^1(Y, \mathcal{L}^{-1})$ which determines the class of the above extension. It is easy to determine how the coefficients a_i and b_i change after replacing U_i by U_j. We find

$$a_i = g_{ij}(a_j - 2c_{ij}),\ b_i = g_{ij}^2 b_j - g_{ij}a_j c_{ij} - c_{ij}^2.$$

Assume $\operatorname{char}(k) \neq 2$

We may replace t_i by $t_i + \frac{1}{2}a_i$ to assume that $a_i = 0$. Then $c_{ij} = 0$ and

$$\mathcal{A} \cong \mathcal{O}_Y \oplus \mathcal{L} \text{ as an } \mathcal{O}_Y\text{-Module},$$

$$\mathcal{A}_i \cong A_i[T]/(T^2 + b_i) \text{ as an } A_i\text{-algebra}$$

Since $b_i = g_{ij}^2 b_j$, the family $\{b_i\}$ defines a non-zero section b of the line bundle $\mathcal{L}^{-2}$. We immediately check that f is separable in this case. Conversely, a line bundle $\mathcal{L}$ and a section $b \in \Gamma(\mathcal{L}^{-2})$ define a separable double cover $f: X \to Y$.

Proposition 0.1.1. Let $\mathcal{L}$ be a line bundle on Y, $b \in \Gamma(\mathcal{L}^{-2}) \setminus \{0\}$ and $f: X \to Y$ the corresponding double cover. Then f is unramified exactly over the complement of the support of the divisor $\operatorname{div}(b)$ of the section b. If Y is smooth, then $x \in X$ is smooth if and only if $f(x)$ is a smooth point of the subscheme $\operatorname{div}(b)$. In this case X is normal if and only if $\operatorname{div}(b)$ is reduced.

PROOF. It is easily verified by local computations.

Assume $\operatorname{char}(k) = 2$.

With the above notation

$$a_i = g_{ij}a_j,\ b_i = g_{ij}^2 b_j + g_{ij}a_j c_{ij} + c_{ij}^2.$$

Thus $\{a_i\}$ form a section of the line bundle $\mathcal{L}^{-1}$ and $\{1, a_i, b_i\}$ form a section s of a vector bundle V of rank 3 with transition matrices of the form

$$\begin{bmatrix} 1 & 0 & 0 \\ 0 & g_{ij}^{-1} & 0 \\ c_{ij}^2 & g_{ij}^{-1}c_{ij} & g_{ij}^{-2} \end{bmatrix}$$

It is isomorphic to an extension

$$0 \to \mathcal{L}^{-2} \to V \to \mathcal{L}^{-1} \oplus \mathcal{O}_Y \to 0.$$

Moreover, s maps to (a,1) under the natural map $\varphi: \Gamma(V) \to \Gamma(\mathcal{L}^{-1}) \oplus \Gamma(\mathcal{O}_Y)$.

Conversely, if V is a rank-3 vector bundle as above and $s \in \Gamma(V)$ such that $\varphi(s) = (a,1)$, we can construct a double cover $f: X \to Y$. A triple $(\mathcal{L}, V, s)$ consisting of an invertible sheaf $\mathcal{L}$, a vector bundle V, and its section s as above will be called an **admissible triple**. It is said to be **splittable** if c_{ij} can be chosen to be zero, or, equivalently, V (resp. $\mathcal{A}$) splits into the sum of line bundles $\mathcal{L}^{-2} \oplus \mathcal{L}^{-1} \oplus \mathcal{O}_Y$ (resp. $\mathcal{L} \oplus \mathcal{O}_Y$). Of course, this condition is satisfied if $H^1(Y, \mathcal{L}^{-1}) = 0$.

Proposition 0.1.2. Let $f: X \to Y$ be the double cover defined by an admissible triple $(V, \mathcal{L}, s)$. Then f is inseparable if and only if $\varphi(s) = (0,1)$. If this is not the case, then f is a separable double cover unramified exactly outside of the support of div(a), where $\varphi(s) = (a,1)$. If Y is smooth and f is inseparable (resp. separable), there exists a section

$$\alpha(f) \in H^0(Y, \mathcal{L}^{-2} \otimes \Omega_Y^1)$$

such that the inverse images of its zeroes (resp. those of them which lie on div(a)) are the singularities of X.

PROOF. The first assertion is checked immediately. Let us check the second one. We know that X is locally given by an equation of the form $y^2 + ay + b = 0$. Since Y is smooth, the singular points of X are found by differentiation. We obtain that X is singular at a point x_0 with $f(x_0) = t_0$ if and only if

$$a(t_0) = 0 \ , \ y\,da + db = 0 \text{ at } t_0.$$

Since we know the way in which y, a, and b are transformed by changing local coordinate system, we easily verify that $\{y\,da + db\}$ is glued together to form a section α of $\mathcal{L}^{-2} \otimes \Omega_Y^1$.

Corollary 0.1.1. Let $f: X \to Y$ and $\alpha(f)$ be as in the previous proposition and C be a smooth irreducible curve on Y. Then $f^{-1}(C) = 2C'$ for some curve on X if and only if the restriction of $\alpha(f)$ to C is zero.

Recall that a closed point x of a Cohen-Macaulay scheme Z is said to be **Gorenstein** if the dualizing sheaf ω_Z is invertible at x. Z is called Gorenstein if every point of it is Gorenstein.

Proposition 0.1.3. Let $f: X \to Y$ be a double cover and $\mathcal{L}$ the corresponding invertible sheaf. Then X is Gorenstein and its canonical sheaf is given by

$$\omega_X \cong f^*(\omega_Y \otimes \mathcal{L}^{-1}).$$

PROOF. Locally X is a hypersurface in $Z_i = U_i \times \mathbb{A}^1$ given by an equation

$$x_i^2 + a_i x_i + b_i = 0 .$$

The schemes Z_i are glued together to form a scheme Z over Y by

$$x_i = g_{ij} x_j + c_{ij} ,$$

which define a structure of an affine bundle over Y of rank 1. In particular, we see that X is Gorenstein (by the adjunction formula for the dualizing sheaf).

Assume that f corresponds to a splittable admissible triple. Then Z has the structure of the line bundle $\mathbb{V}(\mathcal{L}) = \mathrm{Spec}(\mathrm{Sym}(\mathcal{L}))$, the total space of the line bundle $\mathcal{L}$. Let π be the corresponding projection $\mathbb{V}(\mathcal{L}) \to Y$. One checks immediately that

$$\pi^*(\mathcal{L}) \cong \mathcal{O}_{\mathbb{V}(\mathcal{L})}, \qquad \omega_{\mathbb{V}(\mathcal{L})} \cong \pi^*(\omega_Y \otimes \mathcal{L})$$

Let $i: X \hookrightarrow \mathbb{V}(\mathcal{L})$ be the canonical closed embedding. By adjunction

$$\omega_X = i^*(\omega_{\mathbb{V}(\mathcal{L})} \otimes \pi^*(\mathcal{L}^{-2})) \cong i^*(\pi^*(\omega_Y \otimes \mathcal{L}) \otimes \pi^*(\mathcal{L}^{-2})) \cong f^*(\omega_Y \otimes \mathcal{L}^{-1}).$$

In the general case, one can give another proof of the same formula by using Grothendieck's duality formula (**[Lip]**):

$$\omega_X \cong [\mathcal{H}om_{\mathcal{O}_Y}(\mathcal{A}, \omega_Y)]^{\sim}$$

(one regards the sheaf *Hom* as an $\mathcal{A}$-Module and takes the associated sheaf on X). Since $\mathcal{A}$ is an extension

$$0 \to \mathcal{O}_Y \to \mathcal{A} \to \mathcal{L} \to 0,$$

we obtain an extension for $f_*\omega_X$:

$$0 \to Hom_{\mathcal{O}_Y}(\mathcal{L},\omega_Y) \to f_*\omega_X \to Hom_{\mathcal{O}_Y}(\mathcal{O}_Y,\omega_Y) \to 0.$$

On the other hand,

$$f_*(f^*(\omega_Y \otimes \mathcal{L}^{-1}) \cong \omega_Y \otimes \mathcal{L}^{-1} \otimes \mathcal{A}$$

is an extension

$$0 \to \omega_Y \otimes \mathcal{L}^{-1} \to f_*(f^*(\omega_Y \otimes \mathcal{L}^{-1}) \to \omega_Y \to 0$$

which is easily seen to be isomorphic to the preceding extension. This shows that

$$f_*(f^*(\omega_Y \otimes \mathcal{L}^{-1}) \cong f_*\omega_X.$$

Since f is finite, this gives us the required isomorphism.

If char(k) $\neq$ 2 or f is inseparable with splittable admissible triple, we may assume that $a = \{a_i\} = 0$ in the above and $b = \{b_i\}$ defines a section of $\mathcal{L}^{-2}$. We will call W = div(b) the **branch divisor** of f:X $\to$ Y. Its inverse image in X is a Cartier divisor locally given by the equation $x_i^2 = 0$. The corresponding reduced divisor R is called the **ramification divisor**. If y is a generic point of a reduced irreducible component of R , then f*(W) is of multiplicity 2 at y. If 2 is invertible, f is unramified (or étale) outside R. If f is inseparable, f is a principal homogeneous fibration with respect to the finite group scheme μ_2 outside R.

If X and Y are both regular, then

$$f^*(W) = 2R, \text{ in Pic}(X).$$

If W is smooth, for every irreducible component W_i of W, we have

$$R_i^2 = \tfrac{1}{2}W_i^2, \text{ where } R_i = (f^*(W_i))_{red}.$$

Recall also that

$$\mathcal{O}_Y(W) \cong \mathcal{L}^{-2},\ \mathcal{O}_X(R) \cong \pi^*(\mathcal{L}^{-1}),$$

and we can write the formula for ω_X as

$$K_X = f^*(K_Y) + R.$$

If char(k) = 2 is not invertible on Y and f is separable, then we will call $W = \mathrm{div}(a)$ the **branch divisor** of $f{:}X \to Y$. Its inverse image $f^*(W)$ in X is a Cartier divisor locally given by the equation $x_i^2+b_i = 0$. The corresponding positive divisor $R = f^*(W)$ is called the **ramification divisor**. If R_i is an irreducible component of R , then R_i is reduced or of multiplicity 2 at its generic point. In the first case, the restriction map

$$f|R_i\colon R_i \to W_i = f(R_i)$$

is an inseparable double cover. In the second case, its restriction to the reduced subscheme $(R_i)_{red}$ is of degree 1.

If X and Y are both regular, then

$$R_i^2 = 2W_i^2 \text{ if } R_i \text{ is reduced}$$

$$R_i^2 = \tfrac{1}{2}W_i^2 \text{ otherwise.}$$

Recall also that

$$\mathcal{O}_Y(W) \cong \mathcal{L}^{-1}, \qquad \mathcal{O}_X(R) \cong \pi^*(\mathcal{L}^{-1}),$$

and we can write the formula for ω_X again as

$$K_X = f^*(K_Y) + R .$$

Proposition 0.1.4. Let $f{:}X \to Y$ be a separable double cover of a smooth surface Y. Let C be a smooth curve on Y. Assume that C has transversal intersection with the branch divisor W of f at a nonsinglar point y_0. Suppose that X is nonsingular over y_0. Then $f^{-1}(C)$ is smooth at $f^{-1}(y_0)$.

PROOF. Local computation.

Recall that one can define the Betti numbers $b_i(Z)$ and the Euler characteristic $e(Z)$ of every scheme Z of finite type over a field k by using l-adic étale cohomology **[Mi 2]**. Namely, we put

$$b_i(Z) = \dim H^i_{et}(Z, \mathbb{Q}_l),$$

$$e(Z) = \sum_i (-1)^i b_i(Z).$$

If Z is smooth and projective of dimension d, then

$$e(Z) = c_d(Z),$$

where $c_d(Z)$ denotes the top Chern number of Z.

For example, if Z is a smooth projective curve of genus g, then

$$e(Z) = 2-2g.$$

Proposition 0.1.5. Let f: X → Y be a double cover of smooth surfaces. If f is separable, then

$$e(X) = 2e(Y) - (2K_Y \cdot D + 4D^2), \text{ if } \mathrm{char}(k) = 2$$

$$= 2e(Y) - e(R), \text{ if } \mathrm{char}(k) \neq 2,$$

where $\mathcal{L} = \mathcal{O}_X(-D))$ and R is the ramification curve.

If f is inseparable, then

$$e(X) = e(Y).$$

PROOF. We use the Noether formula:

$$12\chi(\mathcal{O}_X) = K_X^2 + e(X), \quad 12\chi(\mathcal{O}_Y) = K_Y^2 + e(Y).$$

Let

$$0 \to \mathcal{O}_Y \to f^*\mathcal{O}_X \to \mathcal{L} \to 0$$

be the canonical sequence. Then

$$\chi(\mathcal{O}_X) = \chi(\mathcal{O}_Y) + \chi(\mathcal{L}).$$

By Riemann-Roch

$$\chi(\mathcal{L}) = \tfrac{1}{2}(D^2 + D \cdot K_Y) + \chi(\mathcal{O}_Y).$$

We have

$$K_X \sim f^*(K_Y+D),$$

hence

$$K_X \cdot f^*(D) + f^*(D)^2 = f^*(K_Y) \cdot f^*(D) + 2f^*(D)^2 = 2K_Y \cdot D + 4D^2.$$

Thus we obtain

$$e(X) = 12\chi(\mathcal{O}_X) - K_X^2 = 24\chi(\mathcal{O}_Y) + 6(D^2 + D \cdot K_Y) - 2(K_Y+D)^2 =$$

$$= 2(12\chi(\mathcal{O}_Y) - K_Y^2) + 4D^2 + 2K_Y \cdot D = 2e(Y) - (2K_Y \cdot D + 4D^2)$$

Assume now that $\mathrm{char}(\kappa) \neq 2$. Then the ramification curve R is smooth and $R \sim f^*(D)$. Thus

$$-e(R) = K_X \cdot f^*(D) + f^*(D)^2 = 2K_Y \cdot D + 4D^2.$$

If f is inseparable then X and Y are étale homeomorphic (**[Mi 2]**). In particular, they have the same Euler characteristic.

We consider now the special case of **principal coverings**. Let G be a group scheme of order 2 over κ. Up to isomorphism, there are three of them **[Oo]**:

$$G \cong (\mathbb{Z}/2)_\kappa,\ \mu_2 \text{ or } \alpha_2,$$

where

$$(\mathbb{Z}/2)_\kappa = \mathrm{Spec}\ \kappa[t]/(t^2+t),\ \mu(\bar{t}) = \bar{t}\otimes 1 + 1\otimes\bar{t},\ \bar{t} = t \bmod (t^2+1);$$

$$\mu_2 = \mathrm{Spec}\ \kappa[t]/(t^2-1),\ \mu(\bar{t}) = \bar{t}\otimes\bar{t},\ \bar{t} = t \bmod (t^2-1);$$

$$\alpha_2 = \mathrm{Spec}\ \kappa[t]/(t^2),\ \mu(\bar{t}) = \bar{t}\otimes 1 + 1\otimes\bar{t},\ \bar{t} = t \bmod (t^2).$$

Note that the latter scheme is defined only if $\mathrm{char}(\kappa) = 2$, and the first two are non-canonically isomorphic if $\mathrm{char}(\kappa) \neq 2$.

For every scheme Z over κ, we denote by G_Z (or simply by G if no confusion arises) the corresponding group scheme $G\times_\kappa Z$ over Z.

Proposition 0.1.6. Assume $\mathrm{char}(\kappa) \neq 2$ and $\Gamma(\mathcal{O}_Y) = \kappa$. There are natural one-to-one correspondences between the isomorphism classes of the following objects:

(i) unramified double covers $f:X \to Y$;

(ii) invertible sheaves $\mathcal{L}$ on Y such that $\mathcal{L}^2 \cong \mathcal{O}_Y$;

(iii) torseurs over $\mu_{2,Y}$ in the étale topology of Y.

PROOF. Let $f:X \to Y$ be an unramified double cover. With the above notation, this means that $b \in \Gamma(\mathcal{L}^{-2})$ does not vanish on Y, hence $\mathcal{L}^{-2} \cong \mathcal{O}_Y$.

Given an isomorphism $b: \mathcal{O}_Y \to \mathcal{L}^{-2}$, we define the double cover locally by the equations $t_i^2 + b_i = 0$, where $b_i = b|U_i$.

Obviously, $\Gamma(\mathcal{L}^{-1}) = \{0\}$, otherwise the cover splits. Thus $\mathcal{L}$ defines a nonzero element in ${}_2\mathrm{Pic}(Y)$. We have a **Kummer exact sequence** in the étale topology (see **[Mi 2]**):

$$0 \to \mu_2 \to G_m \to G_m \to 0$$

and the corresponding cohomology sequence gives an isomorphism

$$H^1(Y,\mu_2) \cong {}_2\mathrm{Pic}(Y).$$

The group on the left classifies the isomorphism classes of principal homogeneous spaces (torseurs) over μ_2.

It is easy to check that all the above correspondences are bijective on the sets of isomorphism classes.

Proposition 0.1.7. Assume $\mathrm{char}(k) = 2$ and $\Gamma(\mathcal{O}_Y) = k$. Let $F: H^1(Y,\mathcal{O}_Y) \to H^1(Y,\mathcal{O}_Y)$ be the linear map induced by the Frobenius morphism. There are natural one-to-one correspondences between the isomorphism classes of the following objects:

(i) non-trivial unramified double covers $f:X \to Y$;

(ii) non-zero elements of $\mathrm{Ker}(\mathbf{F-id})$;

(iii) torseurs over $(\mathbf{Z}/2)_Y$ in the étale topology of Y.

PROOF. Let $f: X \to Y$ be a unramified double cover and $(V,\mathcal{L},s)$ be the corresponding admissible triple. Since f is unramified, $\mathcal{L}$ has a nowhere vanishing section, hence is isomorphic to $\mathcal{O}_Y$. Thus in our local notations, we may assume $g_{ij} = 1$ and $a_i = a$ is a non-zero constant. Hence, X is given locally by

$t_i^2+at_i+b_i = 0$, where

$$b_i = b_j + ac_{ij} + c_{ij}^2.$$

This shows that $\{c_{ij}\}$ form a cocycle $c \in Z^1(Y,\mathcal{O}_Y)$ such that $\mathbf{F}(c)+ac\in B^1(Y,\mathcal{O}_Y)$. Scaling c by multiplying by $\sqrt{a}^{-1}$, we get an element of $\mathrm{Ker}(\mathbf{F}-\mathbf{id})$ which is obviously nontrivial.

Let c be a non-trivial element of $\mathrm{Ker}(\mathbf{F}-\mathbf{id})$. We use the **Artin-Schreier sequence** in the étale topology of Y:

$$0 \to (\mathbf{Z}/2)_Y \to \mathcal{O}_Y \xrightarrow{\wp} \mathcal{O}_Y \to 0,$$

where $\wp$ is given by $x \to x^2-x$. The corresponding cohomology sequence gives an isomorphism:

$$H^1(Y,\mathbf{Z}/2) \cong \mathrm{Ker}(\mathbf{F}-\mathbf{id}:H^1(Y,\mathcal{O}_Y)\to H^1(Y,\mathcal{O}_Y)).$$

The group on the left classifies torseurs over $(\mathbf{Z}/2)_Y$, i.e. unramified double covers of Y.

Proposition 0.1.8. Assume $\mathrm{char}(\kappa) = 2$ and $\Gamma(\mathcal{O}_Y) = \kappa$. There are natural one-to-one correspondences between the isomorphism classes of the following objects:

(i) inseparable double covers $f:X \to Y$ with splittable admissible triple and no ramification divisor;

(ii) invertible sheaves $\mathfrak{L}$ on Y such that $\mathfrak{L}^2 \cong \mathcal{O}_Y, \mathfrak{L} \ncong \mathcal{O}_Y$;

(iii) torseurs over $\mu_{2,Y}$ in the flat topology of Y.

PROOF. Let $f:X \to Y$ be an inseparable cover and $(V,\mathfrak{L},s)$ be the corresponding admissible triple. Let $\{g_{ij}\}$ be the 1-cocycle representing $\mathfrak{L}$. By Proposition 0.1.1, f is locally given by $t_i^2 + b_i = 0$, where b_i are invertible and $b_i = g_{ij}^2 b_j$. This shows that the 1-cocycle $\{g_{ij}^2\}$ is trivial, i.e. $\mathfrak{L}^2$ is trivial. Obviously, $\mathfrak{L} \ncong \mathcal{O}_Y$.

Conversely, let $\mathfrak{L}$ be an invertible sheaves on Y such that $\mathfrak{L}^2 \cong \mathcal{O}_Y$, $\mathfrak{L} \ncong \mathcal{O}_Y$. If $\{g_{ij}\}$ is the corresponding cocycle, then $g_{ij}^2 = b_i/b_j$ for some $b_i\in\Gamma(\mathcal{O}_Y)^*$ and we define locally X by $t_i^2+b_i = 0$.

We have a Kummer exact sequence in the flat topology:

$$0 \to \mu_2 \to \mathbf{G}_m \to \mathbf{G}_m \to 0$$

The corresponding cohomology sequence gives an isomorphism

$$H^1(Y,\mu_2) \cong {}_2\mathrm{Pic}(Y).$$

The group on the left classifies the isomorphism classes of principal homogeneous spaces (torseurs) over μ_2. This explains the connection between (ii) and (iii). It is easy to check that all the above correspondences are bijective on the isomorphism classes.

Proposition 0.1.9. Assume char(k) = 2 and $\Gamma(\mathcal{O}_Y) = k$. Let $F: H^1(Y,\mathcal{O}_Y) \to H^1(Y,\mathcal{O}_Y)$ be the linear map induced by the Frobenius morphism. There are natural one-to-one correspondences between the isomorphism classes of the following objects:

(i) inseparable double covers $f:X \to Y$ with $\mathfrak{L} \cong \mathcal{O}_Y$;

(ii) non-zero elements of Ker(**F**);

(iii) torseurs over $\alpha_{2,Y}$ in the flat topology of Y.

PROOF. Let $f: X \to Y$ be an inseparable double cover with $\mathfrak{L} \cong \mathcal{O}_Y$. In our local notations, we may assume $g_{ij} = 1$ and $a_i = 0$. Hence X is given locally by $t_i^2+b_i = 0$, where

$$b_i = b_j + c_{ij}^2.$$

This defines an element $c = \{c_{ij}\} \in Z^1(Y,\mathcal{O}_Y)$ such that $F(c) \in B^1(Y,\mathcal{O}_Y)$, hence an element of Ker(**F**) which is easily seen to be non-trivial.

Let c be a non-trivial element of Ker(**F**-**id**). We use the following exact sequence in the flat topology of Y:

$$0 \to \alpha_{2,Y} \to \mathcal{O}_Y \xrightarrow{F} \mathcal{O}_Y \to 0,$$

where **F** is given by $x \to x^2$. The corresponding cohomology sequence gives an isomorphism:

$$H^1(Y,\alpha_2) \cong \mathrm{Ker}(F:H^1(Y,\mathcal{O}_Y)\to H^1(Y,\mathcal{O}_Y)).$$

The group on the left classifies torseurs over $\alpha_{2,Y}$ in the flat topology.

The covers corresponding to G-torseurs, where G is a group scheme of order 2 will be called **principal double G-covers.**

Proposition 0.1.10 (Purity theorem). Let Y be a smooth surface , Y' an open subset whose complement Z is a finite set. Then every principal G-cover $f: X' \to Y'$ is extended uniquely to a principal G-cover $f: X \to Y$

PROOF. We apply the exact sequence of local cohomology

$$H^1_Z(Y,G) \to H^1(Y,G) \to H^1(Y',G) \to H^2_Z(Y,G)$$

and use the fact that a nonsingular point $y \in Z$ does not have non-trivial local principal covers.

Proposition 0.1.11. Let $f:X \to Y$ be an inseparable principal double cover. Then

$$H^0(Y,\mathcal{L}^{-2}\otimes\Omega^1_Y) \neq 0.$$

PROOF. Since X is reduced, this follows immediately from Proposition 0.1.2.

We note that a separable double cover $f:X\to Y$ is always a Galois cover of degree 2. This means that there exists a finite group scheme G of order 2 over k, which acts on X , and such that the orbit space X/G is isomorphic to Y.

If 2 is invertible, $G = \mu_2$ which acts locally by sending y_i to $-y_i$.

If 2 is not invertible, then $G = \mathbf{Z}/2$, which acts locally by

$$t_i \to t_i + a_i.$$

The same is true for any principal double G-cover. We leave to the reader the definition of the corresponding group action.

Finally, we explain the relationship between vector fields and inseparable double covers of surfaces. The proofs, when omitted, can be found in **[R-S 1]**.

Let X be an irreducible normal projective surface over k. A **vector field** is a section D of the tangent sheaf Θ_X. For every open affine subset

U = Spec A of X , a vector field D defines a section of $Der_K(A,A)$, i.e. a derivation of the K-algebra A with values in A. We set

$$A^D = \{ x \in A: D(a) = 0 \}.$$

This is a subring of A. Taking an affine covering $\mathfrak{U} = \{U_i\}_i$ of X, we define the **quotient X^D of X with respect to D** by glueing together Spec $\mathcal{O}_X(U_i)$. It is a normal variety which comes with the natural projection

$$\pi_D: X \to X^D .$$

It is obvious that the operation of factorization of X by a vector field is non-trivial only in the case char$(K) = p > 0$. So, from now on , we assume that this is the case. Since $A^p = \{x^p: x \in A\} \subset A^D$, we obtain that the Frobenius morphism $F: X \to X^p$ factors through π_D

$$X \to X^D \to X^p .$$

In particular, π_D is a purely inseparable finite morphism.

Recall that the space $\Gamma(\Theta_X)$ has two operations: the composition of derivations $D \circ D'$ and the Lie bracket $[D,D'] = D \circ D' - D' \circ D$. It is a Lie p-algebra with respect to the second operation.

A vector field D is said to be **p-closed** if $D^p = \lambda D$ for some $\lambda \in K$. Every finite-dimensional Lie p-algebra contains a non-zero p-closed element. By scaling, we obtain that there exists a non-zero vector field D on X such that

$$D^p = D \text{ or } 0.$$

We say that a p-closed vector field is of **multiplicative type** if $D^p = D$, and **additive type** if $D^p = 0$.

Proposition 0.1.12. The correspondence $D \to X^D$ defines a bijection between the set of non-zero p-closed vector fields (defined up to a multiplicative constant) and finite purely inseparable morphisms of degree p of X onto a normal variety.

Let $x \in X$, we say that D vanishes at x (or, x is a zero of D) if $D(\mathcal{O}_{X,x}) \subset \mathfrak{m}_x$, where $\mathfrak{m}_x$ is the maximal ideal of the local ring $\mathcal{O}_{X,x}$. Let $\{U_i\}$ be an affine covering of X, U_i = Spec A_i .The A_i-modules

$$L_i = \{a \in A_i : a^{-1}D \in \Theta_X(U_i)\}$$

are glued together to form a sheaf of modules $\mathcal{I}$ which is an invertible sheaf of ideals if X is smooth. We can write

$$\mathcal{I} = \mathcal{O}_X(-R)$$

for a unique positive divisor R on X, which is called the **zero divisor** of D. If $f_i = 0$ is a local equation of R in U_i, then $D|U_i = f_iD_i$, where $D_i \in \Theta_X(U_i)$ has only isolated zeroes in U_i. In this way, we see that the the set Z of zeroes of D is the union of a divisor R and a finite set of points. We say that D has **isolated singularities** (resp. **no isolated singularities**) if R = 0 (resp. Z = R).

If x is a smooth point of X and t_1,t_2 are local coordinates at x, then D can be written locally in an affine neighborhood of x in the form

$$D = f_x(a_1\partial/\partial t_1)+a_2\partial/\partial t_2),$$

where $f_x = 0$ is a local equation of R at x. The **multiplicity** of D at x is defined as the number

$$\mathrm{mult}_x(D) = \dim \mathcal{O}_{X,x}/(a_1,a_2).$$

Proposition 0.1.13. Assume X is smooth. Then

$$c_2(X) = e(X) = \sum_{x\in X} \mathrm{mult}_x(D) \ -R\cdot K_X \ -R^2.$$

In particular, if R is smooth, then

$$e(X) = \sum_{x\in X} \mathrm{mult}_x(D) + e(R)$$

PROOF. It follows from the properties of Chern classes (**[Gro 1]**) that

$$c_2(\mathcal{I}\otimes\Theta_X) = \sum_{x\in X} \mathrm{mult}_x(D).$$

It remains to express the left hand side via the Chern classes of Θ_X and $\mathcal{I}$.

Proposition 0.1.14. Let $f:X \to Y$ be a finite inseparable morphism of smooth surfaces of degree p. Then, there exists a vector field D on X without isolated zeroes such that $Y \cong X^D$ and $f = \pi_D$. Moreover,

$$K_X = f^*(K_Y) + (p-1)R,$$

where R is the divisor of zeroes of D.

Corollary 0.1.2. Let $f: X \to Y$ be an inseparable double cover of smooth surfaces corresponding to a splittable admissible triple. Then $Y \cong X^D$ for some vector field of multiplicative type and the divisor of zeroes of D is equal to the ramification divisor of f.

Proposition 0.1.15. Let $\pi_D: X \to X^D$, and x be a smooth point of X. Then $\pi_D(x)$ is a smooth point of X^D if and only if x is not an isolated singularity of D.

Remark 0.1.1. Assume, for example, that $p = 2$ and $D = t_1\partial/\partial t_1 + t_2\partial/\partial t_2$ in a neighborhood of x. This happens if and only if $\mathrm{mult}_x(D) = 1$. Let $A \cong k[[t_1,t_2]] = \mathcal{O}_{X,x}$.we easily find that

$$A^D = k[[t_1^2,t_2^2,t_1t_2]] \cong k[[z,x,y]]/(z^2+xy) ,$$

i.e. $\pi_D(x)$ is an ordinary double point. In general, $\pi_D(x)$ is a normal singularity , which can be resolved by blowing up a tree of nonsingular rational curves. This is easy to see by resolving the indeterminacies of the birational map $\varphi \circ \pi_D$, where $\varphi: Y \to X^D$ is a resolution of the singularity $\pi_D(x)$.

Remark 0.1.2. Assume $p = 2$. Let D be a p-closed vector field on X. It defines a morphism

$$\mu^D: X \otimes_k k[\varepsilon] \to X$$

where $k[\varepsilon] = k[t]/(t^2)$, given locally by the formula $a \to a \otimes 1 + D(a) \otimes \varepsilon$. The scheme $G = \mathrm{Spec}\, k[\varepsilon]$ has the group law of the group scheme $\boldsymbol{\mu}_2$ or $\boldsymbol{\alpha}_2$. It is easily checked that μ^D defines the action of $\boldsymbol{\mu}_2$ if D is of multiplicative type and the action of $\boldsymbol{\alpha}_2$ if D is of additive type on X.

§2. Rational double points.

In this section omitted proofs can be found in **[Pi]** or **[Art 2]**. A **singularity** of a surface X is a closed point $x \in X$ such that the local ring $\mathcal{O}_{X,x}$ is non-regular. Two singularities are said to be (formally) isomorphic if the formal completions of their local rings are isomorphic. The isomorphism class of a singularity can be represented by an ideal in a formal power series ring $k[[t_1,\dots,t_n]]$.

A **resolution of the singularity** $x \in X$ is a birational morphism $\pi: Y \to X$, where Y is smooth and $x \in \pi(Y)$. A resolution is called **minimal** if it cannot be factored through another resolution.

If $x \in X$ is **normal** (i.e. $\mathcal{O}_{X,x}$ is normal), then π is an isomorphism over $U \setminus \{x\}$, where U is an open affine neighborhood of x. The set theoretical fibre $E = \pi^{-1}(x)_{red}$ is a connected curve on X. It is called the **exceptional curve** of the resolution π.

A resolution is minimal if and only if none of the irreducible components of E is an exceptional curve of the first kind (i.e. a smooth rational curve C with $C^2 = C \cdot K_Y = -1$).

Proposition 0.2.1. Let $R_1,\dots,R_n$ be some irreducible components of the exceptional curve E of a resolution π. Then the matrix $(R_i \cdot R_j)$ is negative definite.

Let $E = \Sigma R_i$ be the exceptional curve of a resolution of $x \in X$. A positive divisor $Z = \Sigma m_i R_i$ is called an **exceptional cycle**.

The **genus** of a normal singularity $x \in X$ is defined to be $\dim_k (R^1\pi_* \mathcal{O}_Y)_x$. If $\pi(Y)$ is affine and π is an isomorphism outside $\pi^{-1}(x)$, then the genus is equal to $\dim_k H^1(Y, \mathcal{O}_Y)$.

A singularity is said to be **rational** if its genus is equal to 0.

Proposition 0.2.2. Let $\pi: Y \to X$ be a resolution of a normal singularity $x \in X$. The following properties are equivalent:

(i) x is a rational singularity;

(ii) $H^1(Z, \mathcal{O}_Z) = 0$ for every exceptional cycle Z;

(iii) $p_a(Z) := 1 + \frac{1}{2} Z \cdot (Z + K_Y) \leq 0$ for every exceptional cycle Z;

(iv) the homomorphism $\mathrm{Pic}(Z) = H^1(Z, \mathcal{O}_Z^*) \to \mathbf{Z}^r, \mathcal{I} \to (\ldots, \deg(\mathcal{I} \otimes \mathcal{O}_{R_i}), \ldots)$ is bijective for every exceptional cycle $R = \Sigma m_i R_i$.

(v) the canonical maps $H^i(X, \mathcal{O}_X) \to H^i(Y, \mathcal{O}_Y)$ are bijective.

An exceptional cycle $Z = \Sigma m_i R_i$, $m_i > 0$, is called a **fundamental cycle** if

a) $Z \cdot E_i \leq 0$ for every $i = 1, \ldots, n$;

b) Z is minimal among all exceptional cycles satisfying a).

Proposition 0.2.3. A fundamental cycle Z exists and satisfies $p_a(Z) \geq 0$. A singularity is rational if and only if $p_a(Z) = 0$. In this case the multiplicity of (X,x) is equal to $-Z \cdot Z$.

Recall that the **multiplicity** of (X,x) is the multipicity of the maximal ideal of the local ring $\mathcal{O}_{X,x}$ **[A-M]**.

A rational singularity of multiplicity 2 is called a **rational double point**.

Proposition 0.2.4. Let $\pi: Y \to X$ be a minimal resolution of a singularity $x \in X$ and $E = R_1 + \ldots + R_n$ its exceptional curve. Then x is a rational double point if and only if $E_i \cong \mathbb{P}^1$, $R_i^2 = -2$, $i = 1, \ldots, n$.

PROOF. Suppose that x is a rational double point. Let $Z = \Sigma m_i R_i$ be the fundamental cycle. Then, by Proposition 0.2.3, $Z^2 = -2$, $p_a(Z) = 0$. Hence

$$Z \cdot K_Y = \Sigma k_i (R_i \cdot K_Y) = p_a(Z) - 2 - Z^2 = 0.$$

By Proposition 0.2.2, $H^1(R_i, \mathcal{O}_{R_i}) = 0$. Thus $R_i \cong \mathbb{P}^1$. Since π is a minimal resolution, $R_i^2 < -1$, hence, $p_a(R_i) = 0$ implies that $R_i \cdot K_Y \geq 0$. Together with the above equality, this implies

$$R_i^2 = -2,\ R_i \bullet K_Y = 0.$$

Conversely, if $R_i \cong \mathbb{P}^1$, $R_i^2 = -2$ for all i, then $R_i \bullet K_Y = 0$ for all i. Hence, $Z \bullet K_Y = 0$ for every exceptional cycle $Z = \Sigma m_i R_i$. By Proposition 0.2.1, $Z^2 < 0$, and, since Z^2 is even, $Z^2 \leq -2$. Thus $p_a(Z) = 1+\frac{1}{2}(Z \bullet K_Y + Z^2) \leq 0$. By Proposition 0.2.2, x is a rational singularity. Taking the fundamental cycle Z, we have $p_a(Z) = 0$. This implies that $Z^2 = -2$, i.e. x is a double point (Proposition 0.2.3).

Proposition 0.2.5. Let $E = \Sigma R_i$ be the exceptional curve of a minimal resolution of a rational double point. Then $R_i \bullet R_j = 1$, $i \neq j$, $R_i \cap R_j \cap R_k = \emptyset$, $i \neq j \neq k$. Define Γ to be the graph obtained by assigning a vertex to each R_i and joining two of them by an edge if $R_i \bullet R_j = 1$. Then Γ is one of the following graphs:

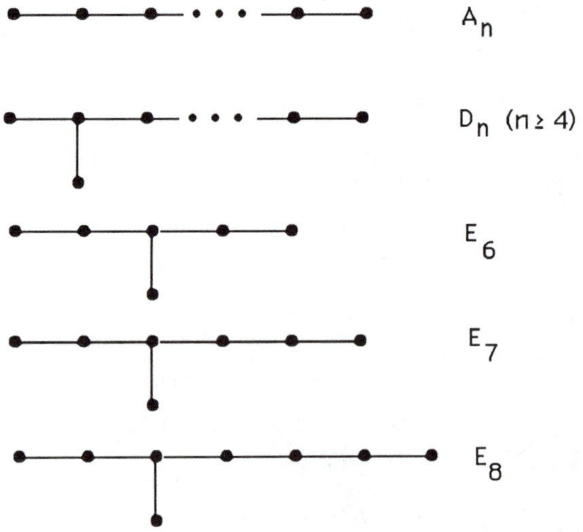

PROOF. The matrix $(R_i \bullet R_j)$ is negative definite and has -2 as diagonal entries. Since E is connected, it is also indecomposable (i.e. cannot be written as a non-trivial block-matrix). All such matrices are classified in the theory of Lie algebras (they are the Coxeter matrices of simple Lie algebras). The corresponding graph is the so-called **Dynkin graph** associated to the Coxeter matrix. See **[Bou 2]**, or Chapter 2.

Remark 0.2.1. Let M be the subgroup of Pic(Y) spanned by the classes α_i = $[R_i]$ of the irreducible components of the exceptional curve of a minimal resolution $\pi: Y \to X$ of a rational double point. The intersection form on Pic(Y) equips M with a structure of an integral lattice. The classes α_i form a root base in M of finite type (see Chapter II). The class of the fundamental cycle Z is a maximal root α_{max} with respect to this basis: for every root $\alpha_i = \Sigma n_i \alpha_i$, $n_i \geq 0$, in M, $\alpha_{max} - \alpha = \Sigma m_i \alpha_i$, where $m_i \geq 0$. We have

$$Z = R_1 + \ldots + R_n \text{ if } \Gamma \text{ is of type } A_n,$$

$$Z = R_1 + R_2 + 2R_3 + \ldots + 2R_{n-1} + R_n \text{ if } \Gamma \text{ is of type } D_n,$$

$$Z = 2R_1 + R_2 + 2R_3 + 3R_4 + 2R_5 + R_6 \text{ if } \Gamma \text{ is of type } E_6,$$

$$Z = 2R_1 + 2R_2 + 3R_3 + 4R_4 + 3R_5 + 2R_6 + R_7 \text{ if } \Gamma \text{ is of type } E_7,$$

$$Z = 3R_1 + 2R_2 + 4R_3 + 6R_4 + 5R_5 + 4R_6 + 3R_7 + 2R_8 \text{ if } \Gamma \text{ is of type } E_8.$$

Here we order $R_1, \ldots, R_n$ in such a way that $R_2, \ldots, R_n$ corresponds to the upper vertices ordered from the left to the right, R_1 corresponds to the most left vertex, if Γ is of type A_n, and to the lower vertex otherwise.

A rational double point will be called **a point of type A_n, D_n, E_6, E_7 or E_8** if the corresponding graph Γ is of the same type.

Proposition 0.2.6. Let x be a rational double point. Then x is isomorphic to one of the following singularities:

Type	Name	Equation
A_n	A_n	$z^{n+1} + xy = 0$, $n \geq 1$.
D_n	D_n	$z^2 + x(y^2 + x^n) = 0$, $n \geq 4$
E_6	E_6	$z^2 + x^3 + y^4 = 0$
E_7	E_7	$z^2 + xy^3 + x^3 = 0$

E_8	E_8	$z^2+x^3+y^5 = 0.$
	$\mathrm{char}(k) \neq 2, 3, 5$	
A_n	A_n	$z^{n+1}+xy = 0, \quad n\geq 1.$
D_{2n}	$D_{2n}^{(0)}$	$z^2+x^2y+xy^n = 0, \ n\geq 2$
	$D_{2n}^{(r)}$	$z^2+x^2y+xy^n+xy^{n-r}z=0, n\geq 2, \quad r=1,\dots,n-1$
D_{2n+1}	$D_{2n+1}^{(0)}$	$z^2+x^2y+zy^n=0, \quad n\geq 2$
	$D_{2n+1}^{(r)}$	$z^2+x^2y+zy^n+xy^{n-r}z=0, n\geq 2, \quad r=1,\dots,n-1$
E_6	$E_6^{(0)}$	$z^2+x^3+y^2z = 0$
	$E_6^{(1)}$	$z^2+x^3+y^2z+xyz = 0$
E_7	$E_7^{(0)}$	$z^2+xy^3+x^3 = 0$
	$E_7^{(1)}$	$z^2+xy^3+x^3+x^2yz = 0$
	$E_7^{(2)}$	$z^2+xy^3+x^3+y^3z = 0$
	$E_7^{(3)}$	$z^2+xy^3+x^3+xyz = 0$
E_8	$E_8^{(0)}$	$z^2+x^3+y^5 = 0$
	$E_8^{(1)}$	$z^2+x^3+y^5+xy^3z = 0$
	$E_8^{(2)}$	$z^2+x^3+y^5+xy^2z = 0$
	$E_8^{(3)}$	$z^2+x^3+y^5+y^3z = 0$
	$E_8^{(4)}$	$z^2+x^3+y^5+xyz = 0$
	$\mathrm{char}(k) = 2$	
A_n	A_n	$z^{n+1}+xy = 0, \quad n\geq 1.$
D_n	D_n	$z^2+x(y^2+x^n)=0, \quad n\geq 4$
E_6	$E_6^{(0)}$	$z^2+x^3+y^4 = 0$
	$E_6^{(1)}$	$z^2+x^3+y^4+x^2y^2 = 0$

E_7	$E_7^{(0)}$	$z^2+xy^3+x^3 = 0$
	$E_7^{(1)}$	$z^2+xy^3+x^3+x^2y^2 = 0$
E_8	$E_8^{(0)}$	$z^2+x^3+y^5 = 0$
	$E_8^{(1)}$	$z^2+x^3+y^5+x^2y^3 = 0$
	$E_8^{(2)}$	$z^2+x^3+y^5+x^2y^2 = 0$
	char(k) = 3	
A_n	A_n	$z^{n+1}+xy = 0$, $n\geq 1$.
D_n	D_n	$z^2+x(y^2+x^n)=0$, $n\geq 4$
E_6	E_6	$z^2+x^3+y^4 = 0$
E_7	E_7	$z^2+xy^3+x^3 = 0$
E_8	$E_8^{(0)}$	$z^2+x^3+y^5 = 0$
	$E_8^{(1)}$	$z^2+x^3+y^5+xy^4 = 0$
	char(k) = 5	

PROOF. See **[Art 7]**.

Remark 0.2.2 . We see from the list above that all rational double points can be realized as the singularities of double covers of smooth surfaces. In the case A_n and char(k) = 2, where this is not obvious, we take an isomorphic singularity given by an equation:

$$z^2+xz+y^{n+1} = 0.$$

In fact, every normal double separable cover in characteristic 2 with smooth branch locus has singularities only of type A_n.

Let $f(x,y)\in k[[x,y]]$ and $f(x,y) = 0$ be the corresponding one-dimensional singularity. Assume char(k) $\neq$ 2. We will call it a **simple singularity** if $z^2+f(x,y) = 0$ is a rational double point. It is said to be of type a_n, d_n, e_n ($e_n^{(r)}$) if the singularity $z^2+f(x,y) = 0$ is of type A_n, D_n, E_n ($E_n^{(r)}$) respectively.

Let Y be a nonsingular surface and $f:X \to Y$ be a double cover of X. Assume that $char(K) \neq 2$. Let $x \in X$ be a point of the ramification divisor R. Then x is a rational double point of X if and only if f(x) is a simple singularity of the branch curve .

A singular point of type A_1 is also called an **ordinary double point** (or a **node**). The exceptional curve of its minimal resolution consists of an irreducible smooth rational curve R with $R^2 = -2$. Such a curve will be called a **nodal curve**.

Proposition 0.2.7. Let $f: X \to Y$ be an inseparable double cover of a smooth surface Y and $\alpha(f)$ be the section of $\mathcal{L}^{-1} \otimes \Omega_Y^{\ 1}$ defined in Proposition 0.1.2. Assume that $\alpha(f)$ has only simple isolated zeroes. Then all singularities of X are ordinary double points.

PROOF. The assertion is local; we may assume that X is given by an equation $z^2+f(x,y) = 0$. It follows from the definition of $\alpha(f)$ that a point P of X is singular if and only if f(P) is a zero of the differential form $df(x,y) = f_x'dx+f_y'dy$. Adding a constant to z, we may assume that $f(P) = (0,0)$. The zero of $\alpha(f)$ at f(P) is simple if the ideal generated by f_x' and f_y' is equal to the maximal ideal $\mathfrak{m}_{Y,f(x)}$. Changing local parameters at f(x), we may assume that $f_x' = y$, $f_y' = x$ and $f = xy + g^2$ for some regular function g on Y. This shows that our singularity is formally isomorphic to the singularity $z^2+xy = 0$.

Proposition 0.2.8. A rational singularity $x \in X$ is Gorenstein if and only it is a rational double point. In this case, for any minimal resolution $\pi:Y \to X$:

$$\pi^*(\omega_X) \cong \omega_Y.$$

Proposition 0.2.9. Let $p: Y \to X$ be a finite surjective morphism onto a surface X whose singularities are rational double points. Let $\pi:X' \to X$ be a minimal resolution of Y, $Y' = (Y \times_X X')_{red}$, $\pi':Y' \to Y$, $p':Y' \to X'$ the corresponding projections. Then

(i) p' is finite,

(ii) $p'_*(\mathcal{O}_{Y'})$ is locally free on X',

(iii) $\pi'_*(\mathcal{O}_{Y'}) = \mathcal{O}_Y$, $R^1\pi'_*(\mathcal{O}_{Y'})=0$.

PROOF. See **[GS-V]**.

Let (X,x) be a normal surface singularity, $A = \tilde{\mathcal{O}}_{X,x}$ be the henselization of its local ring, $Z = \operatorname{Spec} A$, $U = Z\setminus\{x\}$, $j: U \hookrightarrow Z$ is the natural inclusion. Let G be a finite group scheme over κ and $f: Y \to U$ a principal G-cover of U. Let $\mathfrak{a} = f_*\mathcal{O}_Y$; then $j_*\mathfrak{a} = B^\sim$ for some finite henselian local A-algebra B. Let $Z' = \operatorname{Spec} B$; the canonical map $\bar{f}: Y \to Z$ extends f to a finite morphism. Let (Y,y) be a singularity such that $\tilde{\mathcal{O}}_{Y,y} = B$. We will call it the **principal local G-cover** of the singularity of (X,x). Such a cover is called **maximal** if its restriction over U cannot be factored non-trivially through a principal G'-cover of U, where G' is a factor group of G.

The next propositions can be derived from **[Art 7]**.

Proposition 0.2.10. Assume $\operatorname{char}(\kappa) = 0$ (or > 5), then every rational double singularity admits a principal unramified G-cover which is a nonsingular point. Moreover, G is isomorphic to a subgroup of $SL(2,\kappa)$ which is a cyclic group of order $n+1$ if (X,x) is of type A_n, a binary dihedral group of order $4n$ if (X,x) is of type D_n, binary tetrahedral group of order 24 if (X,x) is of type E_6, a binary octahedral group of order 48 if (X,x) is of type E_7 and a binary icosahedral group of order 120 if (X,x) is of type E_8.

Proposition 0.2.11. Let (X,x) be a singularity of type A_n. Then there exists a maximal principal G-cover of (X,x) which is a nonsingular point and $G \cong \mu_{n+1}$. In particular, it is unramified over U if $(n+1, \operatorname{char}(\kappa)) = 1$ and purely inseparable if $n+1 = \operatorname{char}(\kappa)$.

PROOF. Define the action of μ_{n+1} on $\operatorname{Spec} \kappa[[u,v]]$ by

$$u \to u \otimes t,\ v \to v \otimes t^n,$$

where μ_{n+1} = Spec $k[t]/(t^{n+1}-1)$ and check that the ring of invariants is generate by the elements $x= u^{n+1}, y = v^{n+1}$ and $z = uv$, satisfying the relation

$$z^{n+1} = xy.$$

That is,the quotient is a singularity of type A_n. It is easy to see that the action of μ_{n+1} on Spec $k[[u,v]]\setminus(u,v)$ is free.

Proposition 0.2.12. Assume char(k) = 2 . Let (X,x) be a singularity of type D_4 . Then there exists a maximal principal G-cover of (X,x) which is a nonsingular point and $G \cong \alpha_2$ (resp. $\mathbb{Z}/2$) if (X,x) is of type $D_4^{(0)}$ (resp. $D_4^{(1)}$).

PROOF. Recall (Proposition 0.2.5) that (X,x) is isomorphic to a singularity

$$z^2+x^2y+xy^2 = 0 \text{ (type } D_2^{(0)}) \text{ or } z^2+xyz+x^2y+xy^2 = 0 \text{ (type } D_2^{(1)}).$$

Assume we are in the first case. Let $U_0 = U\setminus\{x = 0\}$, $U_1 = U\setminus\{y = 0\}$. Consider the cocycle $a = \{a_{01}\} = \{z/xy\}$ of $\mathcal{O}_U$ in the covering $U = U_0 \cup U_1$. We have

$$a^2 = z^2/x^2y^2 = x^{-1}-y^{-1},$$

this shows that a^2 is trivial. Now, we can define an α_2-cover of U locally by

$$u_0^2 = x^{-1},\ u_1^2 = y^{-1},\ u_0 = u_1 + a_{01}.$$

Setting

$$v_0 = zu_0^2+u_0y,\ v_1 = zu_1^2+u_1y,$$

we see that

$$y = v_0^2,\ z = v_0u_0^{-2}+u_0^{-1}v_0^2,\ x = u_0^{-1} \quad \text{over } U_0,$$

$$x = v_1^2,\ z = v_1u_1^{-2}+u_1^{-1}v_1^2,\ y = u_1^{-1} \quad \text{over } U_1.$$

Thus our singularity has a principal α_2-cover which is a nonsingular point.

Assume we are in the second case. In the above notation, we have

$$a^2+a = x^{-1} - y^{-1}.$$

This defines a $\mathbb{Z}/2$-cover which is a nonsingular point, given locally by

$$u_0^2+u_0 = x^{-1},\ u_1^2+u_1 = y^{-1},\ u_0 = u_1 + a_{01}.$$

Proposition 0.2.13. Assume char(k) = 2 . Let (X,x) be a singularity of type D_5 . Then there exists a maximal principal G-cover of (X,x) which is a singular point of type A_1 . Moreover, $G \cong \boldsymbol{\alpha}_2$ (resp. $\mathbf{Z}/2$) if (X,x) is of type $D_5^{(0)}$(resp. $D_5^{(1)}$).

PROOF. Recall (Proposition 0.2.5) that (X,x) is isomorphic to a singularity

$$z^2+x^2y+zy^2 = 0 \text{ (type } D_5^{(0)}) \text{ or } z^2+xyz+x^2y+zy^2 = 0 \text{ (type } D_5^{(1)}).$$

Assume we are in the first case. Making the substitutions

$$y = u^2,\ z = uv,\ x = v+uw,\ w^2 = uv,$$

we see that our singularity is doubly covered by a point of type A_1. It defines a principal $\boldsymbol{\alpha}_2$-cover of U, the corresponding action of $\boldsymbol{\alpha}_2$ is given by

$$u \to u\otimes 1+v\otimes t, v \to v\otimes 1,\ w \to w\otimes 1+v\otimes t,$$

where $\boldsymbol{\alpha}_2$ = Spec $k[t]/(t^2)$.

Assume we are in the second case. Let (Y,y) be the ordinary double point given by the equation $w^2 = uv$. We define the action of the group scheme $G = \mathbf{Z}/2$ on on (Y,y) as the induced action from the following action on $k[[u,v,w]]$:

$$u \to \frac{u}{1+w+v},\quad v \to \frac{u+v}{1+w+v},\quad w \to \frac{w+u}{1+w+v}.$$

It is immediately checked that this action is free outside the locus u = 0, w+v = 0. Hence it defines an action on (Y,y) which is free outside the origin. Making the substitutions

$$x = \frac{u+vw+uv}{1+w+v},\quad y = \frac{v^2+w^2}{1+w+v},\quad z = \frac{u(v+w)}{1+w+v},$$

we see that (X,x) is isomorphic to the quotient of (Y,y) by the action above. The corresponding projection (Y,y) → (X,x) induces a principal $(\mathbf{Z}/2)$-cover of the complement of the origin. This proves the proposition.

S3. Del Pezzo surfaces.

Recall that a subvariety $X \subset \mathbb{P}^n$ is said to be **non-degenerate** if it is not contained in a proper linear subspace of $\mathbb{P}^n$.

Proposition 0.3.1. Let X be a non-degenerate subvariety of $\mathbb{P}^n$. Then

$$\deg(X) \geq \operatorname{codim} X + 1.$$

PROOF. This is well-known, see for example **[G-H 1]** p.173 or **[Nag]**.

Proposition 0.3.2. Let X be a non-degenerate surface of degree n in $\mathbb{P}^{n+1}$. Then X is isomorphic to one of the following surfaces:

(i) $\mathbb{P}^2$;

(ii) Veronese surface in $\mathbb{P}^5$;

(iii) a rational scroll.

PROOF. Again this is well-known. See ,for example,**[G-H 1]**,p. 525 (p=0) or **[Nag]** ($p \geq 0$).

Recall that a **rational scroll** $S_{n,k}$ is the image of the surface $\mathbb{F}_n = \mathbb{P}(\mathcal{O}_{\mathbb{P}^1} \oplus \mathcal{O}_{\mathbb{P}^1}(n))$ under the map $\mathbb{F}_n \to \mathbb{P}^{n+2k+1}$ given by the linear system $|s+(n+k)f|$, where $k \geq 0$ and

$$\operatorname{Pic}(\mathbb{F}_n) \cong \mathbf{Z}s + \mathbf{Z}f \ , \ s \cdot s = -n \ , \ f \cdot f = 0 \ , \ s \cdot f = 1.$$

Proposition 0.3.3. Let X be a non-degenerate surface of degree n in $\mathbb{P}^n$. Then X is one of the following surfaces:

(i) a projection of a surface of degree n in $\mathbb{P}^{n+1}$;

(ii) a cone over a normal elliptic curve in a hyperplane of $\mathbb{P}^n$;

(iii) an anticanonical Del Pezzo surface;

(iv) a quadric $Q \subset \mathbb{P}^3$ reembedded into $\mathbb{P}^8$ by $|\mathcal{O}_Q(2)|$.

PROOF. Again, this is well-known. See, for example,**[Nag]**.

Recall that a **Del Pezzo surface** of degree n ($1 \le n \le 8$) is a surface $X(\Sigma)$ obtained by blowing up a set Σ of 9-n points (maybe infinitely near) of $\mathbb{P}^2$ satisfying the following conditions:

(i) no more than 3 points lie on a line;

(ii) no more than 6 points lie on a conic;

The anticanonical system $|-K_{X(\Sigma)}|$ of $X(\Sigma)$ is the proper transform of the linear system of plane cubic curves passing through Σ. The above conditions guarantee that $|-K_{X(\Sigma)}|$ does not have fixed components and defines a rational map to the space $\mathbb{P}^n$. Its image will be called an **anticanonical Del Pezzo surface**.

Remark 0.3.1. An anticanonical Del Pezzo surface is a two-dimensional **Fano variety**, i.e. a normal variety X with ample anticanonical class $-K_X$. Conversely any two-dimensional Fano variety is isomorphic to an anticanonical Del Pezzo surface or a quadric in $\mathbb{P}^3$.

The proof of the next propositions can be found in **[Dem]** or **[Ma]** (some additional details in the case where char(k) = 2 can be easily reconstructed by using the results from §1).

Proposition 0.3.4. Let $X(\Sigma)$ be a Del Pezzo surface of degree $n \ge 3$. The anticanonical map is a birational morphism onto a normal surface of degree 9-n in $\mathbb{P}^{9-n}$ with at most rational double points as singularities. It is an isomorphism outside the union of the nodal curves on $X(\Sigma)$. Each connected component of this union blows down to a rational double point of X.

Proposition 0.3.5. Let $X(\Sigma)$ be a Del Pezzo surface of degree n = 2. The anticanonical map f is a morphism of degree 2 onto $\mathbb{P}^2$. It factors into a birational morphism $\varphi: X(\Sigma) \to X'$ which is an isomorphism outside the set of nodal curves on $X(\Sigma)$ and a double cover $f': X' \to \mathbb{P}^2$. If $char(k) \ne 2$, the branch curve of f' is a plane curve of degree 4 with simple singularities, the images of the connected components of the set of nodal curves on $X(\Sigma)$. If char(k) = 2, f' is given by an splittable admissible triple $(V,\mathcal{L},s)$, where $\mathcal{L} \cong \mathcal{O}_{\mathbb{P}^2}(-2)$, $V = \mathcal{O}_{\mathbb{P}^2}(4)\oplus\mathcal{O}_{\mathbb{P}^2}(2)\oplus\mathcal{O}_{\mathbb{P}^2}$, $s = (b,a,1)$. If f' is separable, div(b) is a plane quartic, div(a) is a conic. If f' is inseparable, a = 0, div(b) is a plane quartic. In the latter case φ is not an isomorphism.

Proposition 0.3.6. Let $X(\Sigma)$ be a Del Pezzo surface of degree n = 1. The anticanonical linear system is a pencil of curves of arithmetic genus 1 with one base point x_0. The bi-anticanonical map f is a morphism of degree 2 onto a quadric S of rank 3 in $\mathbb{P}^3$. It factors into a birational morphism $\varphi: X(\Sigma) \to X'$ which is an isomorphism outside the set of nodal curves on $X(\Sigma)$ and a double cover $f': X' \to S$. The point x_0 is the preimage of the singular point of S. Let $\pi:\bar{X}' \to X'$ be the blowing up the point $\varphi(x_0)$, and $\pi': S' \cong \mathbb{F}_2 \to S$ be a minimal resolution of S .Then there exists a double cover $\bar{f}': \bar{X}' \to \mathbb{F}_2$, such that $\pi'\circ\bar{f}' = f'\circ\pi$. If char(k) $\neq$ 2, then the branch curve of f' is the union of the proper transform C of a curve of degree 6 cut out by a cubic surface in $\mathbb{P}^3$ not passing through the node of S and the exceptional curve E of π'. The singularities of C are simple. If char(k) = 2, f is given by an admissible triple $(V,\mathcal{L},s)$, where $\mathcal{L} \cong \mathcal{O}_{\mathbb{F}_2}(-3\ell-2E))$, ℓ is a fibre of the canonical projection $\mathbb{F}_2 \to \mathbb{P}^1$, $V = \mathcal{L}^{-2}\oplus\mathcal{L}^{-1}\oplus\mathcal{O}_{\mathbb{F}_2}$, $s = (b,a,1)$. If f is separable , div(a) is the proper transform of a cubic curve on S passing through the node of s and the curve E. If f is inseparable, a = 0 and div(b) is as in the case char(k) $\neq$ 2.

Remark 0.3.2. One can interpret the two previous propositions as follows: The graded anticanonical ring $A = \oplus_{m\geq 0}\ \Gamma(X(\Sigma),\mathcal{O}_{X(\Sigma)}(-mK_{X(\Sigma)}))$ is finitely generated by 4 elements. If n = 2, one can choose 3 generators x_0,x_1,x_2 of degree 1 and one element x_3 of degree 2 .If n = 1,then one can choose 2 elements x_0,x_1 of degree 1, one element x_2 of degree 2 and one element x_3 of degree 3. They satisfy a relation of the form $x_3^2 = f(x_0,x_1,x_2)$ (resp. $x_3^2 = f(x_0,x_1,x_2)$), where f is a homogeneous polynomial of degree 4 (resp. a quasi-homogeneous polynomial of degree 6 . The anticanonical model X of $X(\Sigma)$ is isomorphic to Proj(A), which is embedded naturally into the weighted projective space $\mathbb{P}(1,1,1,2)$ (resp. $\mathbb{P}(1,1,2,3)$) as a hypersurface of degree 4 (resp. 6).

Proposition 0.3.7. Let $X(\Sigma)$ be a Del Pezzo surface of degree $n \geq 1$. Each nodal curve on $X(\Sigma)$ is the proper transform of one of the following curves on $\mathbb{P}^2$:

(i)$\pi^{-1}(p_i)-\pi^{-1}(p_j)$, where p_i is infinitely near to p_j of order 1;

(ii) a line passing through 3 points in Σ;

(iii) a conic passing through 6 points in Σ;

(iv)($n = 1$) a cubic which passes through all points of Σ and has a double point at one of them.

Proposition 0.3.8. Let $f:X(\Sigma) \to X \subset \mathbb{P}^n$ be the anticanonical map of a Del Pezzo surface of degree $n \geq 3$. Then a line on X (resp. a conic) is equal to the image under f of the proper transform under $\pi:X(\Sigma) \to \mathbb{P}^2$ of one of the following curves:

(i) $\pi^{-1}(p_i)\setminus$ nodal components, where $p_i \in \Sigma$;

(ii) a line passing through exactly 2 points of Σ;

(iii) a conic passing through exactly 5 points of Σ

(resp.

(i) a line passing through exactly one point of Σ;

(ii) a conic passing through exactly 4 points of Σ).

Note that every conic on X moves in a pencil.

Corollary 0.3.1 . Let X be a anticanonical model of a Del Pezzo surface $X(\Sigma)$ of degree 4. The following table gives the number of lines, pencils of conics, and the type of the singularities on X

Lines	Pencils of conics	Singularities
16	10	∅
12	8	A_1
9	6	A_1+A_1
8	7	A_1+A_1
8	6	A_2
6	5	$A_1+A_1+A_1$
6	4	A_2+A_1
5	4	A_3
4	5	A_3
4	4	$A_1+A_1+A_1+A_1$
4	3	$A_1+A_1+A_2$
3	3	A_1+A_3
3	2	A_4
2	3	D_4
2	2	$A_1+A_1+A_3$
1	1	D_5

PROOF. This is a classical result due to Du Val **[DuV]** and Timms **[Ti]**(cf. also **[Seg]**). One can prove it by analyzing all possible configurations of the sets Σ and applying the previous proposition. From the point of view of lattices (see Chapter 2) this classification is equivalent to the classification of root bases in the lattice D_5 or subgroups of the Coxeter group of type D_5 (see **[Cox]**).

§4. Symmetric quartic Del Pezzo surfaces.

A Del Pezzo surface X is called **symmetric** if there exists a double cover f: Y → X which is a principal cover over the complement of the union of nodal curves on X. We say such X is of type μ_2, α_2 or $\mathbf{Z}/2$ if the principal cover is of the respective type. The latter two cases occur only in the case where char(k) = 2.

An anticanonical model of a symmetric Del Pezzo surface is said to be a **symmetric anticanonical Del Pezzo surface**. It is easy to see that every principal double cover of the complement of a set of nodal curves on a Del Pezzo surface induces a principal double cover of the complement of the set of singular points of its anticanonical model. The converse is also true.

In this section we will classify symmetric Del Pezzo surfaces of degree 4.

Proposition 0.4.1. A Del Pezzo surface X of degree 4 is symmetric if and only the singular locus of its anticanonical model is one of the following:

<u>Type μ_2</u>: 4 points of type A_1 or 2 points of type A_1 and one point of type A_3;

<u>Type $\mathbf{Z}/2$</u>: one point of type $D_4^{(1)}$ or $D(_5,^{(1)})$(char(k) = 2);

<u>Type α_2</u>: one point of type $D_4^{(0)}$ or $D_5^{(0)}$ (char(k) = 2).

PROOF. Suppose X is symmetric. Let E be the union of the nodal curves on X.

Assume that X is of type μ_2. By Proposition 0.1.8, there exists a reduced curve $W \subset E$ such that its class in Pic(X) is divisible by 2. Since $K_X \cdot E = K_X \cdot W = 0$, $W \sim 2D$, where $K_X \cdot D = 0$. By adjunction, D^2 is even. Thus, $W^2 = 4D^2$ is divisible by 8. By analyzing the table from Corollary 0.3.1, we see that this is possible only if W is the union of 4 disjoint nodal curves contained in E. Moreover, either E = W or E is the union of two disjoint nodal curves and a disjoint exceptional cycle of type A_3. The anticanonical model of X will have the singularities as stated in the proposition.

Assume that X is of type α_2 or $\mathbf{Z}/2$. Let U = X\E. We know that $H^1(U,G) \neq$

0, where $G = \alpha_2$ or $\mathbb{Z}/2$. Since $H^i(X,\mathcal{O}_X) = 0$, $i = 1,2$, we have $H^i(X,G) = 0$ $i = 1,2$ (see §1). The local cohomology sequence

$$H^1(X,G) \to H^1(U,G) \to H^2_E(X,G) \to H^2(X,G)$$

defines an isomorphism

$$H^1(U,G) \to H_E{}^2(X,G).$$

Let X' be an anticanonical model of X, $\mathrm{Sing}(X') = \{x_1,\dots,x_r\}$, $\hat{\mathcal{O}}_{X,x_i}$ be the formal completion of the local ring of the point x_i, $\hat{X}_i = \mathrm{Spec}\ \hat{\mathcal{O}}_{X,x}$, $\hat{U}_i = \hat{X}_i \setminus \{x_i\}$. It is easy to see that

$$H^2_E(X,G) \cong \oplus_i H^1_{\{x_i\}}(X',G)$$

The local cohomology sequence for $\hat{X}_i$ gives an isomorphism

$$H^1_{\{x_i\}}(X,G) \cong H^1_{\{x_i\}}(\hat{X}_i,G) \cong H^1(\hat{U}_i,G).$$

It shows that $H^1(\hat{U}_i,G) \neq 0$ for some point x_i. Thus the singularity x_i admits a local principal G-cover. It follows from Proposition 0.2.9 that it cannot be a point of type A_n. A glance at the table of Corollary 0.3.1 shows that Sing(X) consists of one point of type D_4 or D_5. By Proposition 0.2.10, this point is of type $D_4^{(0)}$ or $D_5^{(0)}$ if $G = \alpha_2$ and of type $D_4^{(1)}$ or $D_4^{(1)}$ if $G = \mathbb{Z}/2$.

Let X be an irreducible quartic surface in $\mathbb{P}^4$ not contained in a hyperplane. By Proposition 0.3.3, X is one of the following surfaces:

(i) A Del Pezzo surface of degree 4;

(ii) A cone over a quartic curve;

(iii) A projection of a Veronese surface;

(iv) A projection of a scroll $S_{0,2}$ or $S_{2,1}$ in $\mathbb{P}^5$.

Lemma 0.4.1. Let $\mathfrak{a}_X$ be the linear system of quadrics in $\mathbb{P}^4$ containing X. Then

$$\dim \mathfrak{a}_X \le 1.$$

The equality holds if X is a Del Pezzo surface.

PROOF. Suppose $\dim \mathfrak{a}_X \ge 2$. Then X is contained in a quartic surface $Q_1 \cap Q_2$, where Q_1 and Q_2 are two linear independent quadrics from $\mathcal{Q}_X$. Since deg(X) = 4, $X = Q_1 \cap Q_2$ and $\mathcal{Q}_X$ is spanned by Q_1 and Q_2. This proves the first assertion.

Let $\mathcal{Q} = |\mathcal{O}_{\mathbb{P}^4}(2)|$ be the complete linear system of quadrics in $\mathbb{P}^4$. By restriction, we have a map

$$r: H^0(\mathbb{P}^4, \mathcal{O}_{\mathbb{P}^4}(2)) \to H^0(X, \mathcal{O}_X(2)).$$

Clearly, $\mathcal{Q}_X = \mathbb{P}(\mathrm{Ker}(r))$. Suppose X is a Del Pezzo surface. Then $|\mathcal{O}_X(2)|$ corresponds to the linear system of plane sextics passing twice through the points p_i whose blow up defines a minimal nonsingular model of X. Counting constants (or, applying Riemann-Roch), we see that it is of dimension 12. Since $\dim |\mathcal{O}_{\mathbb{P}^4}(2)| = 14$, we obtain that $\dim \mathcal{Q}_X \geq 1$.

Lemma 0.4.2. Let X be a quartic anticanonical Del Pezzo surface in $\mathbb{P}^4$. Suppose that the pencil of quadrics containing X is spanned by two quadrics of rank 3. Then X is isomorphic to one of the following surfaces:

(i) $x_0x_1 + x_2^2 = x_3x_4 + x_2^2 = 0$;

(ii) $x_0x_1 + x_2^2 = x_0^2 + x_3x_4 = 0$.

Moreover, X has 4 nodes in case (i) and 2 nodes and a point of type A_3 in case (ii).

PROOF. Let Q_1 and Q_2 be two quadrics of rank 3 containing X. Diagonalizing the equation of Q_1, we may assume that

$$Q_1: x_0x_1 + x_2^2 = 0.$$

Clearly, one can write the equation of Q_2 as follows:

$$Q_2: x_3h(x_0,x_1,x_2) + x_4h'(x_0,x_1,x_2) + q(x_0,x_1,x_2) + q'(x_3,x_4),$$

where h and h' (resp. q and q') are linear (resp. quadratic) forms. By a linear change of variables, we may reduce q' to the form $q' = x_3x_4$, x_3^2 or 0.

Assume that $q' = x_3x_4$. Then, replacing x_4 by $x_4 + h$ and x_3 by $x_3 + h'$, we may assume that $h = h' = 0$. Since $\mathrm{rk}(Q_2) = 3$, $\mathrm{rk}(q) = 1$. Hence $q = (ax_0+bx_1+cx_2)^2 \neq 0$. If $ab + c^2 \neq 0$, we may assume that $ab+c^2 = 1$. Then, by an orthogonal (with respect to Q_1) transformation, we reduce Q_2 to the form $x_2^2+x_3x_4$ to obtain case (i). If $ab+c^2 = 0$, we reduce Q_2 to the form $x_3x_4+x_0^2$ to obtain case (ii).

Assume that $q' = x_3^2$. If $p \neq 2$, then, replacing x_3 by $x_3 + \frac{1}{2}h$, we are reduced to the case where $h = 0$. If $\mathrm{char}(\kappa) = 2$, then h is proportional to h' (otherwise Q_2 can be reduced to the form $x_3x_0 + x_4x_1 + q + x_3^2$, and $\mathrm{rk}(Q_2) > 3$). Let $h = \lambda h'$. Then, replacing x_4 by $x_3 + \lambda x_4$, we may again assume that $h = 0$. Thus

$$Q_2 = x_4h' + q(x_0,x_1,x_2) + x_3^2.$$

In this case, $h' \neq 0$, otherwise X is a cone. Then, as above, we transform orthogonally x_0,x_1,x_2 to obtain $h' = x_0$ or x_2. We can write

$$Q_2 = x_4x_i + x_i l(x_0,x_1,x_2) + q' + x_3^2,$$

where l is a linear form in x_0,x_1 and x_2 and q' is a quadratic form in x_j, $j = 0,1,2, j \neq i$. Replacing x_4 by $x_4 + l$, we reduce Q_2 to the form $x_4x_i + q' + x_3^2$. It is immediately checked that $\mathrm{rk}(Q_2) = 3$ only if $q' = 0$. If

$$Q_1 = x_0x_1 + x_2^2, \; Q_2 = x_4x_0 + x_3^2,$$

then $x_0 = x_2 = x_3 = 0$ is a singular line of X. So, this case is impossible. If

$$Q_1 = x_0x_1 + x_2^2, \quad Q_2 = x_4x_2 + x_3^2,$$

then up to a permutation of the x_i's, we get (ii).

Finally, assume that $q' = 0$. Since X is not a cone, one of the linear forms h and h', say h', is not zero. By an orthogonal change of variables, we may assume that $h' = x_0$ or x_2. Thus, we can write the equation of Q_2 as follows:

$$Q_2\colon x_ix_4 + x_i l(x_0,x_1,x_2,x_3) + q(x_1,x_2) = 0 \quad (i = 0 \text{ or } 2),$$

where l is a linear form in x_0,x_1,x_2,x_3, q is a quadratic form. Replacing x_4 by $x_4 + l(x_0,x_1,x_2,x_3)$, we see that X is a cone. This shows that this case is impossible.

The type of the singularities is checked directly by using Proposition 0.2.6.

A surface X in $\mathbb{P}^4$ which in some coordinate system is given by the equations

$$x_0x_1 + x_2^2 = x_3x_4 + x_2^2 = 0;$$

$$(\text{resp. } x_0x_1 + x_2^2 = x_0^2 + x_3x_4 = 0).$$

is called a **4-nodal quartic** (resp. **degenerate 4-nodal quartic**) and is denoted by $\mathcal{D}_1$ (resp. $\mathcal{D}_1'$) in $\mathbb{P}^4$.

Proposition 0.4.2. Let Q be a nonsingular quadric in $\mathbb{P}^3$ given by the equation $t_0t_1+t_2t_3 = 0$. Let $\mu_2 = \mathrm{Spec}\ k[t]/(t^2-1)$ acts on Q by the formula

$$(t_0,t_1,t_2,t_3) \to (t_0\otimes\varepsilon, t_1\otimes\varepsilon, t_2, t_3), \quad \varepsilon = t \bmod (t^2-1).$$

Then the quotient Q/μ_2 is isomorphic to a 4-nodal quartic surface X. Moreover, the restriction of the factor projection over $X\setminus\mathrm{Sing}(X)$ is a principal μ_2-cover.

PROOF. The linear system of quadrics in $\mathbb{P}^3$ spanned by the quadrics

$$t_0^2 = 0,\ t_1^2 = 0,\ t_0t_1 = 0,\ t_2^2 = 0,\ t_3^2 = 0$$

is μ_2-invariant and has no base points. Its restriction to Q defines a morphism of degree 2 onto a quartic surface in $\mathbb{P}^4$ which factors through Q/μ_2. It is easy to see that the image satisfies the equations:

$$x_2^2 = x_0x_1, \quad x_2^2+x_3x_4 = 0,$$

where x_0,x_1,x_2,x_3,x_4 correspond to the above quadrics taken in the same order.

It is checked directly that μ_2 acts freely except at 4 points (0,0,1,0), (0,0,0,1), (1,0,0,0) and (0,1,0,0). Their images on Q/μ_2 are the four nodes of $\mathcal{D}_1$.

Proposition 0.4.3. Let X be a nondegenerate surface in $\mathbb{P}^4$, X' its minimal resolution. Let $\mathfrak{a}_X$ be the linear system of quadrics containing X. The following properties are equivalent:

(i) $\mathfrak{a}_X$ is a pencil spanned by two quadrics of rank 3 and contains a quadric of rank 4;

(ii) X is a 4-nodal quartic;

(iii) X is a symmetric quartic Del Pezzo surface of type μ_2 with 4 nodes;

(iv) X is a surface of degree 4 with 4 nodes;

(v) X is an anticanonical quartic Del Pezzo surface with 4 ordinary double points;

(vi) X has exactly 4 lines and 4 pencils of conics;

(vii) X' is a Del Pezzo surface of degree 4 obtained from $\mathbb{P}^2$ by blowing up 5 points $p_1,\dots,p_5$ such that p_3 (resp. p_5) is infinitely near to p_2 (resp. to p_4) and the line joining p_1 and p_2(resp. p_1 and p_4) passes through p_3 (resp. p_5).

PROOF.(i) ⇒ (ii) Follows immediately from lemma 0.4.2 and the definition.

(ii) ⇒ (iii). Follows from Proposition 0.4.2.

(iii) ⇒ (iv) Obvious.

(iv) ⇔ (v)If X is not an anticanonical Del Pezzo surface,then its minimal nonsingular model is either a minimal ruled surface, or $\mathbb{P}^2$. Neither of them contains 4 nodal curves coming from the resolution of X.

(v) ⇒ (vi) Follows from Corollary 0.3.1.

(vi) ⇒ (v) We apply Proposition 0.3.4. Since ruled surfaces and cones contain infinitely many lines, and a projection of the Veronese surface contains infinitely many pencils of conics, X must be an anticanonical Del Pezzo surface. By Corollary 0.3.1, it has 4 nodes.

(v) ⇒ (vii) It is easy to see that the only way (up to Cremona transformations of the plane) to obtain a Del Pezzo surface X(Σ) of degree 4 with exactly 4 disjoint nodal curves is to take the set Σ as described in (v).

(vii) ⇔ (i) By lemma 0.4.1, $\mathfrak{a}_X$ is a pencil . We know that hyperplane sections of X correspond to cubic curves in $\mathbb{P}^2$ passing through the points $p_1,\dots,p_5$. Let H_1 be the hyperplane corresponding to the union of the three lines p_1p_2, p_1p_4 and p_2p_4. Let H_2 be the hyperplane corresponding to the line p_1p_2 taken with multiplcity 2 and the line p_1p_4. Let H_3 be the hyperplane corresponding to the union of the line p_2p_3 taken with multiplicity 2 and the line p_1p_4. The quadrics $2H_1$ and H_2+H_3 cut out the same divisor on X, the union of four lines taken with multiplicity 2. Thus the pencil spanned by these quadrics contains a quadric Q such that $X \subset Q$. Obviously, $\mathrm{rk}(Q) = 3$. Replacing the roles of the points p_2 and p_4, we get another quadric of rank 3 containing X. Applying Lemma 0.4.2, we see that $X \cong \mathcal{D}_1$ or $\mathcal{D}_1'$. An easy inspection shows that only $\mathcal{D}_1$ has 4 nodes, namely the points (0,1,0,0,0) ,(0,0,1,0,0), (0,0,0,1,0) ,(0,0,0,0,1).

Corollary 0.4.1. A minimal nonsingular model of a 4-nodal quartic in $\mathbb{P}^4$ is isomorphic to $\mathbb{P}^1 \times \mathbb{P}^1$ blown up at four points, the vertices of a quadrangle of rulings.

PROOF. Let $f: Y = \mathbb{P}^1 \times \mathbb{P}^1 \to \mathbb{P}^4$ be a rational map given by the linear system of curves of degree (2,2) passing through the 4 vertices of a quadrangle of rulings of Y. Let $\pi: Y' \to Y$ be the blowing up of the four base points $p_1,\ldots,p_4$ of this system. It is easy to check that, after blowing up the base points, this system defines a birational morphism f' from Y' to a quartic surface X in $\mathbb{P}^4$ with 4 lines, the images of $\pi^{-1}(p_i)$ and 4 nodes, the images of the sides of the quadrangle. By Proposition 0.4.3, Y is a 4-nodal quartic.

Remark 0.4.1. We can see easily the 4 lines on a 4-nodal surface and the four pencils of conics. Three lines come from the blowing up of the points p_1,p_3 and p_5 and the fourth one from the line joining p_2 and p_4. Three of the pencils come from the pencils of lines in $\mathbb{P}^2$ passing through the points p_1,p_2 and p_4 respectively. The fourth one comes from the pencil of conics which are passing through the points p_2,p_3,p_4 and p_5.

Let X be a 4-nodal quartic in $\mathbb{P}^4$. The four lines on it form a quadrangle with vertices at the nodes. This follows easily from the equations or from Remark 0.4.1. The quadrangle of lines on X will be called the **quadrangle** of X. Its vertices will be called the **vertices** of X.

Remark 0.4.2. One easily checks the following simple properties of a 4-nodal quartic in $\mathbb{P}^4$:

i) Each of the diagonals of the quadrangle of X is the singular line of one of the two quadrics of rank 3 containing X.

ii) A pair of intersecting sides of the quadrangle lies in the same pencil of conics as the opposite pair of sides.

iii) Each side of the quadrangle taken with multiplicity 2 belongs to a pencil of conics on X . The same pencil contains the opposite side taken with multiplicity 2.

Proposition 0.4.2'. Let Q be a singular quadric in $\mathbb{P}^3$ given by the equation $t_0t_1+t_2^2 = 0$. Let μ_2 = Spec $k[t]/(t^2-1)$ acts on Q by the formula

$$(t_0,t_1,t_2,t_3) \to (t_0\otimes\varepsilon, t_1\otimes\varepsilon, t_2, t_3), \quad \varepsilon = t \bmod (t^2-1).$$

Then the quotient Q/μ_2 is isomorphic to a degenerate 4-nodal quartic surface X. Moreover, the restriction of the factor map over X\Sing(X) is a principal μ_2-cover.

PROOF. Similar to Proposition 0.4.2 and left to the reader.

Proposition 0.4.3'. Let X be a surface of degree 4 in $\mathbb{P}^4$ not contained in a hyperplane. Let $\mathfrak{a}_X$ be the linear system of quadrics containing X. The following properties are equivalent:

(i) $\mathfrak{a}_X$ is a pencil spanned by two quadrics of rank 3 which does not contain any quadric of rank 4;

(ii) X is a degenerate 4-nodal quartic;

(iii) X is a symmetric anticanonical quartic Del Pezzo surface of type $\boldsymbol{\mu_2}$ with 2 nodes and one singular point of type A_3;

(iv) X is a quartic surface with 2 nodes and one singular point of type A_3;

(v) X is an anticanonical Del Pezzo surface with 2 nodes and one singular point point of type A_3;

(vi) X contains exactly 2 lines and 2 pencils of conics;

(vii) a minimal resolution X' of X is a Del Pezzo surface of degree 4 obtained from $\mathbb{P}^2$ by blowing up 5 points $p_1,\ldots,p_5$ such that $p_1 \neq p_2$, p_3 is infinitely near to p_2 , the line l_1 joining p_1 and p_2 passes through p_3 , p_4 is infinitely near to p_1, p_5 is infinitely near to p_4 and there exists a line l_2 passing through p_1,p_4 and p_5.

PROOF. Similar to the proof of Proposition 0.4.2 and left to the reader.

Corollary 0.4.1'. A minimal nonsingular model of a degenerate 4-nodal quartic in $\mathbb{P}^4$ is isomorphic to $\mathbb{P}^1 \times \mathbb{P}^1$ blown up at four points $q_1,\ldots q_4$ such that $q_1 \neq q_2$, q_3 (resp. q_4) is infinitely near to q_1 (resp. q_2), q_1 and q_2 lie on the same ruling which does not pass through p_3 and p_4.

PROOF. Similar to the proof of Corollary 0.4.1 and left to the reader.

Remark 0.4.1'. We can easily see the 2 lines and the 2 pencils of conics on a degenerate 4-nodal quartic surface. The two lines come from the blowing up the points p_1 and p_2 (the last blown up component). The two pencils of conics come from the pencils of lines in $\mathbb{P}^2$ passing through the points p_1 and p_2 .

Let X be a degenerate 4-nodal quartic in $\mathbb{P}^4$. The two lines on it will be called the **degenerate quadrangle**. They intersect at the singular point of type A_3 (called **the A_3-vertex**) and each passes through one node (the nodes are called the **simple vertices**).

Remark 0.4.2'. One easily checks the following simple properties of a degenerate 4-nodal quartic in $\mathbb{P}^4$:

i) The line in $\mathbb{P}^4$ joining the two non-exceptional vertices of the degenerate quadrangle of X is the double line of one of the two quadrics of rank 3 containing X. The double line of the other quadric of rank 3 is tangent to the first one at the exceptional vertex.

ii) The planes passing through the non-exceptional vertices of X cut out a pencil of conics on X.

iii) Each line of the degenerate quadrangle taken with multiplicity 2 belongs to a pencil of conics on X. The same pencil contains the other line taken with multiplicity 2.

Lemma 0.4.3. Let X be a quartic anticanonical Del Pezzo surface in $\mathbb{P}^4$. Suppose that the pencil of quadrics containing X is spanned by a quadric of rank 3 and a quadric of rank 4 , and does not contain other singular quadrics. Then X is isomorphic to one of the following surfaces:

(i) $x_0x_1+x_2^2 = 0,\ x_1x_2 + x_3x_4 = 0$;

(ii) $x_0x_1+x_2^2 = 0,\ x_2(x_1+x_2) + x_3x_4 = 0$;

(iii) $x_0x_1+x_2^2 = 0,\ x_3^2 + x_4x_1 + (x_0+\varepsilon x_2)^2 = 0,\ \varepsilon = 0,1(\mathrm{char}\ (k) \neq 2)$;

(iv) $x_0x_1+x_2^2 = 0,\ x_3^2 + x_4x_2+ (x_0+\varepsilon x_1)^2 = 0,\ \varepsilon = 0,1(\mathrm{char}\ (k) \neq 2)$;

(v) $x_0x_1+x_2^2 = 0,\ x_3(x_0+x_3)+x_4(x_2+\varepsilon x_1) = 0,\ \varepsilon = 0,1(\mathrm{char}\ (k) = 2)$;.

(vi) $x_0x_1+x_2^2 = 0,\ x_3(\varepsilon x_2+x_1)+x_4x_0+x_3^2 = 0\ (\mathrm{char}(k) = 2)$.

Moreover, X has the following singularities:

(i) $A_1+A_1+A_2$, (ii) $A_1+A_1+A_1$, (iii) D_4, (iv) A_3 (ε=1), A_1+A_3 (ε = 0), (v) A_3 (ε = 1), A_1+A_3(ε=0), (vi) $D_4^{(1)}$ (ε=1), $D_4^{(0)}$ (ε=0).

PROOF. As in the proof of Lemma 0.4.2, we can write $X = Q_1 \cap Q_2$,

$$Q_1:\ x_0x_1+x_2^2 = 0,$$

$$Q_2:\ x_3h(x_0,x_1,x_2)+x_4h'(x_0,x_1,x_2)+q(x_0,x_1,x_2)+q'(x_3,x_4) = 0,$$

$q'(x_3,x_4) = x_3x_4,\ x_3^2$ or 0. The last case is excluded as in the proof of Lemma 0.4.2. We may assume that $\mathrm{rk}(Q_2) = 4$.

If $q' = x_3x_4$, we are reduced to the case where $h = h' = 0$ and $\mathrm{rk}(q) = 2$. Since Q_2 and Q_1 are the only singular quadrics in the pencil $\mathcal{Q}_X$, the pencil of conics spanned by $q = 0$ and $q_0 = x_0x_1+x_2^2 = 0$ does not contain any singular conic except $q = 0$. Since $q = 0$ consists of two distinct lines, this can happen only in the case where the singular point of $q = 0$ lies on $q_0 = 0$. Let $q = l_1l_2$, where $l_1 \neq l_2$ are linear forms. As in the proof of Lemma 0.4.2, we may assume that either $l_1 = x_1$ or $l_1 = x_2$. Since $l_1 = 0$ and $l_2 = 0$ intersect at $q_0 = 0$, $l_2 = ax_1+bx_2$, in the first case, and $l_2 = bx_1 +ax_2$ or bx_0+ax_2, in the second case. Since $\mathrm{rk}(Q_2) = 4$, $b\neq 0$. Obviously, we may assume that b= 1, a = 0 or 1. Assume $l_1 = x_1, l_2 = ax_1+x_2$. If a = 0, we get (i). If a = 1, we make the change of variables: $x_2 \to x_2-x_1,\ x_0 \to x_0+2x_2+x_1,\ x_1 \to x_1$ to obtain (i). If $l_1 = x_2$ and a = 0, we get (i) again. If a = 1, we get (i).

Assume $q'(x_3,x_4) = x_3^2$. If $\mathrm{char}(k) \neq 2$, we can get rid of x_3h. Assume

$h' \neq 0$. Again, by an orthogonal transformation, we reduce h' to either x_2 or x_1. Let $h' = x_1$. Write q' in the form $x_1(ax_0+bx_1+cx_2) + dx_0^2+ex_0x_2+fx_2^2$. Replacing x_4 by $x_4+ax_0+bx_1+cx_2$, we may assume that $q = x_0^2+ex_0x_1+fx_1^2$. Hence

$$Q_2: x_3^2 + x_4x_1+ dx_0^2+ex_0x_2+fx_2^2 = 0.$$

Since $rk(Q_2) = 4$, $dx_0^2+ex_0x_2+fx_2^2 = (ax_0+bx_2)^2$, where a and b are not both zero. It is verified that X is not normal if $a = 0$. This immediately implies that we can reduce Q_2 to the form

$$Q_2: x_3^2 + x_4x_1 + (x_0+\varepsilon x_2)^2 = 0, \ \varepsilon = 0,1,$$

without changing the equation of Q_1. This leads to case (iii).

Similarly, if $h' = x_2$, we reduce Q_2 to the form

$$Q_2: x_3^2 + x_4x_2 + (\varepsilon x_0+x_1)^2 = 0, \ \varepsilon = 0,1.$$

This gives equations (iv).

Assume now that $char(k) = 2$ and

$$Q_2: \ x_3h(x_0,x_1,x_2)+x_4h'(x_0,x_1,x_2)+q(x_0,x_1,x_2)+x_3^2 = 0.$$

By a linear transformation of x_0,x_1,x_2 preserving Q_1, we may reduce h' to x_2 or x_1.

Let $h' = x_2$. Replacing x_4 by $x_4 + l(x_3,x_0,x_1)$ for some linear form l, we may assume that

$$Q_2: \ x_3(\alpha x_0+\beta x_1)+x_4x_2+ax_0^2+bx_0x_1+cx_1^2+x_3^2 = 0.$$

By scaling, we make α and β equal to 1 or 0.

Let $\alpha = 1$. Replacing x_3 by x_3+bx_1, we get rid of the term bx_0x_1. It is directly verified that $rk(Q_2) = 4$ if and only if $c = a\beta^2$. Thus

$$Q_2: \ x_3(x_0+\beta x_1)+x_4x_2+ax_0^2+a\beta^2x_1^2+x_3^2 = 0.$$

Then, we replace x_3 by $x_3+\gamma(x_0+\beta x_1)$, where $\gamma^2+\gamma = a$, to obtain

$$Q_2: \ x_3(x_0+\beta x_1+x_3)+x_4x_2 = 0.$$

This leads to (v), by the change $x_0 \to x_0+\beta x_1$, $x_2 \to x_2 +\beta x_1$.

The case $\beta = 1$ is reduced to the previous one by replacing x_0 by x_1 and x_1 by x_0. Assume $\alpha = \beta = 0$. Replacing x_3 by $x_3+\sqrt{a}x_0+\sqrt{c}x_1$, we write

$$Q_2:\ x_4x_2+x_3^2+bx_0x_1.$$

Since $bQ_1+Q_2 = x_4x_2+x_3^2+bx_2^2$ is of rank 3, this case is impossible.

The case $h' = x_0$ is dealt with similarly and leads to the equation

$$Q_2:\ x_3(\alpha x_2+\beta x_1)+x_4x_0+x_3^2 = 0.$$

If $\beta = 0$, the line $x_3 = x_0 = x_2 = 0$ is a singular line of X. Since X is normal, this is impossible.

The case $\alpha = \beta = 0$ is exluded as before. This gives us (vi).The type of the singularities is checked directly by using Proposition 0.2.6.

We denote by $\mathcal{D}_2$ (resp. $\mathcal{D}_3$) the quartic surface given by the equations

$$x_0x_1+x_2^2 = 0\ ,\ x_3(x_2+x_3+x_1)+x_4x_0 = 0\ ;.$$

$$(\text{resp. } x_0x_1+x_2^2 = 0,\ x_3(x_1+x_3)+x_4x_0= 0\ ).$$

Proposition 0.4.4. Assume char(K) = 2. Let τ be the involution of the nonsingular quadric $Q{:}t_0t_1+t_2t_3 = 0$ in $\mathbb{P}^3$ which acts by

$$(t_0,t_1,t_2,t_3)\to(t_1,t_0,t_3,t_2).$$

Then the quotient surface $Q/(\tau)$ is isomorphic to the surface $\mathcal{D}_2$. Morever, the restriction of the factor projection over the nonsingular locus of $\mathcal{D}_2$ is a principal $\mathbf{Z}/2$-cover.

PROOF. Similarly to the proof of Proposition 0.4.3, we consider the linear system spanned by the quadrics:

$$t_0^2+t_1^2 = 0,\ t_2^2+t_3^2 = 0,\ t_0t_2+t_1t_3+t_0t_3+t_1t_2 = 0,\ t_0t_2+t_1t_3 = 0, t_0t_1 = 0.$$

This system of quadrics is τ-invariant and defines a morphism $Q \to \mathbb{P}^4$ which factors through Q/τ and has a quartic surface as its image. To find the equation of its image, we notice the following relations between the quadrics:

$$(x_0+x_1)x_4+(x_3+x_2)x_3 = 0\ ,\ x_2^2+x_0x_1 = 0,$$

where x_0,x_1,x_2,x_3,x_4 are the equations of the above quadrics (taken in the same order). Changing x_2 to x_2+x_1, and x_0 to x_0+x_1, we get the equation of the surface $\mathcal{D}_2$.

Proposition 0.4.5. Assume char(k) = 2. Let Q be a nonsingular quadric in $\mathbb{P}^3$ given by the equation $t_0t_1+t_2t_3 = 0$. Let $\boldsymbol{\alpha}_2$ = Spec $k[t]/(t^2)$ acts on Q by the formula

$$t_0 \to t_0\otimes 1+t_2\otimes\varepsilon+t_3\otimes\varepsilon,\ t_1 \to t_1\otimes 1+t_2\otimes\varepsilon+t_3\otimes\varepsilon,$$

$$t_2 \to t_2\otimes 1+t_1\otimes\varepsilon+t_0\otimes\varepsilon,\ t_3 \to t_3\otimes 1+t_1\otimes\varepsilon+t_0\otimes\varepsilon, \qquad \varepsilon = t \bmod (t^2).$$

Then the quotient is isomorphic to a symmetric quartic Del Pezzo surface with a singularity of type $D_4^{(0)}$ given by the following equations:

$$x_0x_1+x_2^2 = 0,\ (x_3+x_1)x_3+x_4x_0 = 0$$

PROOF. Consider the linear system of quadrics spanned by the following $\boldsymbol{\alpha}_2$-invariant quadrics

$$t_0^2+t_1^2= 0,\ t_2^2+t_3^2 =0,\ (t_0+t_1)(t_2+t_3) = 0,\ t_3^2+t_2^2= 0,\ t_1^2 = 0.$$

It defines an isomorphism between the quotient space and a quartic surface in $\mathbb{P}^4$ given by the above equations. We also check that $\boldsymbol{\alpha}_2$ acts freely over the complement of the point (1,1,1,1).

Proposition 0.4.6. Assume char(k) = 2. Let X be a nondegenerate surface in $\mathbb{P}^4$. The following properties are equivalent:

(i) X is isomorphic to the surface $\mathcal{D}_2$ or $\mathcal{D}_3$;

(ii) X is a symmetric anticanonical quartic Del Pezzo surface of type $\boldsymbol{\alpha}_2$ or $\mathbf{Z}/2$ with a singular point of type D_4;

(iii) X is a surface of degree 4 with a singular point of type D_4;

(iv) X is an anticanonical quartic Del Pezzo surface with a singular point of type D_4;

(v) X has exactly 2 lines and 3 pencils of conics;

(vi) A minimal resolution of X is a Del Pezzo surface $X(\Sigma)$, where Σ consists of 4 infinitely near points $p_5 \to p_4 \to p_3 \to p_2$ and a separate point p_1 such that the line passing through p_1 and p_2 passes through p_3.

PROOF. (i) $\Rightarrow$ (ii) Follows from lemma 0.4.3 and Proposition 0.4.4.

(ii) $\Rightarrow$ (iii) $\Rightarrow$ (iv) $\Leftrightarrow$ (v) $\Rightarrow$ (vi), (iv) $\Rightarrow$ (vi) Similar to the proof of Proposition 0.4.3 .

(vi) $\Rightarrow$ (i) It is easy to see that the pencil of conics passing through the points $p_2,\dots,p_5$ and the pencil of lines through p_1 define two pencils of conics on X which are cut out by two families of plane rulings of a quartic of rank 4 containg X. Similarly, the pencil of lines through p_2 defines a pencil of conics on X cut out by one family of plane rulings of a quadric of rank 3 containg X. No more singular quadrics contain X. Otherwise, we have more than three pencils of conics on X, that would contradict Corollary 0.3.1. Applying Lemma 0.4.3, we excude other possibilities for X by analyzing the singularities of X.

Lemma 0.4.3'. Let X be a quartic anticanonical Del Pezzo surface in $\mathbb{P}^3$. Suppose that the pencil of quadrics containing X contains only one singular quadric which is of rank 3. Then X is isomorphic to one of the following surfaces:

(i) $x_0x_1+x_2^2 = 0$, $\varepsilon x_1^2+x_1x_2+x_0x_4+x_3^2 = 0$, $\varepsilon = 1,0$ ($\mathrm{char}(\kappa) \neq 2$);

(ii) $x_0x_1+x_2^2 = 0$, $x_2x_3 + x_1x_2 + x_4x_0 + x_3^2 = 0$ ($\mathrm{char}(\kappa) = 2$);

(iii) $x_0x_1+x_2^2 = 0$, $x_1x_2 + x_4x_0 + x_3^2 = 0$ ($\mathrm{char}(\kappa) = 2$).

Moreover, X has one singular point of type D_5 ($D_5^{(1)}$ in case (ii) and $D_5^{(1)}$ in case (iii)).

PROOF. We use the notation and the arguments of Lemma 0.4.3. Assume

$$Q_2 : q(x_0,x_1,x_2) + x_3x_4 = 0,$$

Since the pencil $\lambda(x_0x_1+x_2^2) + q(x_0,x_1,x_2) = 0$ has a singular member for some λ, the corresponding quadric λQ_1+Q_2 from the pencil is singular. By assumption, this is impossible. Thus, we may assume that

$$Q_2: q(x_0,x_1,x_2) + x_3h + x_4h' + x_3^2 = 0.$$

If char(K) ≠ 2, we get rid of the term x_3h and reduce Q_2 to the form:

$$Q_2 : ax_0^2+bx_0x_1+cx_1^2 + x_4x_2 + x_3^2 = 0.$$

or

$$Q_2 = ax_1^2+bx_2x_1+cx_2^2 + x_4x_0 + x_3^2 = 0.$$

In the first case, we replace Q_2 by $Q_2+\lambda Q_1$ assuming that

$$Q_2: (\alpha x_0+\beta x_1)^2 + \gamma x_2^2 + x_4x_2 + x_3^2 = 0.$$

Then, we replace x_4 by $x_4+\gamma x_2$ to a find a singular quadric in the pencil diffe- from Q_1. Thus, this case is impossible.

In the second case , replacing x_4 by x_4+cx_1 and subtracting $c(x_0x_1+x_2^2)$, we get rid of the term cx_2^2. Now, obviously b≠ 0 and, by scaling, we may assume that b = 1, a = 0 or 1. This leads to (i).

Assume char(K) = 2. Similarly to the above and Lemma 0.4.3, we are reduced to the case where

$$Q_2: ax_1^2+bx_2^2 + x_3(\alpha x_1+\beta x_2) + x_4x_0 + x_3^2 = 0.$$

and $\alpha,\beta \in \{0,1\}$. If $\alpha = 1$, we find that the quadric $(a\beta^2+b)Q_1+Q_2$ is singular. Thus, this case is impossible. If $\alpha = 0$, $\beta = 1$, we get

$$Q_2: ax_1^2+bx_2^2 + x_3x_2 + x_4x_0 + x_3^2 = 0.$$

Changing x_3 to $x_3+\gamma x_2+\sqrt{a}x_1$, where $\gamma^2+\gamma = b$, we reduce Q_2 to the form

$$Q_2: \sqrt{a}x_1x_2 + x_3x_2 + x_4x_0 + x_3^2 = 0.$$

Scaling x_1,x_2,x_0 and x_4, we get (ii). Similarly, the case $\alpha = \beta = 0$ leads to (iii).

The type of the singularities is verified directly by applying Proposition 0.2.6.

Assume that char(K) = 2. We denote by $\mathcal{D}_2'$ (resp. $\mathcal{D}_3'$) the quartic surface given by the equations

$$x_0x_1+x_2^2 = 0,\ x_2x_3 + x_1x_2 + x_4x_0 + x_3^2 = 0;$$

$$(\text{resp. } x_0x_1+x_2^2 = 0,\ x_1x_2 + x_4x_0 + x_3^2 = 0\ ).$$

Proposition 0.4.4'. Let τ be the involution of the singular quadric $t_0t_1+t_2^2 = 0$ given by the formula:

$$(t_0,t_1,t_2,t_3) \rightarrow (t_0,t_0+t_1,t_0+t_2,t_3+t_1+t_2).$$

Then the quotient space $X = Q/(\tau)$ is isomorphic to the surface $\mathcal{D}_3$. Moreover, the restriction of the factor projection over $X\setminus Sing(X)$ is a principal $\mathbf{Z/2}$-cover.

PROOF. Similar to the proof of Proposition 0.4.4. We consider the linear system of quadrics spanned by the following ones:

$$t_0^2 = 0,\ t_1^2+t_2^2 = 0,\ t_0(t_1+t_2) = 0,\ t_0t_3+t_1t_2+t_0t_1 = 0,$$

$$t_3(t_1+t_2)+t_3^2 = 0$$

and check the assertion.

Proposition 0.4.5'. Assume char(k) = 2. Let Q be the singular quadric in $\mathbb{P}^3$ given by the equation $t_0t_1+t_2^2 = 0$. Let $\boldsymbol{\alpha}_2 = \operatorname{Spec} k[t]/(t^2)$ act on Q by the formula

$$(t_0,t_1,t_2,t_3) \rightarrow (t_0\otimes 1,t_1\otimes 1,t_2\otimes 1+t_0\otimes \varepsilon,t_3\otimes 1+t_2\otimes \varepsilon).$$

Then the quotient is isomorphic to the surface $\mathcal{D}_3'$. Moreover, the restriction of the factor projection over $X\setminus Sing(X)$ is a principal $\boldsymbol{\alpha}_2$-cover

PROOF. Similar to the proof of Proposition 0.4.4. We take the linear system of quadrics spanned by the following ones:

$$t_0^2 = 0,\ t_1^2 = 0,\ t_2^2 = 0\ ,\ t_3^2 = 0\ ,\ t_1t_2+t_0t_3 = 0.$$

and check the assertion.

Proposition 0.4.6'. Assume char(k) = 2. Let X be a nondegenerate surface in $\mathbb{P}^4$ and $\mathcal{Q}_X$ be the linear system of quadrics containing X. The following properties are equivalent:

(i) $\mathcal{Q}_X$ is a pencil containing only one singular quadric which is of rank 3;

(ii) X is isomorphic to the surface $\mathcal{D}_2'$ or $\mathcal{D}_3'$;

(iii) X is a symmetric anticanonical quartic Del Pezzo surface of type $\boldsymbol{\alpha}_2$ or $\mathbf{Z}/2$ with a singular point of type D_5;

(iv) X is a surface of degree 4 with a singular point of type D_5;

(v) X is an anticanonical Del Pezzo surface with a singular point of type D_5;

(vi) X contains exactly one line and one pencil of conics;

(vii) A minimal resolution of X is a Del Pezzo surface $X(\Sigma)$, where Σ consists of 5 infinitely near points $p_5 \to p_4 \to p_3 \to p_2 \to p_1$ such that there exists a line passing through p_1, p_2 and p_3.

PROOF. Similar to the proof of Propositions 0.4.6 and 0.4.3.

Finally, a partial explanation for the name symmetric (another reason will be given in the next paragraph) is given by the following result which shows that the automorphism group of symmetric quartic Del Pezzo surfaces is rather large. Note that every generic quartic Del Pezzo surface X has a finite automorphism group isomorphic to $(\mathbf{Z}/2)^4$. This can be seen by diagonalizing the equations of X.

Proposition 0.4.7. Let X be a symmetric quartic Del Pezzo surface. Then its automorphism group is isomorphic to the following one:

$\mathcal{D}_1$: $D_4 \ltimes \mathbf{G}_m^{\ 2}$;

$\mathcal{D}_1'$: $\mathbf{Z}/2 \ltimes (\mathbf{G}_m \ltimes \mathbf{Aff(1)})$;

$\mathcal{D}_2, \mathcal{D}_3$: $\mathbf{Z}/2 \ltimes \mathbf{Aff(1)}^2$;

$\mathcal{D}_2', \mathcal{D}_3'$: $U \ltimes \mathbf{G}_m^{\ 2}$, where U is a connected unipotent group of dimension 3.

PROOF. Assume X = $\mathcal{D}_1$. Then, the subgroup of Aut(X) which leaves each side of the quadrangle invariant is isomorphic to a 2-dimensonal torus $\mathbf{G}_m{}^2$. In fact, the complement of the quadrangle is itself such a torus and its action on X extends its natural action on itself by translations. The factor group is isomorphic to the group of symmetries of the quadrangle, the dihedral group D_4 of order 8. The automorphisms realizing these symmetries are generated by the following ones:

$$x = (x_0,\dots,x_4) \to (x_3,x_2,x_2,x_1,x_0),$$

$$x \to (x_1,x_0,x_2,x_3,x_4),$$

$$x \to (x_0,x_1,x_2,x_4,x_3).$$

Assume $X = \mathcal{D}_1'$. Then, the argument is similar. The complement of the degenerate quadrangle is isomorphic to to the complement of the exceptional section and two fibres of the minimal ruled surface $\mathbb{F}_2$. Its automorphism group is $\mathbf{G}_m \times \mathbf{Aff(1)} \cong \mathbf{G}_m^2 \ltimes \mathbf{G}_a$. The factor group is isomorphic to the group of order 2 generated by the permutation of the lines. The latter is induced by the automorphism

$$x \to (x_0,x_1,x_2,x_4,x_3) \ .$$

Assume $X = \mathcal{D}_2$ (resp. $\mathcal{D}_2'$). It follows from Proposition 0.4.6 (vi) that the connected component $\mathrm{Aut}(X)^\circ$ of $\mathrm{Aut}(X)$ is isomorphic to the group of projective transformations of $\mathbb{P}^2$ preserving a line and fixing two points on it. The factor group $\mathrm{Aut}(X)/\mathrm{Aut}(X)^\circ$ is isomorphic to the group of 2 elements which acts by permuting the two lines on X. The latter transformation is given by:

$$(x_0,x_1,x_2,x_3,x_4) \to (x_0,x_1,x_2,x_3+x_2,x_4)$$

$$(\text{resp. } (x_0,x_1,x_2,x_3,x_4) \to (x_0,x_1,x_2,x_3+x_1,x_4))$$

Assume $X = \mathcal{D}_2'$ or $\mathcal{D}_3'$. It follows from Proposition 0.4.6' (vi) that $\mathrm{Aut}(X)$ is isomorphic to the group of projective transformations of $\mathbb{P}^2$ preserving a line and fixing one point on it.

We summarize the results of this section with the following table:

Name	Equations	Sing(X)	Type	dimAut(X)
$\mathcal{D}_1$	$x_0x_1+x_2^2=0$ $x_3x_4+x_2^2=0$	$A_1+A_1+A_1+A_1$	μ_2	2
$\mathcal{D}_1'$	$x_0x_1+x_2^2=0$ $x_0^2+x_3x_4=0$	$A_1+A_1+A_3$	μ_2	3
$\mathcal{D}_2$	$x_0x_1+x_2^2=0$ $x_3(x_2+x_3+x_1)+x_4x_0=0$	$D_4^{(1)}$	$\mathbb{Z}/2$	4
$\mathcal{D}_2'$	$x_0x_1+x_2^2=0$ $x_3(x_2+x_2)+x_0x_4+x_1x_2=0$	$D_5^{(1)}$	$\mathbb{Z}/2$	5
$\mathcal{D}_3$	$x_0x_1+x_2^2=0$ $x_3(x_1+x_3)+x_4x_0=0$	$D_4^{(0)}$	α_2	4
$\mathcal{D}_3'$	$x_0x_1+x_2^2=0$ $x_3^2+x_0x_4+x_2x_1=0$	$D_5^{(0)}$	α_2	5

§5. Symmetric cubic Del Pezzo surfaces.

Here we will classify symmetric Del Pezzo surface of degree 3. First of all we note that Proposition 0.3.3 immediately implies that every normal cubic surface in $\mathbb{P}^3$ is either a cone or anticanonical Del Pezzo surface. In the following, by a **cubic surface** we will always mean an anticanonical Del Pezzo surface of degree 3.

Proposition 0.5.1. Let X be a cubic surface in $\mathbb{P}^3$. Then X is symmetric if and only if it is a projection of a symmetric anticanonical quartic Del Pezzo surface from one of its nonsingular point.

PROOF. First of all, any cubic anticanonical Del Pezzo surface X is a projection of a quartic anticanonical Del Pezzo surface from a nonsingular point. Indeed, a minimal nonsingular model $X(\Sigma)$ of X is obtained by blowing up 6 points in $\mathbb{P}^2$. Let $\pi: X(\Sigma) \to \mathbb{P}^2$ be the corresponding blowing down. It is the composition of 6 blowings down of exceptional curves of the first kind. Let $\pi': X(\Sigma) \to X'$ be the first blowing down of a curve E_1. Then X' is a nonsingular model of an anti-canonical Del Pezzo quartic surface Y. Its projection from the point $\pi'(E_1) \in Y$ is our surface X. Conversely, a projection of a symmetric anticanonical quartic Del Pezzo surface from its nonsingular point is, obviously, a cubic surface.

Now, if X is a symmetric surface, then the pull-back of its principal G-cover of X\Sing(X) induces a principal G-cover of Y\Sing(X). And the converse is also true.

The above proposition allows us to classify all symmetric cubic surfaces. First of all, we have 3 types of them, according to the type of the principal cover, similar to the corresponding notion for symmetric anti-canonical quartic Del Pezzo surfaces.

Proposition 0.5.2. Let X be a symmetric cubic surface of type μ_2. Then (i) X is isomorphic to one of the following 3 surfaces $\mathfrak{C}_1, \mathfrak{C}_1'$ or $\mathfrak{C}_1''$:

($\mathfrak{C}_1$) $x_0x_1x_2 + x_0x_1x_3 + x_0x_2x_3 + x_1x_2x_3 = 0,$

($\mathfrak{C}_1'$) $x_0x_1x_3 + x_0x_2^2 + x_1x_2^2 = 0,$

($\mathfrak{C}_1''$) $x_0x_1x_3 + x_0x_2^2 + x_1^3 = 0.$

(ii) X has rational double singularities of one of the following types:

($\mathfrak{C}_1$) 4 points of type A_1,

($\mathfrak{C}_1'$) 2 points of type A_1 and one point of type A_3,

($\mathfrak{C}_1''$) 1 point of type A_1 and one point of type A_5.

(iii) a minimal nonsingular model of X is obtained by blowing up 6 points $p_1,\ldots,p_6$ in $\mathbb{P}^2$ in one of the following special position:

($\mathfrak{C}_1$) $p_1,\ldots,p_6$ is the set of vertices of a complete quadrilateral of lines:

($\mathcal{B}_1'$) p_1,p_2,p_3 are the vertices of a triangle of nonconcurrent lines, $p_4 \neq p_2,p_3$ is on the line l joining p_2 and p_3, p_5 (resp. p_6) is infinitely near to p_2 (resp. p_3), l passes through p_5 and p_6;

($\mathcal{B}_1''$) $p_1 \neq p_2$, p_i is infinitely near to p_{i-1} (i=3,4,5), p_6 is infinitely near to p_1 not on the line l joining p_1 and p_2, there exists a smooth conic tangent to l at p_2 which passes through p_3,p_4,p_5.

PROOF. By the previous proposition and the results of §4, we know that X is a projection of a surface Y = $\mathcal{D}_1$ or $\mathcal{D}_1'$ from one of its nonsingular points P. The group of automorphisms of Y acts on Y' = Y\Sing(Y) with one line and the complement to the set of lines as the orbits. Thus there are only four cases to consider:

Y = $\mathcal{D}_1$ (resp.$\mathcal{D}_1'$), P does not (resp. does) lie on the quadrangle of lines.

(i) The reader finds easily the equation of the surface obtained by projection of Y from P. Note that $\mathcal{B}_1$ (resp. $\mathcal{B}_1'$, resp.$\mathcal{B}_1''$) corresponds to the case $\mathcal{D}_1$, P does not lie on the quadrangle (resp. $\mathcal{D}_1$, P lies on the quadrangle or $\mathcal{D}_1'$ and P does not lie on it, resp. $\mathcal{D}_1'$, P lies on the quadrangle).

(ii) We easily observe from the equations that X has 4 singularities of type A_1 (0,0,0,1)((1,0,0,0),(0,1,0,0),(0,0,1,0), in case $\mathcal{B}_1'$; 2 singularities of type A_1 (1,0,0,0),(0,1,0,0) and one (0,0,0,1) of type A_3 in case ($\mathcal{B}_1'$); 1 singularity of type A_1((1,0,0,0)) and 1 of type A_5 ((0,0,0,1)) in case ($\mathcal{B}_1''$).

(iii) We use the notation of the proof of the previous Proposition. Assume Y = $\mathcal{D}_1$. Let p = $\pi(E_1)$. the corresponding blowing down. We can identify p with a point on Y'. If P = $\pi'(E_1)$ does not lie on the quadrangle of Y, then p lies outside the triangle of lines with the vertices p_1, p_2 and p_3 (we use the notation of Proposition 0.4.3). Applying a standard quadratic transformation centered at the points p_2, p_3 and p, we replace π by X' $\to \mathbb{P}^2$ as in (iii) ($\mathcal{B}_1$). If p lies on the quadrangle of X, then we easily get (iii) ($\mathcal{B}_1'$).

Suppose Y is a degenerate 4-nodal quartic surface. Similarly to the above, we easily check that we get (iii) ($\mathcal{B}_1'$), if the centre of the projection lies outside the degenerate quadrangle, and (iii) ($\mathcal{B}_1''$) otherwise.

Remark 0.5.1. The surface $\mathfrak{C}_1$ is the famous **Cayley cubic surface**. It is the projectivization of the affine surface:

$$x^{-1} + y^{-1} + z^{-1} = 1.$$

It also can be characterized as a cubic surface with maximal possible number of nodes.

The proof of the next propositions is similar to the previous one and is omitted.

Proposition 0.5.3. Let X be a symmetric cubic surface of type $\mathbf{Z}/2$. Then
(i) X is isomorphic to one of the following 3 surfaces $\mathfrak{C}_2, \mathfrak{C}_2'$ or $\mathfrak{C}_2''$:

($\mathfrak{C}_2$) $x_0x_3^2+x_1^2x_2+x_1x_2^2+x_1x_2x_3 = 0,$

($\mathfrak{C}_2'$) $x_0x_3^2+x_1^2x_2+x_3x_2^2+x_1x_2x_3 = 0,$

($\mathfrak{C}_2''$) $x_0x_3^2+x_1^3+x_0x_2^2+x_1x_2x_3 = 0.$

(ii) X has one rational double singularity of the following types:

($\mathfrak{C}_2$) $D_4^{(1)}$,

($\mathfrak{C}_2'$) $D_5^{(1)}$,

($\mathfrak{C}_2''$) $E_6^{(1)}$.

(iii) a minimal nonsingular model of X is obtained by blowing up 6 points $p_1,\dots,p_6$ in $\mathbb{P}^2$ in one of the following special positions:

($\mathfrak{C}_2$) $p_1,\dots,p_5$ as in Proposition 0.4.6 and p_6 disjoint from $p_1,\dots p_5$;

($\mathfrak{C}_2'$) $p_1,\dots,p_5$ as in Proposition 0.4.6' and p_6 disjoint from $p_1,\dots p_5$;

($\mathfrak{C}_2''$) $p_1,\dots,p_5$ as in Proposition 0.4.6' and p_6 is infinitely near to p_5.

Proposition 0.5.4. Let X be a symmetric cubic surface of type α_2. Then
(i) X is isomorphic to one of the following 3 surfaces $\mathfrak{C}_3, \mathfrak{C}_3'$ or $\mathfrak{C}_3''$:

($\mathfrak{C}_3$) $x_0x_3^2+x_1^2x_2+x_1x_2^2 = 0,$

($\mathfrak{C}_3'$) $x_0x_3^2+x_1^2x_2+x_3x_2^2 = 0,$

($\mathcal{C}_3$") $x_0x_3^2+x_1^3+x_0x_2^2 = 0$.

(ii) X has one rational double singularity of the following types:

($\mathcal{C}_3$) $D_4^{(0)}$,

($\mathcal{C}_3$') $D_5^{(0)}$,

($\mathcal{C}_3$")$E_6^{(0)}$.

(iii) A minimal nonsingular model of X is obtained by blowing up 6 points $p_1,\dots,p_6$ in $\mathbb{P}^2$ in one of the following special positions:

($\mathcal{C}_3$) $p_1,\dots,p_5$ as in Proposition 0.4.6 and p_6 disjoint from $p_1,\dots p_5$;

($\mathcal{C}_3$') $p_1,\dots,p_5$ as in Proposition 0.4.6' (case D_5) and p_6 disjoint from $p_1,\dots p_5$;

($\mathcal{C}_3$") $p_1,\dots,p_5$ as in Proposition 0.4.6' (case D_5) and p_6 is infinitely near to p_5.

Now, we give another explanation for the name symmetric cubic surface.

Proposition 0.5.5. Assume char(k) ≠ 2. A cubic surface is symmetric if and only if its equation can be written as the determinant of a symmetric 3x3-matrix whose entries are linear forms in homogeneous coordinates.

PROOF. We verify directly that each surface $\mathcal{C}_1,\mathcal{C}_1',\mathcal{C}_1"$ is isomorphic to a surface given by the following equation:

$\mathcal{C}_1$
$$\begin{bmatrix} x_0 & 0 & x_2 \\ 0 & x_1 & x_3 \\ x_2 & x_3 & -x_2-x_3 \end{bmatrix}$$

$\mathcal{C}_1'$:
$$\begin{bmatrix} x_0 & 0 & x_2 \\ 0 & x_1 & x_2 \\ x_2 & x_2 & x_3 \end{bmatrix}$$

$\mathcal{C}_1"$:
$$\begin{bmatrix} x_0 & 0 & x_1 \\ 0 & x_1 & x_2 \\ x_1 & x_2 & x_3 \end{bmatrix}$$

Conversely, if X is given by a determinantal equation, then there exists a web (= a 3-dimensional linear system) $W \cong \mathbb{P}^3$ of conics Q_x in $\mathbb{P}^2$ such that

$$X \cong \{x \in W: rk(Q_x)\} < 3 .$$

All webs of conics are easily classified (see Chapter 6). In fact, they correspond bijectively to pencils of conics; the latter are classified by analyzing all possible configurations of their base points. Doing this, we easily find that there are only three projective classes of webs which give a normal surface X. Each of them is isomorphic to one of the above surfaces.

S6. Prym canonical maps.

Recall that for every smooth (or, any Gorenstein) connected curve C and a non-trivial element $\varepsilon \in {}_2\mathrm{Pic}(C)$, the sheaf $\omega_C(\varepsilon)$ (resp. the map

$$\varphi_\varepsilon: C \to \varphi_\varepsilon(C) = \Gamma \subset \mathbb{P}(H^0(C,\omega_F(\varepsilon))^*) \cong \mathbb{P}^{g-2},$$

where $g = \dim H^1(C,\mathcal{O}_F) = \dim H^0(C,\omega_C)$, is called a **Prym canonical map**. The image Γ of C under this map is called a **Prym canonical model** of (C,ε). The linear system $|\omega_C(\varepsilon)| = |K_C+\varepsilon|$ is called a **Prym canonical linear system**.

We now recall the basic properties of Prym canonical maps. Proofs are elementary, hence omitted or briefly sketched.

Lemma 0.6.1. Let C be a smooth curve of genus $g \geq 3$. A point $p \in C$ is a base-point of $|K_C+\varepsilon|$ if and only if $|p+\varepsilon| \neq \emptyset$. In particular,if C is not hyperelliptic, $|K_C+\varepsilon|$ is base-point free.

The last statement can be strenghtened. Let $\pi:\bar{C} \to C$ be the etale double cover defined by ε (assume that $\mathrm{char}(k) \neq 2$).

Lemma 0.6.2. $|K_C+\varepsilon|$ is base-point free if and only if $\bar{C}$ is not hyperelliptic.

PROOF. Since $\pi^*(K_C+\varepsilon) = K_{\bar{C}}$, the canonical system of $\bar{C}$ is invariant with respect to the involution associated to the cover π. If $\bar{C}$ is hyperelliptic, then its canonical map $f:\bar{C} \to \mathbb{P}^1$ defines a double cover $\bar{f}:C \to \mathbb{P}^1$ and a double cover $\bar{\pi}:\mathbb{P}^1 \to \mathbb{P}^1$ such that $\bar{\pi}\circ f = f\circ\pi$. Hence C is hyperelliptic. It is easy to see that the ramification divisor of $\bar{f}$ consists of the base points of $|K_C+\varepsilon|$.The converse statement is proved similarly.

Lemma 0.6.3. Let C be a smooth connected curve of genus $g \geq 3$. Then one of the following holds:

(i) deg $\varphi_\varepsilon = 4$, $g = 3$;

(ii) deg $\varphi_\varepsilon = 2$, $p_a(\Gamma) = 0$, $|K_C+\varepsilon|$ has two simple base-points;

(iii) deg $\varphi_\varepsilon = 2$, $p_a(\Gamma) = 1$, $|K_C+\varepsilon|$ has no base-points;

(iv) deg $\varphi_\varepsilon = 1$, $|K_C+\varepsilon|$ has no base points.

PROOF. The degrees of φ_ε and of Γ are related by the following formulas:

$$\deg(\varphi_\varepsilon)\deg(\Gamma) \leq 2g-2,$$

$$\deg(\Gamma) \geq 2g-2.$$

where the equality holds in the first one if and only if $|K_C+\varepsilon|$ is base-point free.

An easy computation leads to cases (i) - (iv) above as well as

(v) deg $\varphi_\varepsilon = 3$, $g = 4$;

(vi) deg $\varphi_\varepsilon = 3$, $g = 3$;

(vii) deg $\varphi_\varepsilon = 1$, $|K_C+\varepsilon|$ has base-points.

Case (v): φ_ε is a triple cover of a conic Γ in $\mathbb{P}^2$. Let $p\in\Gamma$ and $\varphi_\varepsilon^{-1}(p) = p_1+p_2+p_3$. By Riemann-Roch, $|p_1+p_2+p_3+\varepsilon| \neq \emptyset$. Let $q_1+q_2+q_3 \in |p_1+p_2+p_3+\varepsilon|$. Then, $|K_C-p_1-p_2-p_3| = |K_C+\varepsilon-q_1-q_2-q_3|$.

Case (vi): φ_ε is a triple cover of $\mathbb{P}^1$, $\deg\varphi_\varepsilon \deg(\Gamma) = \deg(K_C+\varepsilon) - 1$. Hence $|K_C+\varepsilon|$ has exactly one base-point which contradicts Lemma 4.4.2.

Case (vii): By Lemma 0.6.3, C is hyperelliptic .If p and q are two points of C such that $p\sim q+\varepsilon$, then the linear system $|K_C+\varepsilon-p-q|$ is a multiple of the g_2^1 of C, hence of degree > 1.

The next lemma follows simply from the preceding proof:

Lemma 0.6.4. Let $|K_C+\varepsilon|$ be a Prym canonical map on a smooth curve C of genus $g \geq 3$. Assume it has base-points. Then, it has exactly two simple base-points p and q such that $p = q+\varepsilon$. Moreover, φ_ε is a double cover onto a rational curve of degree g-2 in $\mathbb{P}^{g-2}$.

Lemma 0.6.5. Same notation as in Lemma 0.6.4. Assume that C is neither hyperelliptic nor superelliptic (i.e. a double cover of an elliptic curve). Then φ_ε is an isomorphism unless there exist two points p and q on C such that $|p+q+\varepsilon| \neq \emptyset$.

As a corollary, we obtain

Proposition 0.6.1. Suppose C does not have a g_4^1 (i.e. a pencil of divisors of degree 4). Then φ_ε is an isomorphism onto a curve of degree 2g-2 in $\mathbb{P}^{g-2}$.

Proposition 0.6.2 . Let C be a smooth curve of genus 4. Assume that C is neither hyperelliptic nor superelliptic and does not have vanishing theta constants (i.e. $h^0(\mathfrak{L}) \leq 1$ for any invertible $\mathfrak{L}$ with $\mathfrak{L}^2 = \omega_C$). Then φ_ε maps C onto a plane sextic curve passing twice through 6 points $p_1,\dots,p_6$ (possibly infinitely near). An anti-canonical model of the blow-up of these points is a symmetric cubic Y . The image of C in Y is cut out by a quadric and is a canonical model of C. Conversely, any representation of a canonical model of C as an intersection of a quadric and a symmetric cubic is obtained as above for a unique $\varepsilon \in {}_2\mathrm{Pic}(C)$.

PROOF. This is a classical result due to Wirtinger and Coble. For a modern proof see **[Rec,Cat]**.

Remark 0.6.1. Using the classification of symmetric cubics, Proposition 0.5.2 (especially (iv)), we see three possible forms of Prym canonical models of C. The one corresponding to the Cayley cubic is called a **Wirtinger sextic**. We refer to **[Co 3]**, Chapter V, §50 for the geometry of such curves.

S7. The Picard scheme.

As usual we denote by Pic(X) the Picard group of a scheme X. Its elements are the isomorphism classes of invertible sheaves on X or the classes of Cartier divisors modulo linear equivalence. Equivalently, it is the group of torseurs (principal homogeneous spaces) of the multiplicative group scheme $\mathbf{G}_m$ over X with respect to different Grothendieck topologies (see **[Mi 2]**). Thus we have some canonical isomorphisms:

$$\mathrm{Pic}(X) \cong H^1(X,\mathcal{O}_X^*) \cong H^1_{Zar}(X,\mathbf{G}_m) \cong H^1_{et}(X,\mathbf{G}_m) \cong H^1_{Zar}(X,\mathbf{G}_m).$$

Let f: X → S be a proper morphism of schemes of finite type. The **relative Picard functor** of X over S is defined to be the sheaf $\mathcal{P}ic_{X/S}$ in the flat or étale topology of S associated to the presheaf

$$S' \to \mathrm{Pic}(X \times_S S').$$

In particular,

$$\mathcal{P}ic_{X/S}(S') = H^0(S', R^1_{et} f'_* \mathbf{G}_m),$$

where $f'\colon X' = X \times_S S' \to S'$ is the second projection.

The canonical homomorphism

$$\mathrm{Pic}(X') \to \mathcal{P}ic_{X/S}(S')$$

is the boundary homomorphism $H^1 \to E_1^{0,1}$ in the Leray spectral sequence:

$$H^q(S', R^p_{et} f'_* \mathbf{G}_m) \Rightarrow H^{p+q}_{et}(X', \mathbf{G}_m)$$

Proposition 0.7.1. Assume $f_*'\mathcal{O}_{X'} \cong \mathcal{O}_{S'}$.

(i) The canonical homomorphism $\mathrm{Pic}(X') \to \mathcal{P}ic_{X/S}(S')$ defines an injective homomorphism

$$\alpha: \mathrm{Pic}(X')/f'^*\mathrm{Pic}(S') \to \mathcal{P}ic_{X/S}(S').$$

(ii)The homomorphism α is bijective if $H^2_{et}(S', \mathbf{G}_m) = 0$.

(iii)The homomorphism α is bijective if $X'(S') \neq 0$.

PROOF. (i) and (ii) follow easily from the above Leray spectral sequence. (iii) is deeper and can be found in **[Gro 2]**.

The group $H_{et}^2(S', \mathbf{G}_m)$ is denoted by $\mathrm{Br}(S')$ and is called the **Brauer** (cohomological) **group** of S'. We refer to **[Gro 3]** for the theory of these groups.

Example 0.7.1. Let S be the spectrum of a field κ and X be a complete integral scheme over κ. We have

$$\mathcal{P}ic_{X/S}(S) \cong \mathrm{Pic}(\bar{X})^G,$$

where $\bar{X} = X \otimes_\kappa \bar{\kappa}$, $\bar{\kappa}$ is the separable algebraic closure of κ, $G = \mathrm{Gal}(\bar{\kappa}/\kappa)$. The canonical homomorphism

$$\mathrm{Pic}(X) \to \mathrm{Pic}(\bar{X})^G$$

is injective (since $\mathrm{Pic}(\kappa) = 0$), and is surjective if $X(\kappa) \neq \emptyset$ or $\mathrm{Br}(\mathrm{Spec}\ \kappa) = 0$. The group $\mathrm{Br}(\mathrm{Spec}(\kappa))$ is the Brauer group $\mathrm{Br}(\kappa)$ of the field κ as it is defined in **[Bou 1]**. For example it is zero if $\kappa = \bar{\kappa}$, or κ is finite , or κ is the field of rational functions of an algebraic curve over an algebraically closed field.

Theorem 0.7.1. (Representability). Let $f: X \to S$ be a proper flat morphism of finite type of noetherian schemes. The functor $\mathcal{P}ic_{X/S}$ is represented by a separated group scheme $\mathbf{Pic}_{X/S}$ of locally finite type over S in one of the following cases:

(i) f is projective with geometrically integral fibres;

(ii)S is the spectrum of a field .

PROOF. (i) **[Gro 2]**,(ii)**[Art 3, Mur]**.

The group scheme $\mathbf{Pic}_{X/S}$ is said to be the **Picard scheme** of X over

S.

Let S = Spec k, where k is a field and G be a commutative group scheme locally of finite type over k. We denote by G° the identity component of G, the smallest connected open group subscheme of G. This is an algebraic group scheme over k. The quotient group G/G° is an etale group scheme over k. It is constant if $k = \bar{k}$. We denote by G^τ the open subgroup of G which is the inverse image of the torsion subgroup of G/G°.

Let S be an arbitrary scheme, and G be an abelian group functor on the category of S-schemes whose fibres over any $s \in S$ are representable by a scheme $\mathbf{G}_s$ of locally finite type over k(s). We define the subfunctor G° (resp. G^τ) as follows: for every S-scheme S' $G^\circ(S')$ (resp.$G^\tau(S')$) is the subgroup of G(S') of elements which induce an element of $\mathbf{G}_s^\circ(s')$ (resp. $\mathbf{G}_s^\tau(s')$) for every point $s' \in S'$ lying over a point $s \in S$. If G is representable by a scheme **G** over S, we use $\mathbf{G}^\circ$ and $\mathbf{G}^\tau$ to denote the subschemes of G representing the subfuctors G° and G^τ.

In view of Theorem 0.7.1, the previous terminology is applied to the Picard functor $Pic_{X/S}$ and the Picard scheme $\mathbf{Pic}_{X/S}$ (if it exists) for every proper flat scheme X over S .

Theorem 0.7.2 (finiteness). Let X be a proper scheme over a field k.
(i) $\mathbf{Pic}_{X/k}{}^\tau$ is a group scheme of finite type over k.
(ii) $\mathbf{Pic}_{X/k}{}^\tau/\mathbf{Pic}_{X/k}{}^\circ$ is a finite group scheme .
(iii)$(\mathbf{Pic}_{X/k}/\mathbf{Pic}_{X/k}{}^\circ)(\bar{k})$ is a finitely generated abelian group.

PROOF. See **[Gro 2]**.

For every group functor G on the category S-schemes, one can define the Lie algebra functor Lie(G). Its value on an S-scheme S' is equal to $Ker(G(S' \times_S I_S) \to G(S'))$, where I_S is the scheme of dual numbers over S. The restriction of Lie(G) to the Zariski topology of S is a quasi-coherent sheaf $\mathcal{L}ie(G)$ and Lie(G) is represented by the affine S-scheme $Spec(Sym(\mathcal{L}ie(G)))$.

Proposition 0.7.2. Let $f\colon X \to S$ be a proper flat morphism. Then

$$\mathcal{L}ie(\mathcal{P}ic_{X/S}) \cong R^1f_*\mathcal{O}_X.$$

PROOF. See **[Gro 2]**.

Assume now that $S = \operatorname{Spec} k$, where k is a field and X is a proper k-scheme. We know that $\mathbf{Pic}_{X/k}{}^{\circ}$ is an algebraic group scheme over k with the tangent space at the origin isomorphic to the space $H^1(X,\mathcal{O}_X)$ and $(\mathbf{Pic}_{X/k}{}^{\circ})_{red}$ is an algebraic group over k.

Proposition 0.7.3. (i) If X is smooth, $(\mathbf{Pic}_{X/k}{}^{\circ})_{red}$ is an abelian variety of dimension $h^1(\mathcal{O}_X)$;
(ii) $\mathbf{Pic}_{X/k}{}^{\circ} = (\mathbf{Pic}_{X/k}{}^{\circ})_{red}$ if $H^2(X,\mathcal{O}_X) = 0$ or $\operatorname{char}(k) = 0$.

PROOF. (i) See **[Gro 2]**. (ii) See **[Mu 1]**.

By Theorem 0.7.1, $\mathbf{Pic}_{X/k}{}^{\circ}$ is an algebraic commutative group scheme and $(\mathbf{Pic}_{X/k}{}^{\circ})_{red}$ is a commutative algebraic group. Thus it is an extension of an abelian variety A by an affine commutative algebraic group G. The latter is isomorphic to an extension of an algebraic torus T by a unipotent group U (see **[Ser 1]**). Let n be an integer coprime to $\operatorname{char}(k)$. Then

$${}_nA(\bar{k}) \cong (\mathbf{Z}/n\mathbf{Z})^{2g} \text{ where } g = \dim A,$$

$${}_nT(\bar{k}) \cong (\mathbf{Z}/n\mathbf{Z})^{t}, \text{ where } t = \dim T,$$

$${}_nU(\bar{k}) = (0).$$

The Kummer exact sequence for $\bar{X} = X\otimes_k\bar{k}$ (see **[Mi 2]**)

$$0 \to \mu_n \to \mathbf{G}_m \to \mathbf{G}_m \to 0$$

easily shows that

$${}_n((\mathbf{Pic}_{X/k}{}^{\circ})_{red})(\bar{k}) \cong H^1(\bar{X},\mu_n).$$

This immediately implies:

Proposition 0.7.4. Let $b_1(\bar{X})$ be the first Betti number of $\bar{X}$. Then

$$b_1(X) = 2g + t.$$

In particular, if X is smooth

$$\dim(\mathbf{Pic}_{X/K}{}^{\circ})_{red} = \tfrac{1}{2}b_1(X),$$

and

$$b_1(X) \leq 2\dim H^1(X,\mathcal{O}_X)$$

with equality holding if and only if $\mathbf{Pic}_{X/K}{}^{\circ} = (\mathbf{Pic}_{X/K}{}^{\circ})_{red}$.

Assume X is smooth and proper over an algebraically closed field K. We refer to **[Fu]**) for the notions of algebraic and numerical equvalence of divisors on X

Proposition 0.7.5.
(i)$(\mathbf{Pic}_{X/K}{}^{\circ})_{red}(K)$ = the group of divisor classes of algebraically equivalent to zero on X.
(ii)$(\mathbf{Pic}_{X/K}{}^{\tau})_{red}(K)$ = the group ofdivisor classes of numerically equivalent to zero on X.

PROOF. See **[Gro 2]**.

The algebraic group $(\mathbf{Pic}_{X/K}{}^{\circ})_{red}$ is called the **Picard variety** of X. The factor group

$$\mathrm{Num}(X) = (\mathbf{Pic}_{X/K})_{red}(K)/(\mathbf{Pic}_{X/K}{}^{\tau})_{red}(K)$$

$$(\text{resp. } \mathrm{NS}(X) = (\mathbf{Pic}_{X/K})_{red}(K)/(\mathbf{Pic}_{X/K}{}^{\circ})_{red}(K))$$

of classes of divisors on X modulo numerical (resp. algebraic) equivalence will be called the **Picard lattice** (resp. the. **Neron-Severi group)** of X. It follows from Theorem 0.7.2 that NS(X) is a finitely generated abelian group and

$$\mathrm{Num}(X) = \mathrm{NS}(X)/\mathrm{Torsion}.$$

Example 0.7.2. Let X be a complete algebraic curve over a field k and $X = X_1 \cup \ldots \cup X_k$ be its decomposition into irreducible components, and x_i be the generic point of X_i. Let m_i be the multiplicity of the component X_i. It is defined to be the length of the local ring $\mathcal{O}_{\bar{X},\bar{x}_i}$, where $\bar{x}_i$ is a point of $\bar{X} = X \otimes_k \bar{k}$ over x_i. We know that $\mathbf{Pic}_{X/k}$ exists and is reduced. The algebraic group $\mathbf{Pic}_{X/k}{}^{\circ}$ is called the (generalized) **Jacobian variety** of X and is denoted by $J(X)$. If $X(k') \neq \emptyset$ or $Br(k') = 0$,

$$J(X)(k') = \mathrm{Ker}(d{:}\mathrm{Pic}(X \otimes_k k') \to \mathbf{Z}^k),$$

where d is the degree homomorphism,

$$\mathcal{L} \to (\deg(\mathcal{L}|X_1)/m_1, \ldots, \deg(\mathcal{L}|X_k)/m_k).$$

(Recall that the **total degree** of $\mathcal{L}$ is defined by $\deg(\mathcal{L}) = \chi(\mathcal{L}) - \chi(\mathcal{O}_X)$ and is equal to $\Sigma \deg(\mathcal{L}|X_i)$).

We also have

$$J(X) = \mathbf{Pic}_{X/k}{}^{\tau}.$$

For every vector $v \in \mathbf{Z}^k$, we set

$$\mathrm{Pic}(X \otimes_k k')^v = d^{-1}(v).$$

The functor $k' \to \mathrm{Pic}(X \otimes_k k')^v$ is representable by an algebraic scheme $\mathbf{Pic}_{X/k}{}^{v}$ over k which is a principal homogeneous space over k with respect to $\mathbf{Pic}_{X/k}{}^{\circ}$.

The canoncal inclusion $X' = X_{red} \hookrightarrow X$ induces a homomorphism of group schemes:

$$\mathbf{Pic}_{X/k} \to \mathbf{Pic}_{X/k}.$$

whose kernel is a unipotent algebraic group. If $X = X_{red}$, the computation of its Picard scheme is based on the analysis of the normalization map:

$$p: X' = \coprod X_i' \to X$$

we have the exact sequence of group schemes:

$$0 \to \mathbf{G}_m^{k-1} \to p_*\mathbf{G}_m/\mathbf{G}_m \to \mathrm{Pic}_{X/k} \to \mathbf{Pic}_{X'/k} = \prod \mathbf{Pic}_{X_i'/k} \to 0.$$

For example, if X is geometrically irreducible and has only δ ordinary double points as singularities, the kernel of $\mathrm{Pic}_{X/k} \to \mathrm{Pic}_{X'/k}$ is an algebraic torus of dimension δ.

Bibliographical notes to Chapter 0.

Most of the material from this chapter is rather well-known, though sometimes it is hard to find the needed references. For example, the general nonsense about double covers from §1 can be found in **[B-P-vdV]** for the case p = char(k) ≠ 0. The case p = 2 is dealt in many different papers, for example **[B-M 2, R-S 1, Ek 1]**.

The theory of rational double points from §2 goes back to **[DuV]**. The modern treatment and rather complete theory in case p = 0 is due to M. Artin **[Art 2]**, E. Brieskorn **[Br]** and G. Tyurina **[Tyu]**. An exposition of this theory can be found in **[SS]**. In the case p ≠ 0 (especially p = 2), the theory is still rather rudimentary. The paper of M. Artin **[Art 7]** is the main contribution to the subject.

The material of §3 is rather well known. It is almost as old as algebraic geometry itself. We refer to **[Dem, Ma]** for the most complete exposition of the theory. Note that our definition of Del Pezzo surfaces differs slightly from the standard one: a surface of degree n in $\mathbb{P}^n$.

The classification of intersection of two quadrics in $\mathbb{P}^4$ (p ≠ 2) is due to C. Segre **[Seg]**. The classification of symmetric quartic surfaces in $\mathbb{P}^4$ from §4 follows easily from this in this case. The general definition and the classification of symmetric quartics in all characteristics seem to be new.

The classification of symmetric cubic surfaces from §5 is well known in the case p ≠ 2. One of them, the Cayley cubic, occurs in many situations in algebraic geometry. A modern treatment of this classification can be found in **[Cat]**. The generalization of this classification to the case p = 2 seems to be new.

There is nothing new in §6, everything can be found for example in **[G-H 1]**.

The theory of Picard functors and schemes from §7 is due to A.Grothendieck. The corresponding references can be found in the section.

Chapter I

ENRIQUES SURFACES: GENERALITIES

§1. Classification of algebraic surfaces.

Let k be an algebraically closed field of arbitrary characteristic p. In this section we recall the main results of the classification of nonsingular projective surfaces over k. We refer to **[Mu 2,B-M 1,B-M 2]** for the proofs of all the assertions peculiar to the case of positive characteristic and to general textbooks **[B-P-vdV,Bea 2,G-H 1]** for the case of characteristic zero.

We denote by $\equiv$ (resp. $\sim$) the numerical (resp. linear) equivalence of divisors. We denote by $b_i(Z)$ (resp. $e(Z)= \Sigma(-1)^i b_i(Z))$) the Betti numbers (resp. the Euler-Poincaré characteristic) of an algebraic variety Z computed in étale or classical topology.

Let $\kappa(X)$ denote the **Kodaira dimension** of a projective nonsingular surface X over k, i.e. the maximal number κ (resp. $-\infty$) such that

$$\dim |mK_X| = P_\kappa(m) \text{ for } m >> 0 \text{ (resp. } = 0, \text{ for any } m>0),$$

where $P_\kappa(m)$ is a polynomial of degree κ.

Recall also that a surface X is said to be **minimal** if every birational morphism $f:X \rightarrow X'$ onto a nonsingular surface X' is an isomorphism. Equivalently, this means that X does not contain exceptional curves of the first kind.

Theorem 1.1.1. Let X be a nonsingular projective surface over k with $\varkappa(X) \neq -\infty$. Then there exists a unique birational morphism $f: X \to X'$, where X' is a minimal surface satisfying one of the following properties:

(i) $\varkappa(X') = 2$, $K_{X'}^2 > 0$;

(ii) $\varkappa(X') = 1$, $K_{X'}^2 = 0$, $K_{X'} \not\equiv 0$;

(iii) $\varkappa(X') = 0$, $K_{X'} \equiv 0$.

If $\varkappa(X) = -\infty$, then X is a ruled surface (i.e. birationally isomorphic to the product $C \times \mathbb{P}^1$ for some curve C).

Enriques surfaces belong to case (iii) above.

Theorem 1.1.2. Let X be a minimal nonsingular projective surface over k with $\varkappa(X) = 0$. Then X belongs to one of the following classes of surfaces:

(a) $b_2(X) = 22$, $K_X \sim 0$ (K3-surfaces);

(b) $b_2(X) = 10$ (Enriques surfaces);

(c) $b_2(X) = 6$ $K_X \sim 0$ (abelian surfaces);

(d) $b_2(X) = 2$, $K_X \nsim 0$ (hyperelliptic surfaces).

Note that the condition $K_X \equiv 0$ obviously implies the minimality of a surface. Thus, an **Enriques surface** is characterized and defined by the conditions:

(i) $K_X \equiv 0$; (ii) $b_2(X) = 10$.

or, equivalently:

(i)' X is a minimal surface, (ii)' $\varkappa(X) = 0$; (iii)' $b_2(X) = 10$.

First of all, the value of $b_2(X)$ for a minimal surface X of Kodaira dimension 0 determines the values of $b_1(X)$ and $\chi(X, \mathcal{O}_X)$.

Proposition 1.1.1. Let X be a minimal surface with $\varkappa(X) = 0$. Then

(i) $b_1(X) = 0$, $\chi(X,\mathcal{O}_X) = 2$ if X is a K3-surface;

(ii) $b_1(X) = 0$, $\chi(X,\mathcal{O}_X) = 1$ if X is an Enriques surface;

(iii) $b_1(X) = 4$, $\chi(X,\mathcal{O}_X) = 0$ if X is an abelian surface;

(iv) $b_1(X) = 2$, $\chi(X,\mathcal{O}_X) = 0$ if X is a hyperelliptic surface.

PROOF. Suppose $b_1(X) \neq 0$. Let $f: X' \to X$ be an étale cyclic principal cover of degree n corresponding to a non-trivial element of $H^1_{et}(X,\mu_n)$ ((n,p) = 1). Then $K_{X'} = f^*(K_X)$, hence, $K_{X'} \equiv 0$ and $\varkappa(X') = 0$. Since $b_i(X) \leq b_i(X')$ (the maps $f^*: H^i(X,\mathbb{Q}_l) \to H^i(X',\mathbb{Q}_l)$ are injective) and

$$e(X') = ne(X),$$

we get from Theorem 1.1.2, that X is neither a K3-surface nor an Enriques surface. It also implies that $e(X) = 2-2b_1(X)+b_2(X) = 0$, if $b_1(X) \neq 0$. The assertion about $\chi(X,\mathcal{O}_X)$ follows immediately from the Noether formula

$$12\chi(X,\mathcal{O}_X) = c_2(X) + K_X^2 .$$

Corollary 1.1.1. Let F be an Enriques surface. Then

$$\dim H^i(F,\mathcal{O}_F) \leq 1 , i = 1,2.$$

The equality holds if and only if $K_F \sim 0$.

PROOF. Since $K_X \equiv 0$,

$$\dim H^2(F,\mathcal{O}_F) = \dim H^0(F,\mathcal{O}_F(K_F)) \leq 1$$

and the equality holds if and only if $K_F \sim 0$. It remains to use that $\chi(F,\mathcal{O}_F) = 1$.

An Enriques surface F is said to be **classical**, if

$$K_F \not\sim 0 .$$

For classical Enriques surface

$$H^1(F,\mathcal{O}_F) = H^2(X,\mathcal{O}_X) = 0.$$

For non-classical Enriques surface

$$H^1(F,\mathcal{O}_F) = H^2(F,\mathcal{O}_F) \cong k.$$

Theorem 1.1.3. Assume char(k) $\neq$ 2. Then every Enriques surface is classical.

PROOF. Assume first that char (k) = 0 or , by the Lefschetz principle, $k = \mathbb{C}$. Then, by the Hodge theory

$$0 = b_1(F) = 2\dim H^1(F,\mathcal{O}_F).$$

Thus, F is classical.

Assume now that F is not classical, i.e. $K_F \sim 0$. Let

$$F: H^1(F,\mathcal{O}_F) \to H^1(F,\mathcal{O}_F)$$

be the Frobenius map. Since $H^1(F,\mathcal{O}_F)$ is one-dimensional, only two cases may occur : a)$F \equiv 0$, b) $F-\lambda 1 \equiv 0$ for some $\lambda \in k$, $\lambda \neq 0$. It is explained in Chapter 0, §1 (where only the case p = 2 was considered, the general case being very similar), that there exists a cyclic p-cover $\pi: Y \to F$ which is purely inseparable in the first case and unramified in the second case. Also, we know that Y is Gorenstein and $K_Y = \pi^*(K_F) \sim 0$. This implies that

$$\chi(Y,\mathcal{O}_Y) \leq \dim H^0(Y,\mathcal{O}_Y) + \dim H^2(Y,\mathcal{O}_Y) = 2.$$

On the other hand, the sheaf $\pi_*\mathcal{O}_Y$ has a composition series of length p with factors isomorphic to $\mathcal{O}_F$. This implies that $\chi(Y,\mathcal{O}_Y) = p\chi(F,\mathcal{O}_F) = p$. Hence p = 2.

§2. The Picard group.

We use the notation of §7 of Chapter 0.

Theorem 1.2.1. Let F be an Enriques surface. Then

$$(\mathbf{Pic}^{\circ}_{F/k})_{red} = 0$$

and $\mathbf{Pic}^{\tau}_{F/k}$ is a finite group scheme of order 2. Moreover, let $F: H^1(F,\mathcal{O}_F) \to H^1(F,\mathcal{O}_F)$ be the Frobenius map. Then

$$\mathbf{Pic}^{\tau}_{F/K} \cong (\mathbf{Z}/2)_K \text{ if } F \text{ is classical,}$$

$$\cong \boldsymbol{\alpha}_2 \text{ if } F \text{ is not classical, } \mathbf{F} \text{ is the trivial map,}$$

$$\cong \mu_2 \text{ if } F \text{ is not classical, } \mathbf{F} \text{ is a non-trivial map.}$$

PROOF. Since $b_1(F) = 0$, $(\mathbf{Pic}^{\circ}_{F/K})_{red}$ is an abelian variety of dimension 0. Thus, it is trivial. If F is classical, $H^1(F,\mathcal{O}_F) = 0$. Hence $\mathbf{Pic}_{F/K}$ is a discrete group scheme. In particular, $\mathbf{Pic}^{\tau}_{F/K} = (\mathrm{Tors}(\mathrm{Pic}(F))_K$. Let $\mathcal{L} \in \mathrm{Tors}(\mathrm{Pic}(F))$. By Riemann-Roch, $h^0(\mathcal{L})+h^2(\mathcal{L}) \geq 1$. This implies that either $\mathcal{L} \cong \mathcal{O}_F$ or $\mathcal{L} \cong \mathcal{O}_F(K_F)$, hence, $\mathrm{Pic}^{\tau}(F) \cong \mathbf{Z}/2$.

Assume now that F is not classical. The same argument as above shows that $\mathrm{Pic}^{\tau}(F) = 0$. Thus, $G = \mathbf{Pic}^{\tau}_{F/K}$ consists of one point and the tangent space at it is one-dimensional. It is known (**[Ga]**) that every every finite group scheme G over an algebraically closed field K with $G(K) = \{1\}$ is isomorphic as a scheme to the spectrum of a truncated polynomial ring. Since , in our case, G has one-dimensional tangent space, we obtain that

$$G \cong \mathrm{Spec}\; K[t]/(t^n),$$

for some $n > 0$. Assume $n > 2$. Then, the morphism

$$\mathrm{Spec}\; K[t]/(t^2) \to \mathbf{Pic}_{F/K}$$

defined by a nonzero tangent vector can be extended to a morphism

$$\mathrm{Spec}\; K[t]/(t^3) \to G \hookrightarrow \mathbf{Pic}_{F/K}.$$

By (**[Mu 1]**,Lecture 27) this implies that the first Bockstein operation

$$\beta_1 : H^1(X,\mathcal{O}_X) \to H^2(X,\mathcal{O}_X)$$

is not bijective. However, this is impossible. Indeed, β_1 is the composition of the maps:

$$H^1(F,\mathcal{O}_F) \to H^1(F,\mathcal{O}_F) \otimes H^1(F,\mathcal{O}_F),\; x \to x \otimes x$$

and the cup-product:

$$H^1(F,\mathcal{O}_F) \otimes H^1(F,\mathcal{O}_F) \to H^2(F,\mathcal{O}_F).$$

Since $K_F \sim 0$, the latter coincides with the map

$$H^1(F,\mathcal{O}_F) \otimes H^1(F,\Omega^2_F) \to H^2(F,\Omega^2_F)$$

which is an isomorphism by Serre's duality.

We know that $\mathbf{Pic}^{\tau}_{F/K}$ is an infinitesimal group scheme of order 2. There are only two of them: μ_2 and α_2 (Chapter 0, §1).

It is easily verified that the Frobenius map on the tangent space is non-zero in the case μ_2 and is zero in the case α_2.

Corollary 1.2.1. Let F be an Enriques surface. Then

$$2K_F \sim 0.$$

PROOF. We know that $K_F \equiv 0$ for any surface of Kodaira dimension 0. Thus, K_X is an element of $\mathrm{Pic}^{\tau}_{F/K}$ which is zero if F is not classical, or of order 2 otherwise.

From now on we call a non-classical Enriques surface a **μ_2-surface** (resp. a **α_2-surface**) if $\mathbf{Pic}^{\tau}_{F/K} \cong \mu_2$ (resp. α_2). By analogy with abelian varieties, a μ_2-surface (resp. an α_2-surface) is also called **ordinary** (resp. **supersingular**) Enriques surface.

Denote by $\rho(X)$ (or ρ if no confusion arises) the **Picard number** of X, the rank of the Néron-Severi group NS(X).

Assume $k = \mathbf{C}$. We have a natural homomorphism

$$c_1\colon \mathrm{Pic}(X) \to H^2(X,\mathbf{Z}),$$

the first Chern class. It factors through NS(X) and defines an injective homomorphism

$$c_1\colon \mathrm{NS}(X) \to H^2(X,\mathbf{Z})$$

which is compatible with the intersection pairing on NS(X) and the cup-product on $H^2(X,\mathbf{Z})$. By Lefschetz, c_1 induces an isomorphism

$$\mathrm{Tors}(\mathrm{NS}(X)) \cong \mathrm{Tors}(H^2(X,\mathbf{Z})).$$

We also have the Hodge decomposition

$$H^2(X,\mathbb{C}) = H^{2,0}(X)\oplus H^{1,1}(X)\oplus H^{0,2}(X)$$

and by Lefschetz, the image of the homomorphism c_1 belongs to $H^{1,1}(X)\cap H^2(X,\mathbb{Z})$. Thus, it is equal to $H^2(X,\mathbb{Z})$ if $H^{2,0}(X) = 0$. The latter happens if and only if $H^2(X,\mathcal{O}_X) = 0$. In particular, we obtain in the case of an Enriques surface F.

Proposition 1.2.1. Let F be an Enriques surface over $k = \mathbb{C}$. Then $NS(F) \cong H^2(F,\mathbb{Z})$,

$$\mathrm{Num}(F) \cong H^2(F,\mathbb{Z})/\mathrm{Tors} \cong \mathbb{Z}^{10}$$

and the intersection pairing on Num(F) is a perfect duality of free $\mathbb{Z}$-modules (Poincaré duality).

We will now show that the same result is true in the case of ground fields of arbitrary characteristic. If not stated otherwise, all references can be found in **[Mi 2]**.

Recall that

$$\mathrm{Pic}(X) \cong H^1(X,\mathcal{O}_X^*) \cong H^1_{fl}(X,\mathbf{G}_m) \cong H^1_{et}(X,\mathbf{G}_m),$$

where $\mathbf{G}_m$ is the sheaf in the flat (étale) topology of X represented by the multiplicative group scheme.

The exact Kummer sequence in the flat topology:

$$0 \to \mu_n \to \mathbf{G}_m \to \mathbf{G}_m \to 0, \quad n \in \mathbb{Z},$$

defines an exact sequence

$$0 \to \mathrm{Pic}(X)/n\mathrm{Pic}(X) \to H^2_{fl}(X,\mu_n) \to {}_nH^2_{fl}((X,\mathbf{G}_m) \to 0$$

Since $\mathrm{Pic}^\circ(X) = \mathbf{Pic}^\circ_X(k)$ is n-divisible, we have

$$\mathrm{Pic}(X)/n\mathrm{Pic}(X) \cong NS(X)/nNS(X).$$

Taking $n = l^k$, where l is a prime number, and passing to the projective limit we get an exact sequence

$$0 \to NS(X)\otimes\mathbb{Z}_l \to H^2_{fl}(X,\mathbb{Z}_l[1]) \to T_l(\mathrm{Br}(X)) \to 0,$$

where

$$Br(X) = H^2_{fl}(X, \mathbf{G}_m) \cong H^2_{et}(X, \mathbf{G}_m)$$

is the Brauer group of X and

$$T_l(Br(X)) = \lim.proj_i \, {}_{l^i}Br(X).$$

Since it is known that ${}_nBr(X)$ is finite for every n, the l-primary component Br(X)(l) of Br(X) is isomorphic to a group $(\mathbf{Q}_l/\mathbf{Z}_l)^{t_l} \oplus A$, for some $t_l \geq 0$ and a finite group A This implies that

$$T_l(Br(X) \cong \mathbf{Z}_l^{t_l}.$$

and we obtain:

Proposition 1.2.2. For every prime l

$$Tors(NS(X) \otimes \mathbf{Z}_l) \cong Tors(H^2_{fl}(X, \mathbf{Z}_l[1]))$$

$$\rho(X) = rk_{\mathbf{Z}_l} H^2_{fl}(X, \mathbf{Z}_l[1]) - t_l$$

where $T_l(Br(X) \cong \mathbf{Z}_l^{t_l}$. In particular, if $l \neq p$,

$$\rho(X) = b_2(X) - t(X),$$

where $t(X) = t_l$ is independent of l.

Remark 1.2.1. It follows from the theory of crystalline cohomology that

$$b_2(X) = \rho(X) + t_p + 2h,$$

for some $h \geq 0$ computed in terms of the cohomology group with coefficients in the sheaf of Witt rings or in terms of the height of the formal Brauer group of X (see **[11]**). The homomorphism

$$c_1: NS(X) \otimes \mathbf{Z}_l \to H^2(X, \mathbf{Z}_l)$$

defined above is called the **cycle map** (or, the **first Chern class homomorphism**). It is compatible with the canonical pairings on both the groups

$$NS(X) \otimes \mathbf{Z}_l \times NS(X) \otimes \mathbf{Z}_l \to \mathbf{Z}_l$$

$$H^2(X, \mathbf{Z}_l) \times H^2(X, \mathbf{Z}_l) \to \mathbf{Z}_l .$$

The first one is induced by the intersection of divisors

$$NS(X) \times NS(X) \to \mathbf{Z}.$$

The second one is induced by the $\cap$-product

Corollary 1.2.2. Assume Br(X) is finite. Then

$$\rho(X) = b_2(X).$$

Moreover, the pairing on $\mathrm{Num}(X) \cong \mathbf{Z}^{\rho}$ induced by the intersection of divisors is a perfect duality of free $\mathbf{Z}$-modules.

PROOF. The first assertion follows immediately from the equality $t(X) = 0$. We also obtain that

$$\mathrm{Num}(X) \otimes \mathbf{Z}_l = NS(X) \otimes \mathbf{Z}_l/\mathrm{Tors} \cong H^2(X, \mathbf{Z}_l)/\mathrm{Tors} .$$

By the duality for the flat cohomology of a surface (**[Mi 1, Ber]**), the pairing on the latter group is a perfect duality of free $\mathbf{Z}_l$-modules. Thus, the discriminant of the symmetric bilinear form on Num(X) is an l-adic unit for every prime l. This, of course,implies that it is a unit in $\mathbf{Z}$, i.e. the bilinear form is unimodular.

Remark 1.2.2. If $l \neq p = \mathrm{char}(k)$, then the flat cohomology is equal to the etale cohomology, so we need the former one only in the case $l = p$ (see [Mi 2]).

Theorem 1.2.2. Let ρ be the Picard number of an Enriques surface F. Then

$$\rho = b_2(F) = 10,$$

and the intersection pairing

$$\mathrm{Num}(F) \times \mathrm{Num}(F) \to \mathbf{Z}$$

is a duality of free $\mathbf{Z}$-modules.

PROOF. We have already shown this in the case $k = \mathbf{C}$.

Assume first that F does not have nonzero regular vector fields. It will be shown later in Corollary 1.4.1 that in this case F is liftable to characteristic 0. It is known that the Picard number does not decrease under specialization (**[Gro 4]**). This proves that $\rho(F) = 10$. By the Hodge Index theorem the quadratic space $NS(X)\otimes\mathbb{Q}$ is indefinite. Since every indefinite rational quadratic form of rank ≥ 5 represents 0 (**[Ser 4]**), there exists a divisor class D on F with $D^2 = 0$. It will be proven in Chapter 3 that this implies the existence of a genus 1 fibration on F. In Chapter 5 we will prove that its jacobian fibration is a rational surface and deduce from this that Br(F) is finite. It remains to apply Corollary 1.2.1.

Assume $p = 2$ and F has a regular nonzero vector field D. Let $\pi_D: F \to F^D$ be the corresponding factorization map (see Chapter 0, **S**1). Since π_D is purely inseparable,

$$b_i(F) = b_i(F^D)\ ,\ e(F) = e(F^D) = 12.$$

Let $\varphi: X \to F^D$ be a minimal resolution of singularities and $\pi_{D'}: F' \to X$ the map obtained by resolution of the indeterminancies of the rational map $\varphi^{-1}\circ\pi_D$. It follows from Proposition 0.1.14 that $h^0(nK_{F'}) \geq h^0(nK_X)$ for every $n \geq 0$. Thus,

$$\varkappa(F) = \varkappa(F') \geq \varkappa(X).$$

Since $b_1(F') = b_1(X) = 0$, applying the classification of algebraic surfaces, we find that X is birationally a rational surface, or an Enriques surface, or a K3-surface.

Assume X is a rational surface. Then F^D is a rational surface and,

$$\rho(F) \geq \rho(F^D) = b_2(F^D) = 10.$$

Assume X is birationally an Enriques surface or a K3-surface. Let $f: X \to X'$ be a birational morphism to its minimal model. Then $K_X \equiv A$, where A is the exceptional divisor of f. Applying the formula $K_{F'} = \pi_D'^*(K_X)+R$ from Proposition 0.1.14, we see that A is contained in the exceptional divisor of φ. Since φ is a minimal resolution, f must be an isomorphism, i.e. X is a minimal model.

If X is an Enriques surface, then $e(X)=e(F^D)=12$. This shows that $X = F^D$. Thus we may apply Propositions 0.1.14 and conclude that D has no zeroes.

This contradicts Proposition 0.1.13.

Assume X is a K3-surface. We will see in §4 that only classical or α_2-surfaces may have nonzero vector fields. If F is classical, $h^0(K_F) = 0$. Since $h^0(K_X) = 1$, we see that this case is impossible. Let F be an α_2-surface and $\bar{F} \to F$ be its K3-cover. It is a principal α_2-cover, locally given by $x_i^2+b_i = 0$, where $b_i-b_j \in k(F)^2$ (Proposition 0.1.9). Let η be a regular 1-form on F locally given by db_i. and ω be a non-zero regular 2-form on F. Define a vector field ∂ on F by

$$df \wedge \eta = (\partial f)\omega \text{ for every } f \in k(F).$$

We may assume that $D = \partial$. Since $D(b_i) = 0$, the K3-cover $\bar{F} \to F$ is D-invariant. Its pull-back $\bar{F}'$ on F' is a principal D-invariant α_2-cover of F'. It follows from the theory of inseparable descent (**[Car, Gro 2]**) that $\bar{F}' = X' \times_X F'$ for some principal α_2-cover $X' \to X$. However, it is known that $\mathbf{Pic}_{X/k}$ is reduced (Proposition 1.2.3), hence X' does not have such covers. This contradiction proves the assertion.

Remark 1.2.3. We can avoid reference to the theory of inseparable descent by the corresponding reference to the ordinary theory of descent **[Gro 2]**. In fact, a principal α_2-cover $\bar{F}' \to F'$ corresponds to an element $\mathcal{L} \in \mathrm{Pic}(F' \otimes_k k[\varepsilon])$, where $k[\varepsilon] = k[t]/(t^2)$ is the algebra of dual numbers. The vector field D on F defines a group action of α_2 on the scheme F' (Remark 0.1.3) and hence an involution τ of the scheme $F' \otimes_k k[\varepsilon]$ such that $F' \otimes_k k[\varepsilon]/(\tau) \cong X \otimes_k k[\varepsilon]$. The D-invariance of $\bar{F}' \to F'$ is equivalent to the τ-invariance of $\mathcal{L}$. Applying descent to the pair $(F' \otimes_k k[\varepsilon] \to X \otimes_k k[\varepsilon], \mathcal{L})$, we find an invertible sheaf $\mathcal{L}' \in \mathrm{Pic}(X \otimes_k k[\varepsilon]) \setminus \{0\}$. This contradicts Proposition 1.2.3.

Corollary 1.2.3. Let F be an Enriques surface and D be a vector field on F. The factor surface F^D is rational.

PROOF. Follows from the proof of Theorem 1.2.2.

Finally we include for the comparison the following result about the Picard scheme of a K3-surface.

Proposition 1.2.3. Let X be a K3-surface. Then

$$\mathbf{Pic}_{X/K} \cong (NS(X))_K \cong (Num(X))_K \cong \mathbf{Z}^{\rho},$$

where $\rho \leq 22$.

PROOF. Since $K_X \sim 0$ and $\chi(X,\mathcal{O}_X) = 2$,

$$H^1(X,\mathcal{O}_X) = 0.$$

This shows that the tangent space of $\mathbf{Pic}_{X/K}$ at the origin is trivial. Thus

$$\mathbf{Pic}^{\circ}_{X/K} = 0$$

Let $\mathcal{L} \in \mathrm{Tors}(NS(X))$. By Riemann-Roch,

$$h^0(\mathcal{L})+h^0(\mathcal{L}^{-1}) \geq 2.$$

This shows that $\mathcal{L}$ or $\mathcal{L}^{-1}$ has a non-trivial section. This obviously implies that $\mathcal{L} \cong \mathcal{O}_X$.

Since $b_2(X) = 22$, we get $\rho(X) \leq 22$.

§3. The K3-cover.

Let F be a classical Enriques surface. Then $\omega_F = \mathcal{O}_F(K_F)$ is a non-trivial element of ${}_2\mathrm{Pic}(F)$. As such, it defines a double cover

$$\pi: \bar{F} \to F$$

which is unramified if $p \neq 2$ and is purely inseparable if $p = 2$ (see Chapter 0, §1).

Let F be a μ_2-surface. Then, the Artin-Schreier exact sequence in the etale topology

$$0 \to \mathbf{Z}/2 \to \mathbf{G}_a \xrightarrow{F-1} \mathbf{G}_a \to 0$$

gives

$$H^1_{et}(F,\mathbf{Z}/2) = \mathrm{Ker}(H^1(F,\mathcal{O}_F) \xrightarrow{F-1} H^1(F,\mathcal{O}_F)) \neq 0$$

A non-zero element α of this group defines a separable double cover

$$\pi: \bar{F} \to F$$

(Chapter 0, §1). It is easy to see that it does not depend on the choice of α up to isomorphism.

Finally, let F be an $\boldsymbol{\alpha}_2$-surface. Then, the exact sequence in the flat topology

$$0 \to \boldsymbol{\alpha}_2 \to \mathbf{G}_a \to \mathbf{G}_a \to 0$$

implies that

$$H^1_{fl}(F,\boldsymbol{\alpha}_2) = \mathrm{Ker}(\mathbf{F}: H^1(F,\mathcal{O}_F) \xrightarrow{F} H^1(F,\mathcal{O}_F)) \neq 0 .$$

A non-zero element α of this group defines a double cover

$$\pi: \bar{F} \to F$$

(Chapter 0, §1). It is easy to see that it does not depend on the choice of α up to isomorphism.

Remark 1.3.1. For every finite group scheme G, one has a natural isomorphism

$$\mathrm{Hom}_{gr-sch}(G^*,\mathbf{Pic}_{X/k}) \cong H^1_{fl}(X,G),$$

where $G^*=\mathrm{Hom}_{gr-sch}(G,\mathbf{G}_m)$ is the Cartier dual group scheme (see **[Ray 3]**). This gives another explanation of the above construction of the principal double cover of an Enriques surface X. The corresponding element of $H^1_{fl}(X,G)$ defining the double cover is associated to the natural embedding of the dual group scheme G^* into the Picard scheme. Note also, that we have an isomorphism

$$G^*(k) \cong \mathrm{Ker}(f^*: \mathrm{Pic}(X) \to \mathrm{Pic}(Y)) ,$$

where $f:Y \to X$ is a principal G-cover.

We will call the double cover $\pi:\bar{F} \to F$ the **K3-cover** of F.

The following result shows that $\bar{F}$ is always "K3-like".

Proposition 1.3.1. Let $\pi:\bar{F} \to F$ be the K3-cover of an Enriques surface F. Then X is a Gorenstein surface satisfying

$$H^1(\bar{F},\mathcal{O}_{\bar{F}}) = 0, \; \omega_{\bar{F}} \cong \mathcal{O}_{\bar{F}}.$$

Moreover, $\bar{F}$ is nonsingular if $p \neq 2$ or F is a μ_2-surface, hence is a K3-surface.

PROOF. We have an exact sequence

$$0 \to \mathcal{O}_F \to \pi_*\mathcal{O}_{\bar{F}} \to \mathcal{L} \to 0$$

where $\mathcal{L} = \omega_F$ if F is classical, and $\mathcal{L} = \mathcal{O}_F$. if F is non-classical (see Chapter 0,§1). This implies that

$$\chi(\bar{F},\mathcal{O}_{\bar{F}}) = \chi(F,\pi_*\mathcal{O}_{\bar{F}}) = 2\chi(\bar{F},\mathcal{O}_{\bar{F}}) = 2.$$

On the other hand

$$\omega_{\bar{F}} = \pi^*(\omega_F \otimes \mathcal{L}^{-1}) \cong \pi^*(\mathcal{O}_F) \cong \mathcal{O}_{\bar{F}}.$$

The assertion about $H^1(\bar{F},\mathcal{O}_{\bar{F}})$ follows immediately by inspecting the following exact cohomology sequence

$$0 \to H^0(F,\mathcal{O}_F) \to H^0(\bar{F},\mathcal{O}_{\bar{F}}) \to H^0(F,\mathcal{L}) \to H^1(F,\mathcal{O}_F) \to H^1(\bar{F},\mathcal{O}_{\bar{F}}) \to H^1(F,\mathcal{L}) \to$$
$$\to H^2(F,\mathcal{O}_F) \to H^2(\bar{F},\mathcal{O}_{\bar{F}}) \to H^2(F,\mathcal{L}) \to 0.$$

Theorem 1.3.1. Let $\pi:\bar{F} \to F$ be the K3-cover of an Enriques surface and $\bar{F}'$ be a minimal nonsingular model of $\bar{F}$. Then $\bar{F}'$ is a K3-surface or a rational surface. The first case occurs if and only if $\bar{F}$ has only double rational points as singularities. The second case occurs if and only if $\bar{F}$ is non-normal or has one normal singularity of genus 1 besides double rational points.

PROOF. Applying Proposition 0.2.2 (v) we immediately obtain that $\bar{F}$ has only rational singularities (hence double because $\bar{F}$ is Gorenstein) if and only if $\bar{F}'$ is a K3-surface.

Next we observe that

$$b_1(\bar{F}) = b_1(F) = 0$$

because $\pi: \bar{F} \to F$ is inseparable. Thus it suffices to show that $|nK_{\bar{F}'}| = 0$ for each $n > 0$, where $\bar{F}$ is a nonsingular minimal model of F.

Case 1: $\bar{F}$ has only isolated non-rational singularities.

Then $\bar{F}$ is normal (since it is always Cohen-Macaulay as a double cover of a smooth surface). Let $p:\bar{F}' \to \bar{F}$ be a minimal resolution of singularities of $\bar{F}$. Since $\omega_{\bar{F}} \cong \mathcal{O}_{\bar{F}}$, the canonical class $K_{\bar{F}'}$ has a representative D which is concentrated at the exceptional curve of the resolution. If $|nK_{\bar{F}'}| \neq \emptyset$ for some $n > 0$, one can find a rational function f such that $(f) + nD \geq 0$. Since f is regular outside isolated normal singularities of $\bar{F}$, it is regular everywhere. Thus f is constant, hence nD and D are effective. However, the Leray spectral sequence for p shows that

$$H^0(\bar{F},\ R^1p_*\mathcal{O}_{\bar{F}'}) \cong H^2(\bar{F},\mathcal{O}_{\bar{F}}) \cong k,\ H^2(\bar{F}',\mathcal{O}_{\bar{F}'}) = h^0(K_{\bar{F}'}) = 0$$

$$H^1(\bar{F}',\mathcal{O}_{\bar{F}'}) = \{0\}.$$

By Castelnuovo's criterion **[B-M 2]**, $\bar{F}$ is rational. Moreover, we see that it has one singularity of genus 1 besides double rational points. The converse assertion is proven in the similar way.

Case 2: $\bar{F}$ is not normal. Let $\bar{F}'$ be the normalization of $\bar{F}$ and $p: \bar{F}' \to \bar{F}$ be the natural projection. It follows from the duality theory (cf. §1 of Chapter 0) that

$$p_*(\omega_{\bar{F}'}) \cong \mathcal{H}om_{\mathcal{O}_{\bar{F}}}(p_*(\mathcal{O}_{\bar{F}'}),\omega_{\bar{F}}) \cong \mathcal{H}om_{\mathcal{O}_{\bar{F}}}(p_*(\mathcal{O}_{\bar{F}'}),\mathcal{O}_{\bar{F}}).$$

We have a canonical exact sequence:

$$0 \to \mathcal{O}_{\bar{F}} \to p_*(\mathcal{O}_{\bar{F}'}) \to \mathcal{F} \to 0,$$

where $\mathcal{F}$ is supported on some curves. Dualizing, we obtain that $p_*(\omega_{\bar{F}'})$ is a proper Ideal in $\mathcal{O}_{\bar{F}}$. Let

$$\omega_{\bar{F}'} \to \mathcal{O}_{\bar{F}'}$$

be the homomorphism corresponding to the composition of inclusions

$$p_*(\omega_{\bar{F}'}) \hookrightarrow \mathcal{O}_{\bar{F}} \hookrightarrow p_*(\mathcal{O}_{\bar{F}'}).$$

It defines a homomorphism

$$\omega_{\bar{F}'} \to \mathcal{O}_{\bar{F}'}$$

which must be an inclusion on the open subset U of nonsingular points of $\bar{F}'$. This shows that $\omega_{\bar{F}'}|U$ is an Ideal in $\mathcal{O}_{\bar{F}'}$. The rest of the proof is similar to Case 1 . We prove that $|nK_{\bar{F}''}| = \emptyset$ for each $n > 0$, where $\bar{F}''$ is a minimal nonsingular model of $\bar{F}'$.

Recall that a variety V is said to be **unirational** if there exists a dominant rational map f: $\mathbb{P}^n \to V$. If one can find such f to be a purely inseparable map of degree p = char(k) then V is said to be a **Zariski variety**.

Lemma 1.3.1. Assume p = 2 and let X be a K3-surface with $b_2(X) = \rho(X) = 22$. Then X is a Zariski surface.

PROOF. See **[R-S 2]**.

Corollary 1.3.1. Assume p = 2. Then F is unirational if and only if it is not a μ_2-surface. Moreover F is a Zariski surface unless a minimal nonsingular model of its K3-cover $\bar{F}$ is a K3-surface.

PROOF. First of all a μ_2-surface is not unirational. Indeed, F has a unramified cover of degree 2 (the K3-cover). According to a result of **[Cr]**, the fundamental group of a unirational surface over a field of characteristic p has trivial p-Sylov subgroup.

Assume F is not a μ_2-surface. Let $\pi: \bar{F} \to F$ be the K3-cover. If $\bar{F}$ is rational we are done. In fact in this case F is a Zariski surface since π is an inseparable map of degree 2.Thus we may assume that $\bar{F}$ is not a rational surface. Then its minimal nonsingular model $\bar{F}'$ is a K3-surface satisfying the assumption of the lemma. In fact, $b_2(\bar{F}) = b_2(F) = 10$ because π is inseparable and hence the minimal resolution of $\bar{F}$ is obtained by blowing up 12 nodal curves whose classes in Num($\bar{F}'$) are linearly independent. Together with the inverse images of Num(F) they generate Num($\bar{F}'$). This shows that $\rho(\bar{F}') = b_2(\bar{F}') = 22$. Thus $\bar{F}$ is unirational and we are done.

§4. Differential invariants.

Let X be a smooth projective variety of dimension n over κ. Recall that the **de Rham cohomology** $H^i_{DR}(X)$ is defined as the hypercohomology of the de Rham complex

$$\Omega^*_X: 0 \to \mathcal{O}_X \to \Omega^1_X \to \dots \to \Omega^n_X \to 0.$$

Let

$$b_i^{DR}(X) = \dim_\kappa H^i_{DR}(X).$$

be the **de Rham Betti numbers** and

$$h^{p,q}(X) = \dim_\kappa H^q(X,\Omega^p_X).$$

be the **Hodge numbers**.

The first spectral sequence of hypercohomology

$$E_1^{p,q} = H^q(X,\Omega^p_X) \Rightarrow H^*_{DR}(X)$$

is called the **de Rham-Hodge spectral sequence**.

Lemma 1.4.1. Assume n = 2.

(i) $\Sigma_q(-1)^q h^{p,q}(X)) = \chi(X,\mathcal{O}_X)$, if p = 0 and 2 ;

(ii) $\Sigma_q(-1)^q h^{1,q}(X) = 2\chi(X,\mathcal{O}_X)-c_2(X)$;

(iii) $\Sigma_{p,q}(-1)^{p+q}h^{p,q}(X) = c_2(X)$;

(iv) $\Sigma_i(-1)^i b_i^{DR}(X) = c_2(X)$.

PROOF. (i) follows from Serre duality. (ii) follows from Riemann-Roch. (iii) is deduced from (i) and (ii). Finally (iv) follows from (iii) and the Hodge-de Rham spectral sequence.

Lemma 1.4.2. Assume $p = \mathrm{char}(k) = 0$ or $n \le p$ and X can be lifted to the Witt ring $W_2(k)$. Then the Hodge-de Rham spectral sequence degenerates and

$$b_i^{DR}(X) = \sum_{p+q=i} h^{p,q}(X).$$

PROOF. The case $p = 0$ is reduced to the case $k = \mathbb{C}$, where the assertion follows from the Hodge theory. In the case $p > 0$ the assertion is due to P.Deligne and L.Illusie (see **[D-I, Oe]**.

We also introduce the numbers

$$h^i(\Theta_X) = \dim_k H^i(X,\Theta_X),$$

where $\Theta_X = (\Omega_X^1)^*$ is the tangent bundle of X.

Lemma 1.4.3. Let X be a K3-surface. Then

$$\Theta_X \cong \Omega_X^1$$

and

$$h^{p,q}(X) = h^{q,p}(X) = 1 \text{ if } (p,q) = (0,0),(2,0),(2,2),$$

$$h^{1,1}(X) = 20,$$

$$h^{p,q}(X)) = 0 \text{ otherwise.}$$

PROOF. Since $\omega_X = \Omega_X^1 \cong \mathcal{O}_X$, the natural map

$$\Omega_X^1 \otimes \Omega_X^1 \to \Omega_X^2$$

defines an isomorphism

$$\Omega_X^1 \cong \Theta_X = (\Omega_X^1)^*.$$

This implies that

$$\dim H^q(X,\Omega_X^1) = \dim H^{2-q}(X,(\Omega_X^1)^*) = \dim H^{2-q}(X,\Omega_X^1).$$

By a result of **[R-S 1]**]

$$H^0(X,\Theta_X) = 0$$

Thus

$$\dim H^q(X,\Omega_X^1) = 0 \text{ , } q \neq 1.$$

We also have

$$\dim H^q(X,\Omega_X^0) = \dim H^{2-q}(X,(\Omega_X^2)^*) = \dim H^{2-q}(X,\mathcal{O}_X) = 1, \ q=0 \text{ or } 2.$$
$$= 0 \text{ , otherwise.}$$

It remains to use that $e(X) = 24$.

Proposition 1.4.1. Let F be an Enriques surface. Assume $p \neq 2$.

$$h^{0,0}(F) = h^{2,2}(F) = 1, \ h^{1,1}(F) = 10, \ h^{p,q}(F) = 0 \text{ otherwise;}$$

$$h^1(\Theta_F) = 10 \text{ , } h^i(\Theta_F) = 0 \text{ otherwise,}$$

$$b_i^{DR}(F) = b_i(F).$$

PROOF.Let $\pi: \bar{F} \to F$ be the K3-cover of F. Since it is unramified,

$$\pi^*(\Omega_F^i) \cong \Omega_{\bar{F}}^i.$$

Thus

$$\pi_*(\Omega_{\bar{F}}^i) = \pi_*(\pi^*(\Omega_F^i)) = \Omega_F^i \otimes \pi_*(\mathcal{O}_{\bar{F}}) = \Omega_F^i \oplus \Omega_F^i(K_F)$$

that implies that

$$H^q(\bar{F},\Omega_{\bar{F}}^i) \cong H^q(F,\pi_*\Omega_{\bar{F}}^i)) \cong H^q(F,\Omega_F^i) \oplus H^q(F,\Omega_F^i(K_F)).$$

Hence, by the previous lemma,

$$H^q(F,\Omega_F^i) = 0 \text{ if } q \neq i \text{ . .}$$

Since $e(F) = 12$ and $\chi(F,\mathcal{O}_F) = 1$, we obtain that

$$h^{1,1}(F) = 10.$$

The values of $h^{p,q}(F)$ with $p = 0$ or 2 are obvious.

Let

$$E_2^{p,q} = H^q(F,\Omega_F^p) \Rightarrow H_{DR}^{p+q}(F)$$

be the second spectral sequence associated with the De Rham cohomology (see **[Har 1]**. The corresponding 5-term exact sequence begins as follows:

$$0 \to H^1(F,\mathcal{O}_F) \to H^1_{DR}(F) \to H^0(F,\Omega^1_F) \to H^2(F,\mathcal{O}_F).$$

In our case, it gives an isomorphism

$$H^1_{DR}(F) \cong H^0(F,\Omega^1_F)$$

and by the preceding calculations., we get

$$b_1^{DR}(F) = b_1(F) = 0.$$

The remaining equalities follow from Lemma 1.4.1 and the Poincaré duality for the de Rham and the etale cohomology.

Lemma 1.4.4. Let F be an Enriques surface and $\alpha \in H^0(F,\Omega^1_F)$. Then α is closed (i.e. $d\alpha = 0$).

PROOF. Of course this is well-known if $k = \mathbb{C}$ or $p = 0$ and follows from the degeneracy of the Hodge-de Rham spectral sequence. The proof which follows is due to W.Lang **[La 2]** (see **[Il]** for a less elementary proof). First, this is obvious for a classical Enriques surface F (since $H^0(F,\Omega^2_F) = 0$). So, we may assume that $p = 2$ and F is not classical. In this case $K_F \sim 0$, $\Omega^1_F \cong \Theta_F$ and $H^0(F,\Omega^1_F) = k$. We have to show that the differentiation map

$$d\colon E_1^{1,1} = H^0(F,\Omega^1_F) \to E_1^{2,0} = H^0(F,\Omega^2_F)$$

is trivial.

If this is not true, d is surjective and the differential

$$d\colon H^2(F,\mathcal{O}_F) = E_1^{0,2} \to H^2(F,\Omega^1_F) = E_1^{1,2}$$

is injective. Indeed, $H^4_{DR}(F) \cong k$ and hence the differential $E_1^{1,2} \to E_1^{2,2}$ is zero. Let $a \in E_1^{0,2}$ and let $b \in E_1^{1,0}$. Then $0 = d(a \cup b) = da \cup b \pm a \cup db$ and $da \cup b = \pm a \cup db$.

Let a be a non-zero element in $H^1(F,\mathcal{O}_F)$. As we have seen in the proof of Theorem 1.2.1, its cup-product $a \cap a \neq 0$ in $H^2(F,\mathcal{O}_F)$. Since $p = 2$,

$$d(a \cap a) = a \cap d^*(a) + d^*(a) \cap a = 2(a \cap da) = 0.$$

This contradiction proves the lemma.

Proposition 1.4.2. Let F be an Enriques surface. Assume that $p = 2$. Then the numbers $h^{p,q}(F) = \dim H^q(F,\Omega_F^p)$, $h^i(\Theta_F) = \dim H^i(F,\Theta_F)$ and $b_i^{DR}(F)$ are given by the following table:

type	$h^{1,0}$	$h^{0,1}$	$h^{1,1}$	$h^{2,0}$	$h^{0,2}$	$h^{1,2}$	$h^{2,1}$	$h^{2,2}$	$h^0(\Theta_F)$	$h^1(\Theta_F)$	$h^2(\Theta_F)$	$b_1^{DR}(F)$	$b_2^{DR}(F)$
classical	1	0	12	0	0	1	0	1	$a=?$	$10+2a$	a	1	12
μ_2	0	1	10	1	1	0	1	1	0	10	0	1	12
α_2	1	1	12	1	1	1	1	1	1	12	1	1	12

Moreover $a = 0$ if F is classical and has a regular 1-form with isolated zeroes.

PROOF. Suppose F is classical. We know that $h^{0,1} = h^{2,0} = h^{0,2} = h^{2,2} = 0$. Since F admits a principal μ_2-cover, $h^{1,0} \neq 0$ (Proposition 0.1.11). As in the proof of Proposition 1.4.1 we use the exact sequences

$$0 \to H^1(F,\mathcal{O}_F) \to H^1_{DR}(F) \to H^0(F,\Omega_F^1) \to H^2(F,\mathcal{O}_F).$$

Since every regular 1-form on F is closed (Lemma 1.4.4), we may apply a result of **[Oda]**, which shows that

$$H^1_{DR}(F) \cong DM({}_2\mathbf{Pic}_{F/K}),$$

where DM( ) denotes the Dieudonne module associated to a finite group scheme (see **[Oo]**). In our case, it is one-dimensional. Thus

$$H^1_{DR}(F) \cong H^0(F,\Omega_F^1).$$

This shows that $h^{1,0} = 1$.

Let s be a non-zero section of Θ_F with isolated zeroes. It defines an exact sequence

$$0 \to \mathcal{O}_F \to \Omega_F^1 \to \mathcal{J}_Z\omega_F \to 0,$$

where Z is the subscheme of zeroes of s and $\mathcal{J}_Z$ its ideal sheaf. Note that Z is not empty. In fact, one can prove as in Proposition 0.1.13 that the number of zeroes counted with multiplicities is equal to $e(X) = 12$. Tensoring by ω_F^{-1} and using the isomorphism $\Theta_F \cong \Omega_F^1 \otimes \omega_F^{-1}$, we obtain an exact sequence

$$0 \to H^0(F,\omega_F^{-1}) \to H^0(F,\Theta_F) \to H^0(\mathcal{J}_Z).$$

This obviously implies that $H^0(F,\Theta_F) = 0$.

Suppose F is a μ_2-surface. Then $\Omega_F^2 \cong \mathcal{O}_F$,

$$\Theta_F \cong \Omega_F^1$$

(similar to the proof of Lemma 1.4.3), and

$$h^{1,0} = h^0(\Theta_F) = h^0(\Theta_F) = 0.$$

On the other hand, we know that $h^{0,1} = h^{2,0} = h^{2,0} = h^{2,1} = h^{2,2} = 1$. Since $K_F = 0$, by Serre duality, $h^2(\Theta_F) = h^{1,0} = 0$. The remaining values of $h^{p,q}(F)$ are obtained by applying Lemma 1.4.1.

Suppose F is α_2-surface. Again

$$\Theta_F \cong \Omega_F^1$$

We have seen already in the proof of Theorem 1.2.2 that there exists a nonzero vector field ∂ on F. This shows that $h^0(\Theta_F) \neq 0$. To prove that $h^{1,0} = 1$, we again use the De Rham cohomology. As in the case of classical Enriques surfaces, we see that $H^1_{DR}(F)$ is one -dimensional. Hence $H^1(F,\mathcal{O}_F) \cong H^1_{DR}(F)$ and

$$H^0(F,\Omega_F^1) \hookrightarrow H^2(F,\mathcal{O}_F) \cong K\ .$$

Finally, we have seen already that $b_i^{DR}(F) = 1$ independent of the type of F. By Lemma 1.4.1, we get $b_2^{DR}(F) = 12$. This proves the assertion.

Corollary 1.4.1. Let F be an Enriques surface. Assume that $p \neq 2$ or $h^0(\Theta_F) = 0$ (i.e. F is a μ_2-surface or classical with a regular 1-form with only isolated singularities). Then F can be lifted to characteristic 0.

PROOF. Recall that the latter means the existence of a smooth proper map $f: \mathfrak{F} \to \mathrm{Spec}\, R$ where R is a local integral noetherian ring with the fraction field of characteristic 0 and where the fibre over the closed point of Spec R is isomorphic to F. By general theorems (see **[Gro 2]**) this follows from the vanishing of $H^0(F,\Theta_F)$ which allows one to lift F to a formal scheme over R and the vanishing of $H^2(F,\mathcal{O}_F)$ which allows one to algebraicize a formal lifting). Thus the only problem for us is the case where F is a μ_2-surface. In this case, a formal lifting exists but its algebraization is not clear.

Let F be a μ_2-surface and f: $\bar{F} \to F$ be the K3-cover of F. Since the Frobenius **F** acts trivially on $H^2(F,\mathcal{O}_F) \cong H^1(F,\mathcal{O}_F)$, it also acts trivially on $H^1(\bar{F},\mathcal{O}_{\bar{F}})$. By **[Cr]** $\bar{F}$ is an ordinary K3-surface. Applying a theorem of N.Nygaard **[Ny]** we find a canonical lifting of $\bar{F}$ to characteristic 0. By its definition , the canonical lifting lifts also the involution τ generating the Galois group of f. Since it acts freely on the special fibre, it must act freely on the generic fibre . This shows that a lifting of F exists and can be obtained as a quotient of the lifting of $\bar{F}$.

Remark 1.4.1. Let F be of type α_2. Since $b_1^{DR}(F) \neq h^{1,0}+h^{0,1}$, we may apply Lemma 1.4.2 to obtain that F cannot be lifted to the Witt ring $W_2(k)$.
It is not known whether every α_2-surface can be lifted to characteristic 0. There are examples where such a surface can be lifted to a ramified extension of the Witt ring.

Finally, we give the values of the crystalline cohomology and the cohomology with coefficients in the sheaf of Witt rings for an Enriques surface. It is known that the De Rham cohomology $H^*_{DR}(X/k)$, the Hodge cohomology $H^*(X,\Omega^*_{X/k})$, the Witt cohomology $H^*(X,W\mathcal{O}_X)$ and the crystalline cohomology $H^*(X/W)$ of a proper smooth scheme X over a perfect field k of characteristic p > 0 are closely related by means of the De Rham-Witt complex. We refer for this theory and its application to Enriques surfaces to [Il]. Here we only summarize the results of computations for Enriques surfaces.

Let W = W(k) be the Witt ring of the field k, $W\mathcal{O}_X$ is the sheaf on a scheme X over k defined as a sheaf associated to the presheaf $U \to W\mathcal{O}_X(U) = W(\mathcal{O}_X(U))$ for every open U on X. The cohomology $H^*(X,W\mathcal{O}_X)$ has a natural structure of W-module. Also one can define two k-linear operators **F** and **V** on $H^*(X,W\mathcal{O}_X)$ satisfying:

$$\mathbf{FV} = \mathbf{VF} = \text{multiplication by } p,$$

$$\mathbf{F}a = a^\sigma\mathbf{F},\ a\mathbf{V} = \mathbf{V}a^\sigma, \text{ for any } a \in W,$$

where $(\ )^\sigma$ denotes the lifting of the Frobenius homomorphism $x \to x^p$ to W.

Proposition 1.4.3. Let F be an Enriques surface over k. Then

$$H^0(F, W\mathcal{O}_F) = W,\ H^1(F, W\mathcal{O}_F) = 0,$$

$$H^2(F, W\mathcal{O}_F) = 0 \text{ if } F \text{ is classical},$$

$$= k \text{ with } \mathbf{F}x = x^p,\ \mathbf{V} = 0 \text{ if } F \text{ is an } \mu_2\text{-surface},$$

$$= k \text{ with } \mathbf{F} = \mathbf{V} = 0 \text{ if } F \text{ is an } \alpha_2\text{-surface}.$$

Proposition 1.4.4. Let F be an Enriques surface over k. Then

$$H^0(F/W) \cong H^4(F/W) \cong W;$$

$$H^1(F/W) = 0;$$

$$H^2(F/W) \cong W^{10} \text{ if } p \neq 2,$$

$$\cong W^{10} \oplus k \text{ if } p = 2;$$

$$H^3(F/W) = 0 \text{ if } p \neq 2,$$

$$\cong k \text{ if } p = 2.$$

§5. Riemann-Roch and a vanishing theorem.

Let F be an Enriques surface. Since $K_F \equiv 0$, the **Riemann-Roch formula** on F looks like

$$h^0(\mathcal{O}_F(D)) - h^1(\mathcal{O}_F(D)) + h^2(\mathcal{O}_F(D)) = \tfrac{1}{2}D^2 + 1.$$

By **Serre duality**

$$h^i(\mathcal{O}_F(D)) = h^{2-i}(\mathcal{O}_F(K_F - D)).$$

Since the arithmetic genus $p_a(D)$ of a divisor D is equal to

$$p_a(D) = \tfrac{1}{2}D^2 + 1.$$

we can write the above formula in the form

$$h^0(\mathcal{O}_F(D)) - h^1(\mathcal{O}_F(D)) + h^0(\mathcal{O}_F(K_F - D)) = p_a(D).$$

Note also that , if $D > 0$,$D \neq 0$, then

$$h^0(\mathcal{O}_F(K_F - D)) = h^2(\mathcal{O}_F(D)) = 0.$$

Theorem 1.5.1. Let C be an irreducible curve on F with $C^2 > 0$. Then

$$H^1(F,\mathcal{O}_F(C)) = H^1(F,\mathcal{O}_F(K_F+C)) = 0.$$

PROOF. If $k = \mathbb{C}$, this follows immediately from the Kodaira vanishing theorem. Indeed,

$$H^1(F,\mathcal{O}_F(C)) \cong H^1(F,\mathcal{O}_F(K_F+C')),$$

where $C' = C+K_F \equiv C$ is an ample divisor by the Moishezon-Nakai criterion.

Assume first that $\mathrm{char}(k) = p \neq 2$. The cohomology sequence for the exact sequence

$$0 \to \mathcal{O}_F(-C) \to \mathcal{O}_F \to \mathcal{O}_C \to 0$$

gives easily that

$$H^1(F,\mathcal{O}_F(K_F+C)) \cong H^1(F,\mathcal{O}_F(-C)) = 0.$$

Tensoring the above exact sequence by $\mathcal{O}_F(K_F)$, we obtain similarly

$$H^1(F,\mathcal{O}_F(C)) \cong H^0(C,\mathcal{O}_C(K_F)).$$

Let $\pi\colon \bar{F} \to F$ be the K3-cover of F. The curve $\pi^{-1}(C)$ splits into two connected components if and only if $\mathcal{O}_C(K_F)$ is the trivial element of ${}_2\mathrm{Pic}(C)$ (see Chapter 0,§1), or equivalently $H^0(C,\mathcal{O}_C(K_F)) \neq 0$. Since $C^2 > 0$, $\pi^{-1}(C)$ does not split because of the Hodge Index Theorem.

Assume now $p = 2$. The proof will consist of several steps.

<u>Step 1</u>: $H^1(F,\mathcal{O}_F(K_F+C)) = 0$ if $K_F \nsim 0$.

The proof is the same as in the case $p \neq 2$.

<u>Step 2</u>: $\dim H^1(F,\mathcal{O}_F(C)) \leq 1$ if $K_F \nsim 0$.

The exact sequence

$$0 \to \mathcal{O}_F \to \mathcal{O}_F(C) \to \mathcal{O}_C(C) \to 0$$

and the duality on C gives an isomorphism

$$H^1(F,\mathcal{O}_F(C)) \cong H^1(C,\mathcal{O}_C(C)) \cong H^1(C,\omega_C(K_F)) \cong H^0(C,\mathcal{O}_C(K_F)).$$

Obviously, the dimension of the last space is at most 1.

<u>Step 3</u>: $H^1(F,\mathcal{O}_F(C)) \neq 0$ and $K_F \sim 0$ imply $\dim H^1(F,\mathcal{O}_F(C)) = 1$ and the existence of

a divisor $D \in |C|$ such that $h^0(\mathcal{O}_D) = 2$.

Let $H^0(F,\mathcal{O}_F(C))_0$ be the subspace of $H^0(F,\mathcal{O}_F(C))$ which consists of sections $s:\mathcal{O}_F \to \mathcal{O}_F(C)$ such that the corresponding map $H^1(F,\mathcal{O}_F) \to H^1(F,\mathcal{O}_F(C))$ is trivial. Since $\mathcal{O}_C(C) \cong \omega_C$, we have an exact sequence

$$0 \to \mathcal{O}_F \to \mathcal{O}_F(C) \to \omega_D \to 0$$

where $D = \mathrm{div}(s)$. The corresponding cohomology sequence

$$H^1(F,\mathcal{O}_F) \to H^1(F,\mathcal{O}_F(C)) \to H^1(D,\omega_D) \to H^2(F,\mathcal{O}_F) \to 0$$

shows, by taking $D = C$, that $\dim H^1(F,\mathcal{O}_F(C)) \le 1$. If $H^1(F,\mathcal{O}_F(C)) \neq 0$, then $\mathrm{codim}\ H^0(F,\mathcal{O}_F(C))_0 = \mathrm{codim}\ \mathrm{Ker}(H^1(F,\mathcal{O}_F(C)) \to \mathrm{Hom}(H^1(F,\mathcal{O}_F),H^1(F,\mathcal{O}_F(C))) \le 1$. This implies

$$\dim H^0(F,\mathcal{O}_F(C))_0 \ge \dim H^1(F,\mathcal{O}_F(C)) - 1 = \tfrac{1}{2}C^2+1 > 0.$$

For every $s \in H^0(F,\mathcal{O}_F(C))_0$, by duality,

$$\dim H^0(D,\mathcal{O}_D) = \dim H^1(D,\omega_D) = \dim H^1(F,\mathcal{O}_F(C)) + \dim H^2(F,\mathcal{O}_F) = 2.$$

<u>Step 4</u>: If $K_F \sim 0$, then $H^1(F,\mathcal{O}_F(C)) = 0$.

Assume that $H^1(F,\mathcal{O}_F(C)) \neq 0$. By the previous step, we can choose $D \in |C|$ such that $h^0(\mathcal{O}_D) = 2$. By Chapter 4, Proposition 4.3.1, D is not 1-connected. Thus, $D \sim D_1+D_2$, where D_1 and D_2 are two positive divisors with $D_1 \cdot D_2 \le 0$. By Riemann-Roch,

$$1+\tfrac{1}{2}C^2 = \dim |C| \ge \dim |D_1| + \dim |D_2| \ge \tfrac{1}{2}(D_1^2 + D_2^2) = \tfrac{1}{2}C^2 - D_1 \cdot D_2 .$$

Therefore, $D_1 \cdot D_2 \ge -1$. Since $C^2 > 0$, we may assume that $D_1^2 > 0$. By the Hodge Index theorem, $D_2^2 \le 0$. Thus, $D_1^2 \ge 2+C^2$, and , by Riemann-Roch, $\dim |C| = \dim |D_1|$. This contradicts the irreducibility of C.

<u>Step 5</u>: The assertion is true if $C^2 = 2$.

It will be shown later in Proposition 3.6.1 of Chapter 3 that

$$C \sim E_1+E_2 \text{ or } 2E_1+R + K_F,$$

where E_1 and E_2 are divisors of arithmetic genus 1 with $\dim |E_i| = 0$ and $\dim |2E_i| = 1$, and R is a smooth rational curve. If the first case occurs, we write an exact sequence:

$$0 \to \mathcal{O}_F(E_1) \to \mathcal{O}_F(E_1+E_2) \to \mathcal{O}_{E_1}(E_1+E_2) \to 0$$

Since $E_1 \cdot (E_1+E_2) = 1$ and $h^0(\mathcal{O}_F(E_1)) = 1$, we deduce from Riemann-Roch that

$$H^1(F,\mathcal{O}_F(E_1)) = 0 \ , \ H^1(E_1,\mathcal{O}_{E_1}(E_1+E_2)) = 0.$$

The exact cohomology sequence shows that

$$H^1(F,\mathcal{O}_F(E_1+E_2)) = H^1(F,\mathcal{O}_F(C)) = 0.$$

By Step 1, we are done.

If the second case occurs, we use the exact sequence

$$0 \to \mathcal{O}_F(2E_1+K_F) \to \mathcal{O}_F(2E_1+R+K_F) \to \mathcal{O}_R \to 0$$

To prove our assertion it suffices to show that $H^1(F,\mathcal{O}_F(2E_1+K_F)) = 0$. For this we consider the exact sequence

$$0 \to \mathcal{O}_F(E_1+K_F) \to \mathcal{O}_F(2E_1+K_F) \to \mathcal{O}_{E_1}(2E_1+K_F) \to 0.$$

It is enough to show that

$$H^1(F,\mathcal{O}_F(E_1+K_F)) = H^1(F,\mathcal{O}_{E_1}(2E_1+K_F)) = 0.$$

Since $2E_1$ moves in a pencil and $E_1^2 = 0$, we have

$$\mathcal{O}_{E_1}(2E_1+K_F) \cong \mathcal{O}_{E_1}(K_F).$$

The exact sequence

$$0 \to \mathcal{O}_F(-E_1+K_F) \to \mathcal{O}_F(K_F) \to \mathcal{O}_{E_1}(K_F) \to 0.$$

and the vanishing of $H^1(F,\mathcal{O}_F(-E_1+K_F)) = H^1(F,\mathcal{O}_F(E_1))$ and $H^0(F,\mathcal{O}_F(K_F))$ imply

$$H^1(F,\mathcal{O}_{E_1}(K_F)) = 0.$$

It remain to use that $h^0(\mathcal{O}_F(E_1+K_F)) = 1$ (because $h^0(\mathcal{O}_F(2E_1)) = 2$) and hence , by Riemann-Roch, $H^1(F,\mathcal{O}_F(E_1+K_F)) = 0$.

We finish again by applying Step 1.

<u>Step 6</u>: Assume $K_F \nsim 0$ and $C^2 > 2$, then $H^1(F,\mathcal{O}_F(C)) = 0$.

Let $|C+K_F| = |M| + Z$ be the decomposition of $|C+K_F|$ into its moving part $|M|$ and fixed part Z. Assume that $H^1(F,\mathcal{O}_F(C)) \neq 0$. By step 2, $\dim H^1(F,\mathcal{O}_F(C)) = 1$ and $\dim H^1(F,\mathcal{O}_F(M)) \leq 1$. Since $H^1(F,\mathcal{O}_F(C+K_F)) = 0$ (Step 1),

$$\dim |C+K_F| = \tfrac{1}{2}C^2 = \dim |M| = \tfrac{1}{2}M^2 + \dim H^1(F,\mathcal{O}_F(M)).$$

If $M^2 = C^2$, then $H^1(F,\mathcal{O}_F(M)) = 0$ and $C\cdot Z = Z^2 = 0$. By the Hodge Index theorem, $Z = 0$, i.e. $C+K_F \sim M$ and $H^1(F,\mathcal{O}_F(C)) = H^1(F,\mathcal{O}_F(M+K_F)) = 0$ (Step 1).

If $M^2 = C^2-2$, then $H^1(F,\mathcal{O}_F(M)) \neq 0$. We proceed by induction until we come to the case $M^2 = 0$. Then $C^2 = 2$ and we are done by Step 5.

Corollary 1.5.1. Let C be an irreducible curve on an Enriques surface F with $C^2 > 0$. Then

$$\dim|C| = \dim |C+K_F| = p_a(C)-1 = \tfrac{1}{2}C^2.$$

Remark 1.5.1. If F can be lifted to the Witt ring $W_2(k)$ (see Corollary 1.4.1) then the vanishing of $H^1(F,\mathcal{O}_F(C+K_F)$ follows from a result of M.Raynaud (see **[Oe]**).

S6. Examples.

Here we exhibit some families of classical and non-classical Enriques surfaces in characteristic $p = 2$.

Lemma 1.6.1. Let Y be a surface in $\mathbb{P}^5$ which is a complete intersection of three quadrics. Then $\omega_Y \cong \mathcal{O}_Y$, $H^1(Y,\mathcal{O}_Y) = 0$. In particular, if Y is smooth, it is a K3-surface.

PROOF. The adjunction formula gives immediately $\omega_Y \cong \mathcal{O}_Y$. On the other hand, by Serre's theorem, that $H^1(Y,\mathcal{O}_Y) = 0$.

To construct examples of Enriques surfaces, we can define a fixed-point-free action of a group scheme of order 2 on Y as above. The quotient surface will be the needed example.

Let x_0,x_1,x_2,y_0,y_1,y_2 be homogeneous coordinates in $\mathbb{P}^5$. Let G be a group scheme of order 2 acting on $\mathbb{P}^5$ by the formula

$$(x_i, y_i) \to (x_i, y_i+x_i) \text{ if } G = \mathbf{Z}/2,$$

$$\to (x_i\otimes 1, y_i\otimes 1+x_i\otimes t) \text{ if } G = \text{Spec } k[t]/(t^2) \cong \boldsymbol{\alpha}_2,$$

$$\to (x_i\otimes 1, x_i\otimes t+y_i\otimes(t+1)) \text{ if } G = \text{Spec } k[t,t^{-1}]/(t^2-1) \cong \boldsymbol{\mu}_2.$$

We easily find 12 quadrics in $\mathbb{P}^5$ which are invariant with respect to this action. They are

$$G = Z/2 : x_k^2, x_ix_j, y_k^2+x_ky_k, y_ix_j+y_jx_i \quad 0 \le i<j \le 2, 0 \le k \le 2;$$

$$G = \boldsymbol{\alpha}_2 : x_k^2, x_ix_j, y_k^2, y_ix_j+y_jx_i, 0 \le i<j \le 2, 0 \le k \le 2;$$

$$G = \boldsymbol{\mu}_2 : x_k^2, x_ix_j, y_k^2, y_jx_i+y_jy_i, 0 \le i<j \le 2, 0 \le k \le 2.$$

Let $f: \mathbb{P}^5 \to \mathbb{P}^{11}$ be the corresponding morphism. It factors through the quotient space $\mathbb{P}^5/G$ and embeds it onto a subvariety H of $\mathbb{P}^{11}$. Choose 3 generic hyperplanes L_i in $\mathbb{P}^{11}$ and define Y to be the inverse image of

$$F = H \cap L_1 \cap L_2 \cap L_3.$$

Then Y is an intersection of 3 quadrics in $\mathbb{P}^5$. Since the fixed locus of G in $\mathbb{P}^5$ is one or two planes, dim Sing(H) $\le$ 2. Thus F is smooth. This implies also that the projection $\pi: Y \to F$ is a principal double G-cover. By Lemma 1.5.1 and Proposition 0.1.3, $\pi^*(\omega_F) \cong \mathcal{O}_Y$. Therefore

$$\omega_F \in \text{Ker}(\text{Pic}(F) \to \text{Pic}(Y)) \cong G^*.$$

This implies that $\omega_F \cong \mathcal{O}_F$, if $G \cong \boldsymbol{\alpha}_2$ or $\mathbf{Z}/2$, and $\omega_F^{\otimes 2} \cong \mathcal{O}_F$ if $G = \boldsymbol{\mu}_2$. We also know that $\chi(\mathcal{O}_F) = \frac{1}{2}\chi(\mathcal{O}_Y) = 1$. Thus F is an Enriques surface, which is classical (resp. $\boldsymbol{\mu}_2$-surface, resp. $\boldsymbol{\alpha}_2$-surface) if $G = \boldsymbol{\mu}_2$ (resp. $\mathbf{Z}/2$, resp. $\boldsymbol{\alpha}_2$).

For example, take Y to be a complete intersection of the three quadrics

$$x_0^2+x_0x_1+x_1^2+x_0y_1+x_1y_0+ y_2^2 = 0 ,$$

$$x_1^2 +x_1x_2+x_2^2+x_1y_2+x_2y_1 +y_0^2 = 0 ,$$

$$x_2^2+x_0x_2+x_0^2+x_2y_0+x_0y_2+y_1^2 = 0 .$$

This surface has exactly 12 nodes which have the following coordinares:

$$(1,1,1,1,1.1), \quad (1,t,t,t,t^2,t^2), \quad \text{where } t^3+t^2+1=0,$$

$$(1,t,t^2,0,0,0), \quad \text{where } t^2+t+1 = 0.$$

The quotient surface with respect to the action of α_2 is an Enriques α_2-surface.

Bibliographical notes to Chapter I.

The classification of algebraic surfaces over fields of characteristic zero is mostly due to Italian algebraic geometeres. It has been revived in the sixties in the works of K.Kodaira, I.Shafarevich and his students and O.Zariski. Shafarevich **[AS]** was the first who introduced the notion of what is now called, the Kodaira dimension and arranged the classification according to the values of this invariant.Since then, the place of Enriques surfaces in the classification of all surfaces has become especially clear. We made the references to the modern expositions of the theory of algebraic surfaces in §1. In **[B-P-VdV]** one can find all the results of this chapter in the case $k = \mathbb{C}$.

In **[Mum 2, B-M 1-2]** the classification of algebraic surfaces is extended to the case of positive characteristic.The definition of Enriques surfaces in all characteristic and the theory of the Picard scheme of an Enriques surface over a field of characteris- tic 2 from §2 is given in **[B-M 2]**. The proof given there of the fact that the second Betti number b_2 of an Enriques surface is equal to its Picard number r is different from ours. They prove first that every Enriques surface contains an elliptic or a quasi-elliptic pencil by generalizing the classic reducibility lemma to characteristic 2. Then they used an idea of M.Artin which allows one to show that the difference b_2-r stays the same after passing to the jacobian fibration. Our approach is a mix of this idea and an idea of W.Lang **[La 2]** which allows one to prove the existence of an elliptic pencil without using the reducibility lemma. This work contains another proof of this result which does not use the theory of elliptic fibrations. A proof of Theorem 1.2.2 based on the theory of crystalline cohomology can be found in **[Il]**.

The results from §3 were first obtained by P.Blass **[Bl 2]**. The fact that a μ_2-surface is not unirational was fist noticed by R.Crew **[Cr]**. In the case where k is the algebraic closure of a finite filed, the fact was noticed by T.Katsura **[Kat 1]**.We refer to [La 5] and [Ek 2] for more information on K3-covers of classical Enriques surfaces

§4 is mostly due to W.Lang [**La** 3].We refer to **[Ek 2]**for more information on vector fields on classical Enriques surfaces.

§4 is mostly due to W.Lang **[La 3]**.We refer to **[Ek 2]** for additional information on vector fields on classical Enriques surfaces.

The vanishing theorem from §5 seems to be a new result in the case p = 2. Note that the first counterexamples to the Kodaira-Ramanujam vanishing theorems for nonsingular surfaces in characteristic p > 0 were constructed by W. Lang **[La 1]** and M. Raynaud **[Ra 4]**.

The construction of examples from §6 is due to M.Reid. It is reproduced in **[B-M 2]**. Another example of an Enriques surface in characteristic 2 was considered in **[Bl 1]**.

Chapter II

LATTICES AND ROOT BASES.

§1. Generalities.

A **lattice** is a free abelian group M of finite rank rk(M) equipped with a symmetric bilinear form $\varphi: M\times M \to \mathbf{Z}$. The value of this form on a pair $(x,y)\in M\times M$ will be denoted by $x\cdot y$. We write x^2 to denote $x\cdot x$.

By linearity, φ defines a symmetric bilinear form on the real vector space $M_{\mathbb{R}} = M\otimes_{\mathbf{Z}}\mathbb{R}$ .The **signature** of a lattice M is the signature (t_+,t_-,t_0) of the corresponding real quadratic form $x \to x^2$. Thus. we can speak about positive (negative) **definite, semi-definite and indefinite** lattices. A lattice is called **non-degenerate** if $t_0 = 0$. In this case, we often drop t_0 in the notation of the signature. A lattice of signature (1,rk(M)−1) is said to be **hyperbolic**.

A homomorphism of lattices $f:M \to M'$ is a homomorphism of abelian groups such that $f(x)\cdot f(y) = x\cdot y$ for any $x,y\in M$. An injective (resp. bijective) homomorphism of lattices is called an **embedding** (resp. **isometry**). Two lattices are called **isomorphic** if there exists an isometry from one to another. The set of isometries $\sigma:M \to M$ is a group with respect to composition of maps. It is called the **orthogonal group** of M and is denoted by O(M).

A **sublattice** M' of a lattice M is a subgroup of M equipped with the induced bilinear form. A sublattice M' of M is said to be **primitive** (resp.

of finite index a) if the quotient group M/M' is a free abelian group (resp. a finite group of order a). An element $m \in M$ is called a **primitive element** if the sublattice $\mathbf{Z}m \subset M$ is primitive. The **sum** of two sublattices M_1 and M_2 of M is the minimal sublattice of M containing M_1 and M_2. It is denoted by M_1+M_2. If $x \cdot y = 0$, for any $x \in M_1$, $y \in M_2$, this sum is said to be the **orthogonal sum of sublattices** and is denoted by $M_1 \perp M_2$.

The **orthogonal sum of two lattices** M and M' is the lattice $M \oplus M'$ equipped with the bilinear form $(x,x')\cdot(y,y') = x\cdot y + x'\cdot y'$. Similarly, one defines the orthogonal sum of any finite number of lattices. The **orthogonal complement** of a sublattice M' of a lattice M is the sublattice

$$M'^{\perp} = \{x \in M : x \cdot y = 0 \text{ for all } y \in M'\}.$$

Let M^* denote the abelian group $\mathrm{Hom}_{\mathbf{Z}}(M,\mathbf{Z})$; the map $x \to \varphi(x,\cdot)$ is a homomorphism $i_M : M \to M^*$ of abelian groups. The kernel of i_M is denoted by Rad(M) and is called the **radical** of M. The lattice M is non-degenerate if and only if $\mathrm{Rad}(M) = \{0\}$. The cokernel

$$D(M) = M^*/i_M(M)$$

is denoted by D(M) and is called the **discriminant group** of the lattice M. If M is non-degenerate, D(M) is a finite group and its order is called the **discriminant** of M and is denoted by discr(M). A lattice M is called **unimodular** if i_M is bijective.

If $\underline{e} = \{e_1,\ldots,e_r\}$ is a basis of M , the matrix $G(\underline{e}) = (e_i \cdot e_j)$ is called the **Gram matrix** of M with respect to $\underline{e}$. If M is non-degenerate, then

$$\mathrm{discr}(M) = |\det(G(\underline{e}))| .$$

It is clear that a lattice structure on a free abelian group M is determined by the Gram matrix with respect to some basis of M.

A lattice M is called **even** if $x^2 \in 2\mathbf{Z}$ for every $x \in M$. There is a natural one-to-one correspondence between even lattices and **integral quadratic forms** (or free quadratic Z-modules in sense of **[Ser 4]**. The latter is defined as a free abelian group M together with a map $q: M \to \mathbf{Z}$ such that $q(nx) = n^2 q(x)$ for any $n \in \mathbf{Z}$ and the map $(x,y) \to q(x+y)-q(x)-q(y)$ is a bilinear map from $M \times M$ to $\mathbf{Z}$.

Let n be an integer. The following notations will be used.

$M(n)$: the lattice obtained from the lattice M by multiplying the values of its symmetric bilinear form by n.

$\langle n \rangle$: the lattice $\mathbf{Z}x$ of rank 1 with $x^2 = n$.

$M_n = \{ x \in M : x^2 = n \}$ (the elements of M_0 are called **isotropic vectors**).

M_n' : the subset of primitive vectors from M_n.

M^n : the orthogonal sum of n copies of a lattice M .

Here are some standard lattices we will be dealing with:

$U_{[n]} = \mathbf{Z}e_1 + \ldots + \mathbf{Z}e_n$, where $e_i \cdot e_j = 1 - \delta_{ij}$, δ_{ij} is the Kronecker symbol (when n = 2 it is called the **standard hyperbolic plane** and is denoted by U).

$Q_{p,q,r} = \mathbf{Z}\alpha_0 + \ldots + \mathbf{Z}\alpha_{p+q+r-2}$, where $\alpha_i^2 = -2$, and $\alpha_i \cdot \alpha_j = 1$ or 0 depending on whether α_i is joined to α_j in the following graph $T_{p,q,r}$:

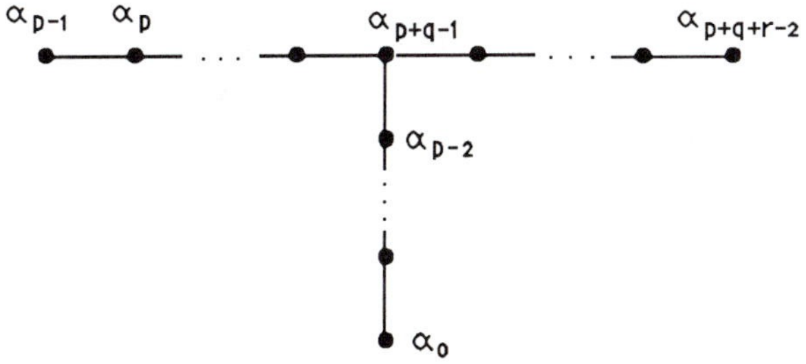

$E_6 = Q_{2,3,3}$, $E_7 = Q_{2,3,4}$, $E_8 = Q_{2,3,4}$, $D_n = Q_{2,2,n-2}$ $(n \geq 4)$, $A_n = Q_{1,1,n}$, $\tilde{E}_6 = Q_{3,3,3}$, $\tilde{E}_7 = Q_{2,4,4}$, $\tilde{E}_8 = Q_{2,3,6}$.

Note that all the above lattices are even. The lattices U and $Q_{p,q,r}$

$(p^{-1}+q^{-1}+r^{-1} < 1)$ are hyperbolic. The lattices A_n, D_n, E_n are negative definite. The lattices $\tilde{E}_n$ are negative semi-definite with radical of rank 1.

$$\mathrm{discr}(Q_{p,q,r}) = |pqr-pq-qr-pr|.$$

Proposition 2.1.1. Let M be an even (resp. non-even) unimodular indefinite lattice of signature (a,b) with $a \le b$. Then $M \cong U^a \oplus E_8{}^{b-a}$ (resp. $M \cong \langle 1\rangle^a \oplus \langle -1\rangle^b$)

PROOF. See **[Ser 4]**, Chapter 5, Theorem 5.

S2. Root bases and their Weyl groups.

Let M be a lattice. A non-empty subset B of M_{-2} is called a **root basis** if $\alpha\cdot\beta \ge 0$ for all $\alpha \ne \beta$ of B and every α is not a linear combination with positive coefficients of elements of $B\setminus\{\alpha\}$. For every $\alpha \in B$ the map

$$s_\alpha: x \to x + (x\cdot\alpha)\alpha$$

is an isometry of M, called the **simple reflection into α** . The subgroup of O(M) generated by all simple reflections s_α is called the **Weyl group** of B and is denoted by $W_B(M)$ (or simply W if no confusion arises). An element $\alpha' \in M_{-2}$ is called a **root** (with respect to B) if $\alpha' = w(\alpha)$ for some $w \in W$ and $\alpha \in B$ (that is, $\alpha' \in W_B$, the orbit of the set B with respect to W). The set of roots is denoted by R(B)(or simply by R).

Let B be an ordered finite root basis. The matrix $(\alpha\cdot\beta)_{\alpha,\beta\in B}$ is called the **Cartan matrix** of B. It can be given geometrically by the **Dynkin diagram** of B which is the graph $\Gamma(B)$ whose set of vertices is the set B and two vertices are joined by an edge if $\alpha\cdot\beta \ge 1$, the edge is labelled by the number $\alpha\cdot\beta - 1$ if it is greater than 1.

A root basis B is said to be **irreducible** if $B \ne B_1 \amalg B_2$, where B_1 and B_2 are root basiss such that $\alpha\cdot\beta = 0$ for any $\alpha \in B_1$, $\beta \in B_2$. Obviously, B is irreducible if and only if its Dynkin diagram is connected.

The basis $\{\alpha_0,\dots,\alpha_{p+q+r-2}\}$ of the lattice $Q_{p,q,r}$ from **S1** and the corresponding graph $T_{p,q,r}$ is an example of an irreducible root basis and its

The basis $\{\alpha_0,\ldots,\alpha_{p+q+r-2}\}$ of the lattice $Q_{p,q,r}$ from **S1** and the corresponding graph $T_{p,q,r}$ is an example of an irreducible root basis and its Dynkin diagram. A root basis in $Q_{p,q,r}$ is called **canonical** if its Dynkin diagram is of type $T_{p,q,r}$.

Let $M_{\mathbb{R}} = M\otimes\mathbb{R}$ be the real inner product vector space associated with M. Let

$$C(B) = \{x\in M_{\mathbb{R}}: x\cdot\alpha_i \geq 0 \text{ for all } \alpha_i\in B\}$$

(called the **fundamental chamber** with respect to B) and

$$K(B) = \bigcup_{w\in W} w(C(B)),$$

called the **Tits cone** with respect to B.

Proposition 2.2.1. (i) K(B) is a convex cone.

ii) W acts properly discontinously in the interior $K(B)^\circ$ of K(B) with fundamental domain equal to C(B);

iii) $w(C(B)^\circ)\cap C(B)^\circ = \emptyset$ for any $w\in W$;

iv) $K(B)^\circ\cap C(B) = \{x\in C(B)$: the stabilizer subgroup W_x is finite and is generated by the s_{α_i}'s such that $x\cdot\alpha_i = 0\}$.

PROOF. See **[Vi 1,Kac]**.

Recall that a **Coxeter group** is a pair (G,S), where G is a group and S is a set of generators of G such that $s^2 = 1$, for any $s\in S$, and these relations together with the relations $(ss')^{\mathrm{ord}(ss')} = 1$ are the defining relations for G (this means that for every group G' and every injective map $f:S \to G'$, there exists a unique homomorphism of groups $f:G \to G'$ extending f). The matrix $(m(s,s'))_{s,s'\in S}$, where $m(s,s') = \mathrm{ord}(ss')$ (or ∞) is called the **Coxeter group** of (G,S). Equivalently, it can be given geometrically, by the **Coxeter graph**, a graph whose set of vertices is the set S and such that two vertices are joined by an edge if $m(s,s')\geq 3$, the edge being labelled by the number $m(s,s')-2$ if it is greater than 1.

Proposition 2.2.2. The pair $(W,\{s_\alpha\}_{\alpha\in B})$ is a Coxeter group. Its Coxeter matrix $(m(s_\alpha,s_\beta))$ is given by

$$m(s_\alpha,s_\beta) = 1, \quad \alpha = \beta$$
$$= 2, \quad \alpha\bullet\beta = 0$$
$$= 3, \quad \alpha\bullet\beta = 1$$
$$= \infty, \quad \alpha\bullet\beta > 1$$

PROOF. See **[Vi 1,Kac]**.

Note that the Coxeter graph of $(W,\{s_\alpha\}_{\alpha\in B})$ is equal to the Dynkin graph $\Gamma(B)$ if and only if $\alpha\bullet\beta \le 1$ for any $\alpha,\beta\in B$.

Let (G,S) be a Coxeter group. For any non-empty subset S' of S the subgroup $G_{S'}$ generated by S' defines a Coxeter group $(G_{S'},S')$ **[Bou 2]**.

Proposition 2.2.3. Suppose that a system of generators S of a Coxeter group (G,S) is divided into two subsets S_1 and S_2 such that the order of any product s_1s_2, where $s_1\in S_1$, $s_2\in S_2$, is even or infinite. Then

$$G = G_1 \ltimes N ,$$

where G_1 is the subgroup generated by S_1 and N is the minimal normal subgroup containing S_2.

PROOF. **[Vi 3]**,Lemma 1.

Corollary 2.2.1. Let B be a root basis. There is a bijective correspondence between the W-orbits of roots and the set of connected components of the graph $\Gamma(B)'$ obtained from the Dynkin diagram $\Gamma(B)$ by deleting the labelled edges.

PROOF. By definition, every root $\alpha = w(\alpha_i)$, where $\alpha_i\in B$. Thus, the reflection $s_\alpha : x \to x + (x\bullet\alpha)\alpha$ is conjugate to the simple reflection s_{α_i}. Conversely, the conjugate $w\circ s_{\alpha_i}\circ w^{-1}$ of a simple reflection is a reflection s_α, where $\alpha = w(\alpha_i)$ is a root. Thus the set of W-orbits of roots is in bijection with the set of conjugacy classes of the set of generators of the Coxeter group $(W, S=\{s_\alpha\}_{\alpha\in B\}$. Let

$S_2 = \{s_\alpha$: α belongs to a connected component $\Gamma(B)'^\circ$ of $\Gamma(B)'\}$, $S_1=S\setminus S_2$. Then it follows from the proposition that none of the roots $\alpha\in\Gamma(B)'^\circ$ is W-equivalent to a root $\beta\notin B_1$. If $\alpha\bullet\beta = 1$, then $s_\beta\circ s_\alpha(\beta) = s_\beta(\beta+\alpha) = \alpha$. Thus the vertices of every connected component of $\Gamma(B)'$ are W-equivalent. This proves the proposition.

We will denote by l: $W(B) \to \mathbf{Z}_{\geq 0}$ the **length function** of the Weyl group. Recall that for every $w\in W(B)$, $l(w)$ is the minimal number q such that w can be written as a product of q simple reflections s_α, $\alpha\in B$.

Proposition 2.2.4. For every $\alpha\in B$ let $A_\alpha = \{x\in M_{\mathbb{R}}: x\bullet\alpha \geq 0\}$. Let $w\in W(B)$. Then:

$$w(C(B)) \subset s_\alpha(A_\alpha) \text{ if and only if } l(s_\alpha\circ w) = l(w)-1.$$

PROOF. See **[Bou 2]**, Chapter V, **§4**, n°4).

Corollary 2.2.2. For every $x\in K(B)$ there exists $w\in W(B)$ such that $w(x)\in C(B)$ and

$$x = w(x) + \sum_{\alpha\in B} m_\alpha \alpha,$$

where $m_\alpha \geq 0$.

PROOF. Let $w\in W(B)$ be such that $w(x)\in C(B)$. Write w as a product of $\kappa = l(w)$ simple reflections $s_{\alpha_1}\circ\ldots\circ s_{\alpha_\kappa}$. Since $l(s_{\alpha_\kappa}\circ w^{-1}) = l(w^{-1})-1$,

$$s_{\alpha_\kappa}(x) = s_{\alpha_\kappa}(w^{-1}(w(x)))\in s_{\alpha_\kappa}(s_{\alpha_\kappa}(A_{\alpha_\kappa})) = A_{\alpha_\kappa},$$

i.e. $s_{\alpha_\kappa}(x)\bullet\alpha_\kappa > 0$ hence $x\bullet\alpha_\kappa < 0$ and $s_{\alpha_\kappa}(x) = x-m_\kappa\alpha_\kappa$ for some $m_\kappa \geq 0$. Applying induction on κ, we finish the proof.

A root $\alpha\in R_B$ is called **positive** (with respect to B) if

$$\alpha\bullet x \geq 0 \text{ for all } x\in C(B).$$

It follows from Proposition 2.2.4 that for every root α either α or $-\alpha$ is positive. Thus, denoting by R_B^+ the set of positive roots, we obtain

$$R_B = R_B^+ \sqcup R_B^-,$$

where

$$R_B^- = \{-\alpha : \alpha \in R_B^+\}.$$

Note also that in the case B is a basis $\{\alpha_1,\dots,\alpha_r\}$ of $M_{\mathbb{R}}$,

$$C(B) = \mathbb{R}_{\geq 0}\alpha_1{}^* + \dots + \mathbb{R}_{\geq 0}\alpha_r{}^*,$$

where $\{\alpha_1{}^*,\dots,\alpha_r{}^*\}$ is the dual basis in $(M_{\mathbb{R}})^*$. In particular,

$$R_B^+ = R_B \cap (\mathbb{Z}_{\geq 0}\alpha_1 + \dots + \mathbb{Z}_{\geq 0}\alpha_r)$$

S3. Root bases of finite and affine type.

Let M be a lattice and B be a root basis in M. We denote by M_B the sublattice of M spanned by the subset B.

A root basis B is said to be of **finite type** (resp. **affine type**) if the sublattice M_B is negative definite (resp. negative semi-definite).

Proposition 2.3.1. Let B be an irreducible root basis. The following properties are equivalent:

(i) B is of finite type;

(ii) $W_B(M)$ is a finite group;

(iii) $K(B) = M_{\mathbb{R}}$;

(iv) B is finite and its Cartan matrix is negative definite;

(v) R(B) is finite ;

(vi) $\Gamma(B) = A_n, D_n$ or E_n for some n.

PROOF. See **[Bou 2]**.

Proposition 2.3.2. Let B be an irreducible root basis. The following properties are equivalent:

(i) B is of affine type;

(ii) there exists a unique vector $f = \Sigma m_i\alpha_i$, $m_i \in \mathbf{Z}_{\geq}$, such that $\mathrm{Rad}(M_B) = \mathbf{Z}f$, and $M_B/\mathbf{Z}f \cong M_{\bar{B}}$ is of finite type, where $\bar{B}$ is the image of B in $M_B/\mathbf{Z}f$.

(iii) every proper subdiagram of the Dynkin diagram $\Gamma(B)$ is of finite type and the Cartan matrix A(B) is singular;

(iv) the Tits cone $K(B) = \{x \in M_{\mathbb{R}}: x \cdot f > 0\} \cup \mathbb{R}f$, for some vector $f \in M_B$;

(v) the Dynkin diagram $\Gamma(B)$ is of type $\tilde{E}_n$ (n=6,7,8), or of type $\tilde{A}_n$, or of type $\tilde{D}_n$, where

$\tilde{D}_n$

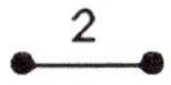

$\tilde{A}_1$

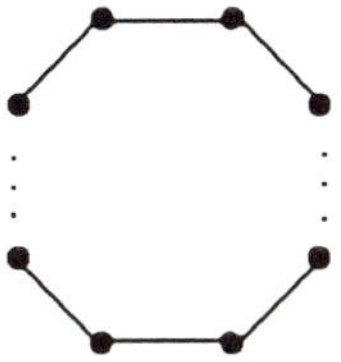

$\tilde{A}_n$, $n > 1$

Here n is equal to the number of vertices minus 1.

PROOF. See **[Kac]**.

Note that in the affine case

$$W_B(M) \cong W_{\bar{B}}(M) \ltimes M_{\bar{B}}{}^*,$$

where $\bar{M} = M/\mathbf{Z}f$ and the homomorphism $M_{\bar{B}}{}^* \to W_B(M)$ is defined by assigning to a linear function $\varphi \in M_{\bar{B}}{}^*$ the transformation $t_\varphi : x \to x + \varphi(\pi(x))f$. The group $W_B(M)$ is isomorphic to the affine Weyl group associated to the root system $\bar{B}$ of finite type (see **[Bou 2,Kac]**). If B is of type $\tilde{A}_n$, $\tilde{D}_n$, or $\tilde{E}_n$, then $\bar{B}$ is of type A_n,D_n, or E_n respectively.

In the case where B is of finite type and $M = M_B$, the Weyl group $W_B(M)$ is "almost" equal to the whole orthogonal group $O(M)$. In fact,

$$O(M) = A(B) \ltimes W_B(M),$$

where $A(B)$ is the group of automorphisms of the Dynkin diagram $\Gamma(B)$ (see **[Bou 2]**). In the affine case, we have a similar result:

$$O(M)' = A(B) \ltimes W_B(M),$$

where $O(M)' = \{\sigma\in O(M):\sigma(f) = f\}$ is a subgroup of index 2 of $O(M)$.

S4. Root bases of hyperbolic type.

A root basis B in a lattice M is said to be of **hyperbolic type** if the sublattice M_B of M is hyperbolic.

Lemma 2.4.1. Let B is a root basis of hyperbolic type. Then there exists a vector $x_0\in C(B)^\circ$ such that $x_0 = \sum_{\alpha\in B} m_\alpha \alpha$, where $m_\alpha > 0$.

PROOF. In fact, this is true for any root basis B such that M_B is indefinite. We refer for the proof to **[Vi 1]** or **[Kac]**, Corollary 4.3.

To simplify notations, we let $M = M_B$. Since $M_{\mathbb{R}}$ is an inner product space with signature $(1,n-1)$, $n = rk(M)$,

$$V = \{ x\in M_{\mathbb{R}}: x\cdot x > 0\} = V^+ \amalg V^-,$$

where $V^\pm$ is a convex cone defined by the property $\pm x\cdot x_0 > 0$ for some fixed vector x_0 with $x_0\cdot x_0 > 0$.

We choose x_0 to be a vector with the property from the previous lemma. By this choice

$$x_0 \in C(B)\cap V^+, \ C(B)\cap V^- = \emptyset.$$

We also have, for every simple reflection s_α, $\alpha\in B$,

$$s_\alpha(x_0)\cdot x_0 = x_0\cdot\alpha + (x_0\cdot\alpha)^2 > 0,$$

and this, of course, implies that $s_\alpha(V^+) \subset V^+$. Hence,

$$w(V^+) \subset V^+, \text{ for any } w \in W_B(M).$$

Let $H(V) = V^+/\mathbb{R}_+ \subset \mathbb{P}(V)$ be the **hyperbolic** (or **Lobachevsky**) **space** of dimension n-1 associated to the inner product space $M_{\mathbb{R}}$. It has a natural structure of Riemannian space of constant negative curvature (if $n \geq 3$).The group of motions of H(V) is the subgroup $O(M_{\mathbb{R}})'$ of the orthogonal group of V which leaves V^+ invariant. Clearly,

$$O(M_{\mathbb{R}}) = O(M_{\mathbb{R}})' \times \{\pm id_V\}.$$

The Weyl group $W = W_B(M)$ is identified with the subgroup of $O(M_{\mathbb{R}})'$ generated by the hyperbolic reflections about the faces of the convex polyhedron (called the **fundamental polyhedron**)

$$P(B) = (C(B) \cap V^+)/\mathbb{R}_+.$$

Clearly

$$W \subset O(M)' := O(M) \cap O(V)'.$$

In particular, W is a discrete subgroup of $O(M_{\mathbb{R}})'$ and acts properly discontinuously in H(V) and the polyhedron P(B) is a fundamental domain for this action.

Proposition 2.4.1. The Tits cone K(B) contains V^+.

PROOF. For every $x \in V^+$, the isotropy subgroup $(W_B)_x$ is finite (because $(\mathbb{R}x)^\perp$ is negative definite). This shows that W_B acts in V^+ properly discontinuously. On the other hand, it is known that K(B) is the maximal subset of V where W_B acts properly discontinuosly **[Kac,Lo 1]**.

A root basis B (or the Weyl group W_B) is said to be **crystallographic** if P(B) is of finite volume. The latter is equivalent to the property that

$$C(B) \subset \bar{V}^+ = \{x \in V^+ : x^2 \geq 0\} \text{ or } K(B)^\circ = V^+.$$

This also implies that B is finite and P(B) is the convex hull of a finite set of points in $\bar{V}^+/\mathbb{R}^+ = H(V) \cup H(V)^\infty$. The elements of such a set (assumed to be minimal with the above property) are called **vertices** of P(B) (they are said to be **proper** if they belong to H(V) and **improper** otherwise). Note that P(B) is compact if and only if it is a convex hull of a finite set of its proper vertices.

Proposition 2.4.2. A root basis B is crystallographic if and only if $W_B(M)$ is a subgroup of finite index in O(M).

PROOF. It follows immediately from the fact that O(M)' acts in H(V) with a fundamental domain of finite volume **[Sie]**.

Let Γ(B) be the Dynkin diagram of B (if it is defined). Obviously, there is a natural bijective correspondence between the following sets:

{non-empty subsets of B },

{full subgraphs of Γ(B)} ,

{principal submatrices of the Cartan matrix} .

A subgraph of Γ(B) is said to be of **finite type** (resp. **affine type**) if the corresponding subset of B is a root basis of finite (resp. affine) type. Its **rank** is defined as the rank of the corresponding Cartan matrix , or geometrically, as the number of vertices minus the number of connected components.

Proposition 2.4.3. A root basis B of hyperbolic type is crystallographic if and only if every connected subgraph of affine type of Γ(B) is contained in a subgraph of affine type of maximal rank (= n-2).

PROOF. See **[Vi 2]**,Theorem 2.3.

Note that the proper (resp. improper) vertices of the fundamental polyhedron P(B) of a crystallographic root basis B correspond bijectively to the subdiagrams of finite type of rank n-1 (resp. of affine type of rank n-2) of Γ(B).

Remark 2.4.1. Notice the discrepancy between our terminology and the terminology of **[Bou 2]**, Chapter V, §4 , Exercises 12-18 or **[Kac]**, §4.10. According to the latter, a root basis B is of hyperbolic type (resp. strictly hyperbolic) if every proper subgraph of $\Gamma(B)$ is of finite or affine (resp. of finite) type. This is equivalent to saying that B is crystallographic (resp. P(B) is compact) and linearly independent (or P(B) is a simplicial polyhedron). The possible Dynkin diagrams satisfying these conditions are:

$$T_{2,3,4}\ ,\ T_{2,4,5}\ ,\ T_{3,3,4}$$

of rank 10,9 and 8 respectively, and

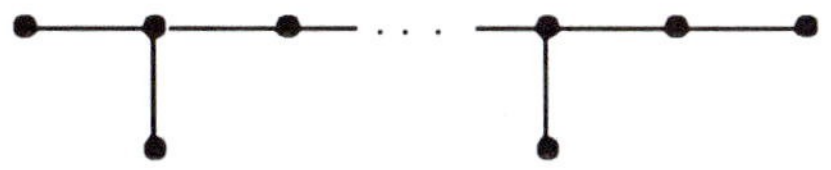

with $n \leq 10$ vertices.

Proposition 2.4.4. Let B be a hyperbolic root basis. Suppose that B' is crystallographic for some subset B' of B. Then B is crystallographic and B = B'.

PROOF. Clearly $W_{B'}(M) \subset W_B(M)$. By Theorem 2.4.1, $W_{B'}(M)$ is of finite index in O(M), hence $W_B(M)$ is of finite index too. Thus, B is crystallographic. If $\alpha \in B \setminus B'$, then $\alpha \cdot \beta \geq 0$ for any $\beta \in B'$. Thus $\alpha \in C(B')$. Since B' is crystallographic, $C(B') \subset V^+$. Hence, $\alpha^2 \geq 0$. This is absurd.

Let M be a lattice. One can define the **Weyl group** (or the **2-reflection group**) W(M) of M as the subgroup of O(M) generated by all reflections $x \to x + (x \cdot e)e$, where $e \in M_{-2}$. Obviously, W(M) is a normal subgroup of O(M)'.

For every $e \in M_{-2}$ let H_e denote the hyperplane $\{x \in M_{\mathbb{R}}: x \cdot e = 0\}$. Let C be a connected component of $V^+\backslash(\cup_{e\in M_{-2}} H_e)$ and $x_0 \in C$. We have

$$M_{-2} = R^+ \sqcup R^-,$$

where $R^{\pm} = \{ e\in M_{-2}: \pm e\cdot x_0 > 0\}$. Also, we have

$$C = \{ x\in V^+: x\cdot e > 0, \text{ for any } e\in R^+\}$$

An element $e\in R^+$ is called **irreducible** if it cannot be represented by a positive linear integral combination of elements from $R^{\pm}\backslash\{e\}$. The image of C in $H(M_{\mathbb{R}})$ is a convex polyhedron P(M) bounded by the hyperplanes H_e corresponding to some irreducible elements from R^+. The set of such elements is a root basis B(M) in M and $W(M) = W_{B(M)}(M)$. Clearly, B(M) is uniquely defined up to a choice of a connected component of $V^+\backslash(\cup_{e\in M_{-2}} H_e)$.

Let

$$A(P(M)) = \{\sigma\in O(M)': \sigma(P(M)) \subset P(M)\}.$$

If B(M) is finite, A(B) is clearly the subgroup of the group of symmetries of the Dynkin diagram $\Gamma(B(M))$ of B(M). If B(M) is linearly independent, then it is equal to the whole group of symmetries. Obviously, both O(M)' and W(M) act transitively on the set of connected components of $V^+\backslash(\cup_{e\in R} H_e)$. Thus

$$O(M)' = A(P(M)) \ltimes W(M)$$

A hyperbolic even lattice is called a **Nikulin lattice** if W(M) is a crystallographic (or, equivalently, W(M) is of finite index in O(M)).

A complete classification of isomorphism classes of Nikulin lattices can be found in **[Ni 2,Ni 3]** (see also a survey of Nikulin's results in **[Do 5]**. If $W_B(M)$ is crystallographic for some root basis B, M must clearly be a Nikulin lattice. Moreover, it follows from the proof of Proposition 2.4.4 that B = B(M) and $W_B(M) = W(M)$ in this case.

Example 2.4.1. A lattice $Q_{p,q,r}$ is a Nikulin lattice if and only if (p,q,r) is one of the following 9 triples

(2,3,7),(2,3,8),(2,3,9),(2,3,10),(2,4,5),(2,4,6),(3,3,4),(3,3,5),(3,3,6).

Only three of them ((2,3,7),(2,4,5) and (3,3,4)) have crystallographic canonical root basis.

§5. The Enriques lattice.

Let F be an Enriques surface and

$$\mathrm{Num}(F) = \mathrm{NS}(F)/\mathrm{Tors}$$

As we know from Chapter 1, §2 , the intersection theory of divisors on F defines a lattice structure on Num(F) which is of rank 10 and unimodular.

Theorem 2.5.1. There is an isomorphism of lattices

$$\mathrm{Num}(F) \cong Q_{2,3,7} \cong U \oplus E_8.$$

PROOF. Since K_F is numerically trivial, for every irreducible curve C on F

$$C^2 = 2p_a(C) - 2 \in 2\mathbf{Z} .$$

This shows that Num(F) is even. Now, the assertion follows from Proposition 2.1.1 since the three lattices are even, unimodular, hyperbolic and of the same rank.

For every $n \geq 4$ we denote the lattice $Q_{2,3,n-3}$ by E_n. In the notation of §1, we have

$$E_4 = A_4,\ E_5 = D_5,\ E_i = E_i,\ i = 6,7,8,\ E_9 = \tilde{E}_8.$$

Let $B_n = \{\alpha_0,\dots,\alpha_n\}$ be the canonical root base , and $B_k = \{\alpha_0,\dots,\alpha_k\}$, $k \leq n$. Clearly

$$E_k \cong (E_n)_{B_k}.$$

We denote by W_n the Weyl group of the canonical root basis in E_n.

The lattice E_{10} is denoted by $\mathbf{E}$ and is called the **Enriques lattice**. Its Dynkin diagram looks like

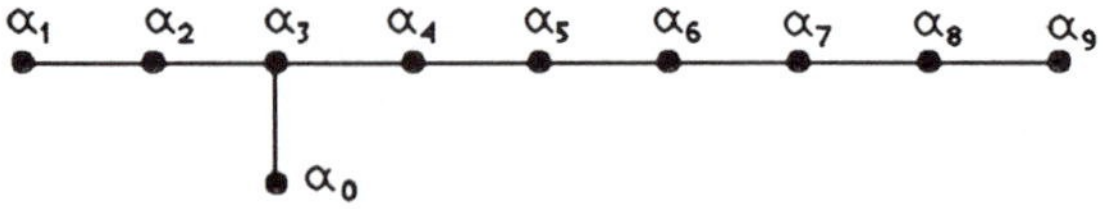

Its only subgraph of affine type is the subgraph with the set of vertices B_9 = $B\setminus\{\alpha_9\}$. Its radical is spanned by the vector

$$f = 3\alpha_0+2\alpha_1+4\alpha_2+6\alpha_3+5\alpha_4+4\alpha_5+3\alpha_6+2\alpha_7+\alpha_8$$

The sublattice $Q_{2,3,5} = E_8$ is embedded naturally into $\mathbf{E}$ as the sublattice spanned by the set B_8. Obviously,

$$E_8^\perp = \mathbf{Z}f+\mathbf{Z}\alpha_9 \cong U .$$

This gives a natural embedding $U \hookrightarrow \mathbf{E}$ such that $\mathbf{E} \cong E_8 \oplus U$.

The fundamental polyhedron P(B) in the 9-dimensional hyperbolic space $H(M_{\mathbb{R}})$ is simplicial. It has 10 vertices P_i corresponding to the subsets $B_i = B\setminus\{\alpha_i\}$, the only one of them which is improper being P_9.

Since B is crystallographic, the Weyl group W_{10} coincides with the Weyl group $W = W(\mathbf{E})$ of the lattice $\mathbf{E}$. Also, the fundamental polyhedron P(B) is a fundamental domain for the action of the group W in the Lobachevsky space $V^+/\mathbb{R}_+$, where

$$V^+ = \{x\in \mathbf{E}_{\mathbb{R}}: x^2 > 0\}.$$

Since $T_{2,3,7}$ does not have nontrivial symmetries we have

$$O(\mathbf{E})' = W.$$

Proposition 2.5.1.

$$E_9 \cong E_8\oplus\langle 0\rangle,\ E_n \cong \mathbf{E} \oplus A_{n-10} \cong E_8\oplus U\oplus A_{n-10} ,\ n \geq 10 .$$

PROOF. Assume $n \geq 9$. We already know that this assertion is true for $n = 10$. Let $\alpha_0,\dots,\alpha_{n-1}$ be the canonical root basis in E_n. We define the isotropic vector $f\in E_n$ by

$$f = 3\alpha_0+2\alpha_1+4\alpha_2+6\alpha_3+5\alpha_4+4\alpha_5+3\alpha_6+2\alpha_7+\alpha_8$$

It is immediately verified that

$$\mathrm{Rad}(E_9) = \mathbf{Z}f.$$

This checks the assertion for $n = 9$.

If $n > 10$, we observe that

$$\mathbf{Z}\alpha_0+\ldots+\mathbf{Z}\alpha_7 \cong E_8,\ \mathbf{Z}f+\mathbf{Z}(\alpha_9+\ldots+\alpha_{n-1}) \cong U,\ \mathbf{Z}\alpha_9+\ldots+\mathbf{Z}\alpha_{n-2} \cong A_{n-10}$$

and

$$E_n = (\mathbf{Z}\alpha_0+\ldots+\mathbf{Z}\alpha_7) \perp (\mathbf{Z}f+\mathbf{Z}(\alpha_9+\ldots+\alpha_{n-1}) \perp (\mathbf{Z}\alpha_9+\ldots+\mathbf{Z}\alpha_{n-2}).$$

This proves the assertion.

Let $H^{1,n} = \langle 1\rangle \oplus \langle -1\rangle^n$ be the standard unimodular non-even hyperbolic lattice of rank $n+1$ and $\{e_0,e_1,\ldots,e_n\}$ be its orthonormal basis, that is

$$e_0^2 = 1,\ e_i\bullet e_j = -\delta_{ij}.$$

Define the vector k_n by

$$k_n = -3e_0+e_1+\ldots+e_n\ .$$

Note that

$$k_n^2 = 9-n.$$

Corollary 2.5.1. The lattices E_n are hyperbolic for $n \geq 10$. The canonical root basis is crystallographic only if $n = 10$.

PROOF. Since $k_n^2 = -(n-9)$ and $H^{1,n}$ is hyperbolic, E_n is hyperbolic for $n \geq 10$. Since the only subdiagram of affine type of $T_{2,3,n}$ $(n \geq 9)$ is $T_{2,3,6}$, the second assertion follows immediately from Proposition 2.4.2.

Proposition 2.5.2.

$$E_n \cong (\mathbf{Z}k_n)^\perp,$$

and a canonical root basis $B = \{\alpha_0,\ldots,\alpha_{n-1}\}$ in E_n can be given by

$$\alpha_0 = e_0-e_1-e_2-e_3,\ \alpha_1 = e_1-e_2,\ \alpha_2 = e_2-e_3,\ \ldots,\alpha_{n-1} = e_{n-1}-e_n.$$

PROOF. This is immediately verified .

Corollary 2.5.2.

$$H^{1,10} \cong \langle -1\rangle \oplus E$$

From now on the Weyl group W_n of the canonical root basis of E_n will be considered either as a subgroup of $O(E_n)$ or as a subgroup of $O(H^{1,n})$.

Corollary 2.5.3. Let E_n be identified with $(\mathbf{Z}k_n)^{\perp} \subset H^{1,n}$. For every $4 \le k < n$,

$$E_{B_k} = (\mathbf{Z}e_{k+1}+\mathbf{Z}e_n)^{\perp} \cap E_n,$$

$$W_{B_k} = W_{e_{k+1}} \cap \ldots \cap W_{e_n},$$

where W_e denotes the isotropy subgroup of a vector $e \in H^{1,n}$.

Remark 2.5.1. Let V be a surface obtained by blowing up n points in $\mathbb{P}^2$ among which some can be infinitely near. Let $\pi: V \to \mathbb{P}^2$ be the blowing down which is decomposed into a composition

$$V = V_n \to V_{n-1} \to \ldots \to V_1 \to V_0 = \mathbb{P}^2,$$

where each $\pi_i: V_i \to V_{i-1}$ is the blowing up of a point $p_i \in V_{i-1}$ (i= 1,..., n). Denote by e_i the class of $(\pi_n \circ \ldots \circ \pi_i)^{-1}(p_i)$ in Pic(V) and by e_0 the class of $\pi^{-1}(l)$, where l is a line in $\mathbb{P}^2$. Then, it is immediately verified that

$$e_0^2 = 1,\ e_i^2 = -1,\ i = 1,\ldots,n,$$

$$\mathrm{Pic}(V) = \mathbf{Z}e_0+\mathbf{Z}e_1+\ldots+\mathbf{Z}e_n \cong H^{1,n},$$

$$k_n = K_V, \text{ the canonical class of } V.$$

An example of a root in Pic(V) with respect to the canonical root basis of $(K_V)^{\perp} \cong E_n$ is the class of a nodal curve on V. However, if the points $p_1,\ldots,p_n$ are taken to be in general position, none of the roots is obtained in this way. We refer to **[D-O]** and Chapter VII for further discussions of this matter.

Proposition 2.5.3. Assume $n \le 10$. Then

$$O(E_n) = W_n \times (\tau),$$

where $\tau^2 = 1$ and $\tau = 1$ (resp. -1) if n = 7,8 (resp. $n \ge 9$). If n = 4,5,6, τ is induced by the symmetry of the Dynkin diagram of the canonical root basis.

PROOF. This follows immediately from §3 ($n \le 9$) and from §4 ($n = 10$).

Corollary 2.5.4. Assume $n \le 10$. Then

$$O(E_n) = \{\sigma\in O(H^{1,n}) : \sigma(k_n) = \pm k_n\}.$$

PROOF. Let $G = \{\sigma\in O(H^{1,n}) : \sigma(k_n) = \pm k_n\}$. By restriction to $E_n = (\mathbf{Z}k_n)^\perp$, we have a homomorphism $G \to O(E_n)$. Its kernel acts identically on the sublattice $E_n \perp \mathbf{Z}k_n$. Since the latter is of finite index in $H^{1,n}$, the kernel is trivial. This allows us to identify G with a subgroup of $O(E_n)$. Obviously, $W_n \subset G$. It remains to apply the previous proposition and verify directly that $\tau\notin G$.

Proposition 2.5.4. Assume $n \le 10$. If $n \ne 7, 8$, W_n has two orbits on the set of canonical root bases. If $n = 7$ or 8, W_n acts transitively on this set.

PROOF. Fix an orthonormal basis in $H^{1,n}$ and the corresponding vector k_n. If $\{\alpha_0,\dots,\alpha_n\}$ is a canonical basis in E_n, we can define a primitive embedding

$$j: E_n \hookrightarrow H^{1,n} \text{ with } j(E_n)^\perp = \mathbf{Z}k_n$$

by sending α_0 to $e_0-e_1-e_2-e_3$, α_i to e_i-e_{i+1}. Conversely, given such an embedding j, it defines a canonical root basis in E_n by taking

$$\alpha_0 = j^{-1}(e_0-e_1-e_2-e_3),\ \alpha_i = j^{-1}(e_i-e_{i+1}),\ i \ne 0.$$

Assume $n \ne 9$. It follows from an analog of Witt's theorem for lattices that $O(H^{1,n})$ acts transitively on primitive embeddings of $E_n \hookrightarrow H^{1,n}$ (**[Ni 1]**, Proposition 1.1.4.1 and Remark 1.14.6). This means that for every two such embeddings $j,j':E_n \hookrightarrow H^{1,n}$ there exists an isometry $\sigma\in O(H^{1,n})$ such that $j = \sigma\circ j'$. Since $j(E_n)^\perp = j'(E_n)^\perp = \mathbf{Z}k_n$, $\sigma(k_n) = \pm k_n$. By Corollary 2.5.4, $\sigma\in O(E_n)$. This shows that W_n has at most two orbits on the set of canonical bases. They are represented by a basis $B = \{\alpha_0,\dots,\alpha_{n-1}\}$ and $-B = \{-\alpha_0,\dots,-\alpha_{n-1}\}$. Since $-1\in W_n$ only for $n = 7$ and 8, we get the assertion of the proposition.

Assume $n = 9$. The assertion easily follows from the corresponding assertion for $E_8 \cong E_9/\mathrm{Rad}(E_9)$. We leave the details to the reader.

Let M be a lattice, an ordered (resp. unordered) r-tuple $(f_1,\ldots,f_r)$, of $r \geq 2$ isotropic vectors in M with $f_i \cdot f_j = 1$, $i \neq j$, will be called an **isotropic r-sequence** (resp. **r-set**). Such a sequence defines an embedding of the lattice $U_{[r]}$ into M by sending its generators to the vectors f_i. Conversely an embedding as above defines an isotropic r-sequence. Fixing a basis of $U_{[r]}$ we get a bijective correspondence between isotropic r-sequences and embeddings $U_{[r]} \hookrightarrow M$.

Lemma 2.5.1. The lattice $U_{[r]}$ is an even lattice of rank r. If $r \neq 1$, it is a hyperbolic lattice and its discriminant group is a cyclic group of order r-1. Every isotropic r-sequence $(f_1,\ldots,f_r)$ in $U_{[r]}$ is a basis. The vector

$$f = f_1+\ldots+f_r$$

satisfies the properties $f^2 = r(r-1)$ and $|f \cdot x| \geq r-1$ for every isotropic vector x in $U_{[r]}$. Conversely, if a vector $f' \in U_{[r]}$ satisfies these properties, then $f' = \pm f$.

PROOF. By definition, $U_{[r]}$ is spanned by an isotropic r-set. Writing down its Gram matrix, we immediately check that this set is linearly independent and the discriminant group is cyclic of order r-1. In the same way, we check the signature of $U_{[r]}$. The assertion about the evenness of $U_{[r]}$ is obvious.

Let f be defined as above. Then $f^2 = r(r-1)$ and $f \cdot f_i = r-1$ for every $i = 1,\ldots,r$. For every isotropic vector $x = a_1f_1+ \ldots +a_rf_r$ in $U_{[r]}$ $f \cdot x \neq 0$ (since $U_{[r]}$ is hyperbolic) and we have

$$|f \cdot x| = (r-1)|a_1+\ldots+a_r| \geq r-1$$

Conversely, let $f' = a_1f_1+\ldots+a_rf_r$ satisfies the two properties of f stated in the lemma. First of all, we note that for every vector x with $x^2 > 0$, $x \cdot f_i > 0$ for one i implies $x \cdot f_j$ for all j. Indeed, otherwise we can find a vector y of the form af_i+bf_j with $a,b > 0$ such that $x \cdot y = 0$. Since $y^2 = 2ab > 0$, this contradicts the hyperbolicity of $U_{[r]}$. Therefore we may assume that

$$f' \cdot f_i = \sum_{j \neq i} a_j > 0 \, , \; i = 1,\ldots,r.$$

Since

$$f'\cdot f_i = (\Sigma a_j) - a_i \geq r-1 \text{ for all } i,$$

by adding, we obtain $\Sigma a_i \geq r$, and the above inequality shows that $a_i \leq 1$. Together this implies $a_i = 1$ for all i, hence $f' = f$.

Proposition 2.5.5. Let $\bar{U}_{[10]}$ be the subgroup of $U_{[10]} \otimes \mathbb{R}$ spanned by $U_{[10]}$ and the vector $\Delta = (f_1+\ldots+f_{10})/3$, where $(f_1,\ldots,f_{10})$ is an isotropic 10-sequence in $U_{[10]}$. The inner product of $U_{[10]} \otimes \mathbb{R}$ induces a lattice structure on $\bar{U}_{[10]}$ and

$$\bar{U}_{[10]} \cong \mathbf{E}.$$

There is a bijective correspondence between the set of canonical root bases in **E** and the set of isotropic 10-sequences. It is given by the formulae

$$\alpha_0 = \Delta - f_1 - f_2 - f_3,\ \alpha_1 = f_1 - f_2,\ \ldots\ ,\ \alpha_9 = f_9 - f_{10}\ ;$$

$$f_1 = 3\alpha_0+3\alpha_1+5\alpha_2+7\alpha_3+6\alpha_4+5\alpha_5+4\alpha_6+3\alpha_7+2\alpha_8+\alpha_9,$$

$$f_i = f_{i-1} - \alpha_i,\ i = 2, \ldots ,10.$$

PROOF. Since Δ^2 and $\Delta\cdot f_i$ are integers, $\bar{U}_{[10]}$ is a lattice. It contains $U_{[10]}$ as a sublattice of index 3. Since the discriminant of $U_{[10]}$ is equal to 9, $U_{[10]}$ is unimodular. It is obviously even ($\Delta^2 = 10$), hyperbolic and of rank 10. Thus it isomorphic to **E**. The last assertion is verified directly by checking the formulae.

Corollary 2.5.5. For every isotropic set $(f_1,\ldots,f_{10})$ of length 10 in **E**, the vector

$$\Delta = (f_1+\ldots+f_{10})/3$$

is defined and satisfies $\Delta^2 = 10$, $\Delta\cdot f_i = 3$. Conversely, if $x \in \mathbf{E}$ satisfies $x^2 = 10$ and $|x\cdot f| \geq 3$ for any isotropic vector $f \in \mathbf{E}$, then $f = \pm\Delta$ for a uniquely defined isotropic set $(f_1,\ldots,f_{10})$.

PROOF. We have to prove only the converse statement. We know that the Enriques lattice **E** contains a sublattice M of index 3 which is isomorphic to $U_{[10]}$. Thus $3x \in M$ and satisfies the properties: $(3x)^2 = 90$ and $|3x\cdot f| \geq 9$ for all isotropic vectors in M. It follows from Lemma 5.2.1 that $\pm 3x = \Sigma f_i$ for some isotropic set $(f_1,\ldots,f_{10})$ in $M \subset E$. It remains to prove the uniqueness of $(f_1,\ldots,f_{10})$. Assume

$$f_1+\ldots+f_{10} = f_1'+\ldots+f'_{10}.$$

Then intersecting the both sides with f_i, we obtain

$$9 = (f_1'\cdot f_i) + \ldots + (f'_{10}\cdot f_i).$$

If $f_j'\cdot f_i > 0$, then $f'_K\cdot f_i > 0$ for all K (see the proof of Lemma 2.5.10). This immediately implies that $f_i = f_K'$ for some K. Subtracting the f_i from the both sides of the above equality and repeating the argument, we prove the assertion.

A vector $e \in H^{1,n}$ is said to be **exceptional** if

$$e^2 = -1,\ e\cdot K_n = -1.$$

We denote by $\mathcal{E}_n$ the set of exceptional vectors in $H^{1,n}$. An ordered set $(e_1,\ldots,e_n)$ of elements of $\mathcal{E}_n$ with $e_i\cdot e_j = 0$, $i\neq j$, is said to be an **exceptional r-sequence**.

Lemma 2.5.2. Assume $n \leq 9$. The group W_n acts transitively on the set of exceptional r-sequences if $r \neq n-1$ If $r = n-1$, W_n has two orbits on this set represented by the sequences $(e_0-e_1-e_2,e_3,\ldots,e_n)$ and $(e_2,e_3,\ldots,e_n)$.

PROOF. See **[Dem,Ma]**.

Proposition 2.5.6. The Enriques lattice contains isotropic r-sequences for every $r \leq 10$ .The group $O(\mathbf{E})$ acts transitively on the set of isotropic r-sequences for every $r \neq 9$. The subgroup $W(\mathbf{E})$ has two orbits on this set, represented by some isotropic sequences $(f_1,\ldots,f_r)$ and $(-f_1,\ldots,-f_r)$.

PROOF. Let $e \in \mathcal{E}_{10}$, then

$$f = e-K_{10} \in \mathbf{E}_0,$$

We denote by $\mathcal{E}_{10}'$ the subset of $\mathcal{E}_{10}$ which consists of vectors e such that $e-K_{10}$ is primitive. Since

$$e \cdot e' = 0 \Leftrightarrow (e-K_{10}) \cdot (e'-K_{10}) = 1,$$

there is a bijective correspondence between the set of exceptional r-sequences of vectors from $\mathcal{E}_{10}'$ and the set of isotropic r-sequences. Clearly, if $r > 1$, every exceptional r-sequence consists of vectors from $\mathcal{E}_{10}'$.

Thus we have to show that $O(\mathbf{E})$ acts transitively on the set of exceptional r-sequences from $\mathcal{E}_{10}'$ for every $r \neq 9$. It follows from Corollary 2.5.7 below that $O(\mathbf{E})$ has only one orbit on the set of primitive isotropic vectors. Thus $O(\mathbf{E})$ acts transitively on the set $\mathcal{E}_{10}'$ and the assertion is true for $r = 1$. Assume $r > 1$. Let $(e_1,\dots,e_r)$ and $(e_1',\dots,e_r')$ be two exceptional r-sequences. By the above there exists $w \in W_{10}$ such that $w(e_r) = e_r'$. Thus we may assume that $e_r = e_r'$. Hence $(e_1,\dots,e_{r-1})$ and $(e_1',\dots,e_{r-1}')$ belong to $(\mathbb{Z} e_r)^{\perp} \cong H^{1,9}$. By Corollary 2.5.3 the isotropy subgroup of e_r is isomorphic to W_9 which is naturally identified with a subgroup of $O(H^{1,9})$. It remains to apply Lemma 2.5.1.

Corollary 2.5.6. Every isotropic r-sequence in $\mathbf{E}$, $r \neq 9$, can be extended to a 10-sequence.

Remark 2.5.2. Let V be a rational surface with $\mathrm{Pic}(V) \cong H^{1,n}$ introduced in Remark 2.5.1. Fix an isomorphism φ: $\mathrm{Pic}(V) \to H^{1,n}$. Every exceptional curve E of the first kind on V is represented in $H^{1,n}$ by an exceptional vector $e \in \mathcal{E}_n$. If $n \leq 9$, and the points $p_1,\dots,p_n$ are taken in general position, every exceptional vector is obtained in this way. If $n = 10$, the same is true only for vectors from $\mathcal{E}_{10}'$. An example of a vector from $\mathcal{E}_{10} \setminus \mathcal{E}_{10}'$ is $5e_0-3e_1-3e_2-e_3-\dots-e_{10}$. If it represents an irreducible curve on V, its image in $\mathbb{P}^2$ is a curve of degree 5 with 2 triple points. Passing a line through these points, we see the contradiction by applying Bezout's theorem. For any n and generic V, only the subset $W_n e_1$ of $\mathcal{E}_n$ is represented by the classes of irreducible exceptional curves of the first kind on V. We refer to **[D-O,Ha ,Nag]** for discussions of this matter in the case of arbitrary V.

We have two expressions for a canonical root basis in **E**. The first one comes from the realizing **E** as a sublattice of $H^{1,n}$(Proposition 2.5.2), the second one from using isotropic 10-sequences (Proposition 2.5.5). Combining them both, we obtain a formula expressing an isotropic 10-sequence in **E** via an orthonormal basis in $H^{1,n}$:

$$f_i = e_i - \kappa_{10},\ i = 1,\dots,10,\ \Delta = (1/3)(f_1+\dots+f_{10}) = 10e_0-3e_1-\dots-3e_{10}.$$

From now on we will always consider the model of **E** given by Proposition 2.5.2. and denote by κ the vector κ_{10}.

Let $\{\alpha_0,\dots,\alpha_9\}$ be a canonical root basis in **E**, $(f_1,\dots,f_{10})$ be the isotropic 10-sequence associated to it, and $\alpha_0=e_0-e_1-e_2-e_3, \alpha_i=e_i-e_{i+1}$, $i\geq 1$, be the corresponding embedding of **E** into $H^{1,10}$.

Proposition 2.5.7. The dual basis $\{\omega_0, \dots ,\omega_9\}$ of $\mathbf{E}^* = \mathbf{E}$ corresponding to the canonical root basis $\{\alpha_0, \dots ,\alpha_9\}$ in **E** is given by

$$\omega_0 = e_0-3\kappa = \Delta = (f_1+\dots+f_{10})/3,$$

$$\omega_1 = e_0-e_1-2\kappa = \Delta-f_1,$$

$$\omega_2 = 2e_0-e_1-e_2-4\kappa = 2\Delta-f_1-f_2,$$

$$\omega_i = e_{i+1}+\dots+e_{10}-(10-i)\kappa = f_{i+1}+\dots+f_{10},\ i = 3,\dots,9.$$

The Gram matrix $(\omega_i\bullet\omega_j)$ of this set is given by

$$\begin{bmatrix}
10 & 7 & 14 & 21 & 18 & 15 & 12 & 9 & 6 & 3\\
7 & 4 & 9 & 14 & 12 & 10 & 8 & 6 & 4 & 2\\
14 & 9 & 18 & 28 & 24 & 20 & 16 & 12 & 8 & 4\\
21 & 14 & 28 & 42 & 36 & 30 & 24 & 18 & 12 & 6\\
18 & 12 & 24 & 36 & 30 & 25 & 20 & 15 & 10 & 5\\
15 & 10 & 20 & 30 & 25 & 20 & 16 & 12 & 8 & 4\\
12 & 8 & 16 & 24 & 20 & 16 & 12 & 9 & 6 & 3\\
9 & 6 & 12 & 18 & 15 & 12 & 9 & 6 & 4 & 2\\
6 & 4 & 8 & 12 & 10 & 8 & 6 & 4 & 2 & 1\\
3 & 2 & 6 & 6 & 5 & 4 & 3 & 2 & 1 & 0
\end{bmatrix}$$

PROOF. Straightforward.

Corollary 2.5.7.

$$E_0' = O(E)\omega_9,$$

$$E_2 = O(E)\omega_8,$$

$$E_4 = O(E)\omega_1 \amalg O(E)(\omega_8+\omega_9),$$

$$E_6 = O(E)\omega_7 \amalg O(E)(\omega_8+2\omega_9),$$

$$E_8 = O(E)(\omega_8+3\omega_9) \amalg O(E)(2\omega_8),$$

$$E_{10} = O(E)\omega_0 \amalg O(E)(\omega_7+\omega_9) \amalg O(E)(\omega_8+4\omega_9)$$

PROOF. Since $C(B)= \mathbb{R}_{\geq 0}\omega_0+\ldots+\mathbb{R}_{\geq 0}\omega_9$ is a fundamental domain for the action of W in V^+, and $O(\mathbf{E}) = W\times\{\pm id_E\}$, every vector $x \in \mathbf{E}_r$ is $O(\mathbf{E})$-equivalent to a vector $x = \Sigma n_i\omega_i$ with $x^2 = r$. Now, the assertion follows by direct computation using the Gram matrix $(\omega_i \bullet \omega_j)$.

The elements $\omega_1,\ldots,\omega_{10}$ of the dual basis of a root basis B are called the **fundamental weights** of B.

Remark 2.5.3. Note that every fundamental weight ω_i can be written as a positive linear rational combination of isotropic vectors in **E**. This is obvious for all $i \neq 2,3$. If $i = 2$, we notice that $f_{i,j} = \Delta-f_i-f_j$, $1 \leq i < j \leq 10$, are isotropic vectors and we can write

$$\omega_2 = \Delta-f_1 = f_{1,2}+f_1 ,$$

$$\omega_3 = 2\Delta-f_1-f_2 = \Delta+f_{1,2}.$$

Finally, we mention the following model of **E** although it will not be used in the future.

Proposition 2.5.8.

$$\mathbf{E} \cong \{x = x_0e_0+\ldots+x_9e_9 \in V^{1,9}: \text{ all } x_i \in \mathbf{Z} \text{ or all } x_i \in \tfrac{1}{2}+\mathbf{Z},\ \ x_0-x_1-\ldots-x_9 \in 2\mathbf{Z}\}$$

PROOF. Let M be the lattice defined on the right side of the quality. The vectors $\frac{1}{2}(e_0+e_1)$ and (e_0-e_1) span the sublattice M' of M isomorphic to U. Its orthogonal complement M" in M is spanned by the vectors

$$\alpha_0 = e_2+e_3,\ \alpha_1 = \tfrac{1}{2}(e_3+\ldots+e_8)-\tfrac{1}{2}(e_2+e_9),\ \alpha_2 = e_2-e_3,$$

$$\alpha_3 = e_3-e_4,\ \ldots\ ,\alpha_7 = e_7-e_8.$$

It is immediately checked that the latter is isomorphic to the lattice E_8 (cf. **[Bou 2]**, Chapter VI, §4, n°10) and $M = M'\perp M"$.

§6. The Reye lattice.

We denote the lattice $Q_{2,4,n-4}$, $n\geq 5$, by R_n. We have

$$R_5 = A_5,\ R_6 = D_6,\ R_7 = E_7,\ R_8 = \tilde{E}_7.$$

Let $B_n = \{\alpha_0,\ldots,\alpha_{n-1}\}$ be a canonical root base in R_n with Dynkin diagram $T_{2,4,n-4}$. Denote by B_r (resp. B_r') the subset $\{\alpha_0,\ldots,\alpha_{r-1}\}$ (resp. $\{\alpha_0,\alpha_2,\ldots,\alpha_r\}$) of B. In this way, for every $r \leq n$ the lattices R_r (resp. E_r) can be identified with the sublattices $(R_r)_{B_r}$ (resp. $(R_r)_{B_r'}$). We denote by W'_r the Weyl group W_{B_r}.

The lattice R_{10} will be used in Part II for the study of generic nodal Enriques surfaces which can be represented by the Reye congruences of lines in $\mathbb{P}^3$. For this reason we call this lattice the **Reye lattice** and denote it by **R**.

First, we note that R_n is an even lattice of rank n and discriminant $|2n-16|$. It is hyperbolic if $n \geq 9$.

Let $B_{10} = \{\alpha_0,\ldots,\alpha_9\}$ be the canonical root basis of R_{10} with the Dynkin diagram of type $T_{2,4,6}$

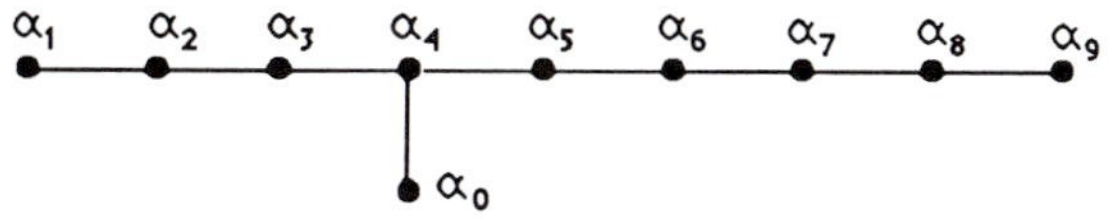

There are two subgraphs of this diagram which are of affine type. They correspond to the subset B_8 (of type $\tilde{E}_7$) and B_9 (of type $\tilde{E}_8$). Since $\Gamma(B_8)$ is not contained in a subgraph of parabolic type of maximal rank (=8), B_{10} is not a crystallographic root basis (Theorem 2.4.2).

Let $Rad(R_{B_9}') = \mathbf{Z}f_1$,$Rad(R_{B_8}) = \mathbf{Z}f_2$. It is easy to check that

$$f_1 = 3\alpha_0+2\alpha_1+4\alpha_2+6\alpha_3+5\alpha_4+4\alpha_5+3\alpha_6+2\alpha_7+\alpha_8,$$

$$f_2 = 2\alpha_0+\alpha_1+2\alpha_2+3\alpha_3+4\alpha_4+3\alpha_5+2\alpha_6+\alpha_7 .$$

Let

$$\mathfrak{r} = f_2 - \alpha_9 .$$

It is immediately verified that $\mathfrak{r} \in \mathbf{R}_{-2}$, and $B = \{\alpha_0,\ldots,\alpha_{10}\}$ is a root basis with Dynkin diagram

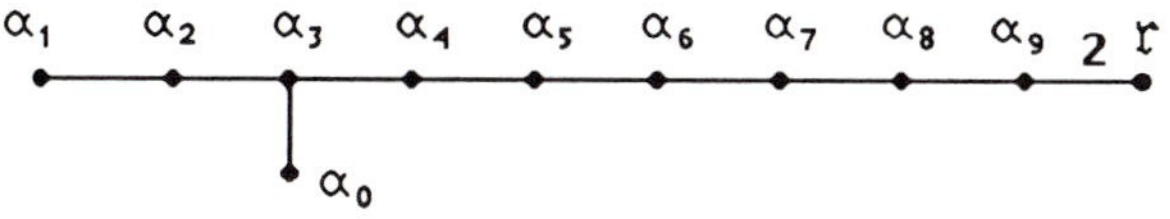

For every finite subset $\{g_1,\ldots,g_k\}$ of a group G we will denote by

$$\overline{\langle g_1,\ldots,g_k\rangle},$$

the minimal normal subgroup of G containing this subset.

Proposition 2.6.1. Let $B = B_{10}$, $\bar{B} = B \cup \{\mathfrak{r}\}$.

(i) B is a crystallographic root basis in $\mathbf{R}$;

(ii) $O(\mathbf{R}) = W_{\bar{B}}(\mathbf{R})\times\{\pm id_{\mathbf{R}}\}$;

(iii) $W_{\bar{B}}(\mathbf{R}) = W_B(R)\ltimes \overline{\langle \mathfrak{r} \rangle}$;

(iv) $\mathbf{R}_{-2} = R(\bar{B}) = R(B)\sqcup W_{\bar{B}}(\mathbf{R})\breve{\mathfrak{r}}$.

PROOF. (i)There are three connected subgraphs of parabolic type in $\Gamma(\bar{B})$ of type $\tilde{E}_7$, $\tilde{E}_8$ and $\tilde{A}_1$. The first two, $\Gamma(B_8)$ and $\Gamma(B_9')$, come from the subgraph $\Gamma(\bar{B})$ which sits naturally in $\Gamma(\bar{B})$. The third one corresponds to the subset $\{\alpha_9, \mathfrak{r}\}$. The subgraph of type $\Gamma(B_9')$ is of maximal rank. The disjoint sum of the other two is of maximal rank too. Thus (i) follows from Theorem 2.4.2.
(ii) Since B is crystallographic, $W_{\bar{B}}(\mathbf{R})$ is the reflection group of $\mathbf{R}$. Thus (ii) follows from the fact that $\Gamma(B)$ does not have symmetries.
(iii)This follows from Theorem 2.2.3.
(iv) This is Corollary 2.2.1.

We will call the root basis $\bar{B} = B\cup\{\mathfrak{r}\}$ the **crystallographic extension** of B.

Here is another model of the lattice $\mathbf{R}$.

Let

$$'H^{1,n} = \langle 2\rangle\oplus\langle -1\rangle^n \quad (n \geq 5)$$

and $e_0, e_1, \ldots, e_n$ be the corresponding basis. Let

$$k_n' = -2e_0+e_1+\ldots+e_n \in {'H}^{1,n}.$$

Proposition 2.6.2.

$$R_n \cong (\mathbf{Z}k_n')^{\perp}.$$

A canonical root basis $B = \{\alpha_0, \ldots, \alpha_{n-1}\}$ in $(\mathbf{Z}k_n')^{\perp}$ can be given as follows:

$$\alpha_0 = e_0-e_1-e_2-e_3-e_4, \quad \alpha_i = e_i-e_{i+1}, \quad i = 1, \ldots, n-1.$$

PROOF. Directly verified.

Remark 2.6.1. The lattices R_n and has an interpretation similar to one for E_n given in Remark 2.5.1. We refer to **[D-O]** or later chapters for this.

Fix a canonical root basis $B_n = \{\alpha_0,\dots,\alpha_{n-1}\}$ in R_n and define the following sublattices of R_n.

$$R_{B_7} \cong E_7\ ,\ R_{B_8}{}' \cong E_8,\ \mathbf{Z}f_2+\mathbf{Z}\alpha_8 \cong U;\ \mathbf{Z}\mathfrak{r}+\mathbf{Z}\alpha_{10}+\dots+\mathbf{Z}\alpha_{n-1} \cong A_{n-9};$$

$$\mathbf{Z}f_1+\mathbf{Z}\mathfrak{r} \cong V(2)\ ,\ \text{where } V=\mathbf{Z}v_1+\mathbf{Z}v_2\ ,\ v_1\cdot v_1=0\ ,v_2\cdot v_2=v_1\cdot v_2=1.$$

Proposition 2.6.3.

(i) $R_n = R_{B_7} \perp \mathbf{Z}f_2+\mathbf{Z}\alpha_8 \perp \mathbf{Z}\mathfrak{r}+\mathbf{Z}\alpha_{10}+\dots+\mathbf{Z}\alpha_{n-1} \cong E_7 \oplus U \oplus A_{n-9}$, $n \geq 9$;

(ii) $R_n = R_{B_8}{}' \perp (\mathbf{Z}f_1+\mathbf{Z}\mathfrak{r}) \perp (\mathbf{Z}\alpha_{10}+\dots+\mathbf{Z}\alpha_{n-1}) \cong E_8 \oplus V(2) \oplus A_{n-10}$, $n \geq 10$.

PROOF. It is directly verified that the sum on the right hand side is orthogonal. Since both sides are lattices of the same discriminant $2n-16$, they must be equal.

Remark 2.6.2. Note that $\mathbf{R}_{B_7} \oplus U_B \subset \mathbf{R}_{B_9} \cong Q_{2,4,5}$. Since both lattices have the same discriminant (=2), they coincide. Thus

$$Q_{2,4,5} \cong E_7 \oplus U, \text{ and } \mathbf{R} \cong Q_{2,4,5} \oplus A_1.$$

In particular, we see that

$$(\mathbf{Z}\alpha)^{\perp} \cong Q_{2,4,5} \text{ for any } \alpha \in W_{\bar{B}}(\mathbf{R})\mathfrak{r}$$

However,

$$(\mathbf{Z}\alpha)^{\perp} \cong E_7 \oplus V(2) \text{ for any } \alpha \in R(B).$$

Corollary 2.6.1. **R** is isomorphic to a sublattice of the Enriques lattice **E** of index 2.

PROOF. Obviously, we can define an embedding $j{:}E_7\oplus A_1 \hookrightarrow E_8$ onto a sublattice of index 2 . Then embed $\mathbf{R} \cong E_7 \oplus A_1 \oplus U$ into $\mathbf{E} \cong E_8 \oplus U$ by $j\oplus id_U$. Or , use the isomorphism $\mathbf{R} \cong Q_{2,4,5} \oplus A_1$ from Remark 2.6.1, and note that $Q_{2,4,5}$ embeds naturally into $\mathbf{E} \cong Q_{2,4,7}$ with orthogonal complement of rank 1 and discriminant 2, i.e. isomorphic to A_1. Hence $\mathbf{R} \cong Q_{2,4,5} \oplus A_1$ embeds into $\mathbf{E}$ as a sublattice of index 2.

Remark 2.6.3. Since $\alpha_i \cdot \mathfrak{r} \equiv 0 \bmod 2$ for $i=0,\dots,9$, we have

$$\mathbf{R} \cong \{x \in \mathbf{E} : x \cdot \alpha \equiv 0 \bmod 2\},$$

where α is any (fixed) vector from $\mathbf{E}_{-2}$. Indeed, it is true if $\alpha = \mathfrak{r}$. On the other hand, $\mathbf{E}_{-2} = \mathrm{O}(\mathbf{E})_{\mathfrak{r}}$. In particular, we have

$$\mathrm{O}(\mathbf{R}) = \{\sigma \in \mathrm{O}(\mathbf{E}): \ \sigma(\alpha) - \alpha \in 2\mathbf{E}\}.$$

A root basis B in $\mathbf{E}$ with Dynkin diagram of type $T_{2,4,6}$ will be called a **Reye root basis**. The corresponding sublattice $\mathbf{E}_B$ is isomorphic to $\mathbf{R}$.

Proposition 2.6.4. Let $(f_1,\dots,f_{10})$ be an isotropic sequence of length 10 in $\mathbf{E}$. Then a Reye root basis $B = \{\alpha_0,\dots,\alpha_9\}$ in $\mathbf{E}$ can be obtained by the formulas

$$\alpha_0 = f_1+f_2+f_3+f_4-\Delta, \ \alpha_1 = f_2-f_1, \ \dots, \ \alpha_9 = f_{10} - f_9.$$

The dual basis $\check{B} = \{\omega_0,\dots,\omega_9\}$ to B in $(\mathbf{R}_B)^* \subset \mathbf{R}_r{}^* = \mathbf{R}_r$ is given as follows:

$$\omega_0 = f_{10} + \tfrac{1}{2}\mathfrak{r}, \ \omega_1 = f_1, \ \omega_2 = f_1 + f_2, \ \omega_3 = f_1 + f_2 + f_3,$$

$$\omega_4 = f_1 + f_2 + f_3 + f_4, \ \omega_5 = f_{6,7} + f_{8,9} + \tfrac{1}{2}\mathfrak{r}, \ \omega_6 = f_{7,8} + f_{9,10},$$

$$\omega_7 = f_{8,9} + \tfrac{1}{2}\mathfrak{r}, \ \omega_8 = f_{9,10}, \ \omega_9 = \tfrac{1}{2}\mathfrak{r},$$

where

$$f_{i,j} = \Delta - f_i - f_j \in \mathbf{R}_0 \text{ for any } i \neq j.$$

Moreover, the Gram matrix $(\omega_i \cdot \omega_j)$ of $\check{B}$ is the following:

$$\begin{bmatrix}
\frac{5}{2} & \frac{3}{2} & 3 & \frac{9}{2} & 6 & 5 & 4 & 3 & 2 & 1 \\
\frac{3}{2} & 0 & 1 & 2 & 3 & \frac{5}{2} & 2 & \frac{3}{2} & 1 & \frac{1}{2} \\
3 & 1 & 2 & 4 & 6 & 5 & 4 & 3 & 2 & 1 \\
\frac{9}{2} & 2 & 4 & 6 & 9 & \frac{15}{2} & 6 & \frac{9}{2} & 3 & \frac{3}{2} \\
6 & 3 & 6 & 9 & 12 & 10 & 8 & 6 & 4 & 2 \\
5 & \frac{5}{2} & 5 & \frac{15}{2} & 10 & \frac{15}{2} & 6 & \frac{9}{2} & 3 & \frac{3}{2} \\
4 & 2 & 4 & 6 & 8 & 6 & 4 & 2 & 2 & 1 \\
3 & \frac{3}{2} & 3 & \frac{9}{2} & 6 & \frac{9}{2} & 2 & \frac{3}{2} & 1 & \frac{1}{2} \\
2 & 1 & 2 & 3 & 4 & 3 & 2 & 1 & 0 & 0 \\
1 & \frac{1}{2} & 1 & \frac{3}{2} & 2 & \frac{3}{2} & 1 & \frac{1}{2} & 0 & -\frac{1}{2}
\end{bmatrix}$$

PROOF. This is verified by direct computations.

Remark 2.6.4. Note that $\mathfrak{r} = \Delta - 2f_{10}$ and

$$\mathfrak{r}\cdot f_i = 1\ ,\ i\neq 10\ ,\ \mathfrak{r}\cdot f_{10} = 3\ ,\ \mathfrak{r}\cdot\Delta = 4\ .$$

Corollary 2.6.2. Let B be a Reye root basis in **E** as above and $\bar{B} = B\cup\{\mathfrak{r}\}$ be the corresponding crystallographic basis in $\mathbf{E}_B$. Then

$$C(B) = \mathbb{R}_{\geq 0}\omega_0 + \dots + \mathbb{R}_{\geq 0}\omega_9$$

$$C(\bar{B}) = \mathbb{R}_{\geq 0}\omega_0 + \dots + \mathbb{R}_{\geq 0}\omega_8 + \mathbb{R}_{\geq 0}\omega_{10} + \dots + \mathbb{R}_{\geq 0}\omega_{17}$$

where $\omega_0,\dots,\omega_9$ are given as above, and

$$\omega_{10} = \omega_0+2\omega_9,\ \omega_{11} = \omega_1+\omega_9,\ \omega_{12} = \omega_2+2\omega_9,\ \omega_{13} = \omega_3+3\omega_9\ ,$$

$$\omega_{14} = \omega_4+4\omega_9,\ \omega_{15} = \omega_5+3\omega_9\ ,\ \omega_{16} = \omega_6+2\omega_9,\ \omega_{17} = \omega_7+\omega_9.$$

PROOF. The first assertion is obvious. To see the second one, it suffices to note that

$$C(\bar{B}) = \{x\in C(B):\ x\cdot\alpha_{10} \geq 0\} =$$

$$= \{\lambda_0\omega_0+\dots+\lambda_9\omega_9:\ \ \lambda_i\geq 0\ \ ,2\lambda_0+\lambda_1+2\lambda_2+3\lambda_3+4\alpha_4+3\lambda_5+2\lambda_6+\lambda_7-\lambda_9\geq 0\}.$$

Corollary 2.6.3. Let $B_9 \subset B$ be the root basis with Dynkin diagram of type $T_{2,4,5}$. Then its dual basis $\{\omega_0',\dots,\omega_8'\}$ in $\mathbf{E}_B'^{*} \subset \frac{1}{2}\mathbf{E}_B'$ is given by

$$\omega_0' = f_{10} + \tfrac{3}{2}\mathfrak{r},\ \omega_1' = f_1 + \tfrac{1}{2}\mathfrak{r},\ \omega_2' = f_1 + f_2 + \mathfrak{r},$$

$$\omega_3' = f_1 + f_2 + f_3 + \tfrac{3}{2}\mathfrak{r},\ \omega_4' = f_{5,6} + f_{7,8} + f_{9,10} + 2\mathfrak{r},$$

$$\omega_5' = f_{6,7} + f_{8,9} + 2\mathfrak{r},\ \omega_6' = f_{7,8} + f_{9,10} + \mathfrak{r},\ \omega_7' = f_{8,9} + \mathfrak{r},$$

$$\omega_8' = f_{9,10}.$$

Proposition 2.6.5. Let $B = \{\alpha_0,\dots,\alpha_9\}$ be a canonical root basis in **R** and $\bar{B} = B\cup\{\mathfrak{r}\}$ be the corresponding crystallographic root basis. Let $W = W_B(R)$. Then

$$WC(\bar{B}) := \bigcup_{w\in W} w(C(\bar{B})) = \{x\in V^+ : x\cdot w(\mathfrak{r}) \geq 0\ ,\ \text{for any } w\in W\ \}.$$

PROOF. Let $x \in A = \{x \in V^+ : x \cdot w(\mathfrak{r}) \geq 0$, for any $w \in W\}$. By Proposition 2.4.1, $C(B) \cap V^+$ is a fundamental domain of W in V^+. Hence there exists $w \in W$ such that $w(x) \in C(B)$. Since also $w(x) \cdot \mathfrak{r} = x \cdot w^{-1}(\mathfrak{r}) \geq 0$, $w(x) \in C(\bar{B})$. This shows that $A \subset WC(\bar{B})$. Conversely, let $x \in WC(\bar{B})$, $x = w(y)$ for some $y \in C(\bar{B})$ and $w \in W$. We have to show that

$$w(y) \cdot \mathfrak{r} \geq 0 \text{ for each } y \in C(\bar{B}) \text{ and } w \in W.$$

We use induction on the length $l(w)$ of w. The assertion is true for $w = 1$,i.e. $l(w) = 0$. Write $w = w' \circ s_i$, where s_i is the simple reflection into some α_i, $i \neq 10$, and $l(w') = l(w)-1$. Hence

$$w(y) \cdot \mathfrak{r} = (w' \circ s_i(y)) \cdot \mathfrak{r} = w'(y+(y \cdot \alpha_i)\alpha_i) \cdot \mathfrak{r} =$$

$$= w'(y) \cdot \alpha_{10} + (y \cdot \alpha_i)(w'(\alpha_i) \cdot \mathfrak{r}) = w'(y) \cdot \mathfrak{r} + (y \cdot \alpha_i)(\alpha_i \cdot w'^{-1}(\mathfrak{r})).$$

By induction $w'(y) \cdot \mathfrak{r} \geq 0$. Since $y \in C(B)$, $y \cdot \alpha_i \geq 0$. We have $w^{-1} = s_i \circ w'^{-1}$ and $l(w'^{-1}) = l(w) - 1$. Thus $w'^{-1}(y) \cdot \alpha_i \geq 0$ for every $y \in C(B)$ **[Bou 2]**,Chapter V,S4, n°4). In particular, $\alpha_i \cdot w'^{-1}(\mathfrak{r}) \geq 0$. This shows that $w(y) \cdot \mathfrak{r} \geq 0$.

S7. The function φ_M.

Let M be a hyperbolic lattice with $M_0 \neq \{0\}$, $M_{>0} = \{x \in M: x \cdot x > 0\}$. Define

$$\varphi_M: M_{>0} \to \mathbf{Z}_{\geq 0}$$

by

$$\varphi_M(x) = \inf_{f \in M_0'} \{(x \cdot f)^2 / x^2\}$$

Proposition 2.7.1. Let $x \in M_{>0}$.

(i) $\varphi_M(x) = (x \cdot f_0)^2 / x^2 > 0$ for some $f_0 \in M_0$;

ii) $\varphi_M(\sigma(x)) = \varphi_M(x)$ for any $\sigma \in O(M)$;

(iii) φ_M extends by linearity to a homogeneous function of degree 0 on the set $\{x \in M_{\mathbb{R}} : x^2 > 0\}$.

PROOF. (i) Since M is hyperbolic, $x\cdot f \neq 0$ for every $x\in M_{>0}$ and $f\in M_0'$. Thus the function $f \to (x\cdot f)^2$ takes values in $\mathbf{Z}_{>0}$, and hence has a positive minimum. The same is true for the function $f \to (x\cdot f)^2/x^2$.
(ii) and (iii) are obvious.

Remark 2.7.1. One can also define the function φ_M for indefinite non-hyperbolic lattices. However, if M is of signature (t_+,t_-) with $t_+ > 1$ and $rk(M) > 5$, then $((\mathbf{Z}x)^{\perp})_0 \neq \emptyset$ for any $x\in M_{>0}$. This shows that $\varphi_M(x) \equiv 0$, hence φ_M has sense for only a few lattices besides the hyperbolic ones.

Problem. Estimate φ_M from above.

Following an idea of E.Looijenga, we will show how to solve this problem for Nikulin lattices.

Let M be a Nikulin lattice with $M_0 \neq \emptyset$ (e.g. $rk(M) \geq 5$), B a crystallographic root basis, $W = W_B(M)$ the reflection group of M , C(B) a fundamental chamber for W, $P(M) = C(B)/\mathbb{R}_+ \subset P(M)$ the corresponding fundamental polyhedron, $\{v_1,\ldots,v_r\}$ (resp. $\{v_1,\ldots,v_k\}$, $k\leq r$) the set of its vertices (resp.improper vertices), representing the W-orbits on the set of vectors in M_0', $v_1,\ldots,v_r \in C(B)\cap M$ the corresponding representatives of these orbits.

Lemma 2.7.1. If $x\in C(M)$, then there exists a vertex v_i such that

$$\inf_{f\in M_0'} \{|x\cdot f|\} = x\cdot v_i .$$

PROOF. Let

$$\inf_{f\in M_0'} \{|x\cdot f|\} = x\cdot f_0$$

for some $f_0\in M_0'$. If $f_0 \neq v_i$, $i = 0,\ldots,k$, then there exists $w\in W$, $w\neq 1$, such that $f_0 = w(v_i)$ for some i. Hence,

$$x\cdot v_i \geq x\cdot f_0 = w^{-1}(x)\cdot v_i.$$

Thus the assertion would follow if we prove that

$$w(x)\cdot v_i \geq x\cdot v_i \quad \text{for any } w\in W.$$

Let $l:W \to \mathbb{Z}$ be the length function on the Coxeter group W. If $l(w) > 1$, then we can write $w = s_\alpha \circ w'$, where $l(w') = l(w)-1$ and $\alpha\in B$. Since the reflection hyperplane H_α separates $C(B)^\circ$ and $w(C(B)^\circ)$, $\alpha\cdot y \leq 0$ for each $y\in w(C(B))$. Hence

$$w'(x)\cdot v_i = (s_\alpha w)(x)\cdot v_i = (w(x)+(w(x)\cdot\alpha)\alpha)\cdot v_i =$$

$$= w(x)\cdot v_i + (w(x)\cdot\alpha)(\alpha\cdot v_i) \leq w(x)\cdot v_i$$

(because $w(x)\cdot\alpha \leq 0$, $\alpha\cdot v_i \geq 0$). Thus the assertion follows by induction on $l(w)$.

Define the function

$$\varphi_i : P(B)^\circ \to \mathbb{R},$$

by

$$\bar{\varphi}_i(\bar{x}) = (x\cdot v_i)^2/x^2, \text{ where } \bar{x} = \mathbb{R}_+ x, \quad x\notin\{v_1,\dots,v_K\}.$$

Lemma 2.7.2. φ_i is a convex function on $P(B)^\circ$.

PROOF. Let $x,y\in C(B)\cap M_{>0}$. Since x and y are proportional or span a plane of signature (1,1) in $M_{\mathbb{R}}$, we obtain $(x\cdot y)^2 \geq (x\cdot x)(y\cdot y)$. Let $\bar{x}$ and $\bar{y}$ denote the images of x and y in $P(B)^\circ$. For any $\lambda,\mu\in\mathbb{R}_+$, we have

$$\varphi_i(\lambda x+\mu y)((\lambda x+\mu y)^2 = ((\lambda x+\mu y)\cdot v_i)^2 =$$

$$= (\lambda^2(x\cdot v_i)^2 + 2\lambda\mu(x\cdot v_i)(y\cdot v_i) + \mu^2(y\cdot v_i)^2)$$

$$\leq \lambda^2\varphi_i(x)x^2 + 2\lambda\mu(\varphi_i(x)\varphi_i(y)(x^2y^2)^{1/2} + \mu^2\varphi_i(y)y^2 \leq$$

$$\leq \max\{\varphi_i(x),\varphi_i(y)\}(\lambda^2x^2+2\lambda\mu(x\cdot y)+\mu^2y^2) =$$

$$= \max\{\varphi_i(x),\varphi_i(y)\}((\lambda x+\mu y)^2.$$

Thus $\varphi_i(\lambda x+\mu y) \leq \max\{\varphi_i(x),\varphi_i(y)\}$.

Theorem 2.7.1. Let $C(B) = \mathbb{R}_{\geq 0}v_1 + \dots + \mathbb{R}_{\geq 0}v_r$. Let

$$A = \max_{j>K}(\min_{j\leq K}(v_j\cdot v_i)^2/(v_j^2)), \qquad B = \min_{\substack{i\leq K\\ j\leq K}}\{v_i\cdot v_j\}.$$

Then

$$\max \varphi_M = \max\{A, \tfrac{1}{2}B\}.$$

PROOF. Let $x \in M_{>0}$. We may assume that $x \in C(M)$. By Lemma 2.7.1,

$$\varphi_M(x) = \min_{j \le K} (x \cdot v_i)^2 / x^2.$$

Let $x = \lambda_1 v_1 + \ldots + \lambda_K v_K + \ldots + \lambda_r v_r = x_1 + x_2$, where $x_1 = \sum_{j \le K} \lambda_j v_j$, $x_2 = \sum_{j > K} \lambda_j v_j$. If $x_1 \ne v_i$ for some i and $x_2 \ne 0$, then by Lemma 2.7.2,

$$\varphi_M(x) \le \min\{(x_1 \cdot v_i)^2 / x_1^2, (x_2 \cdot v_i)^2 / x_2^2\}$$

$$\le \min \{\varphi_M(x_1), (v_j \cdot v_i)^2 / v_j^2\} \le \min\{\varphi_M(x_1), A\}.$$

Thus we may assume that either $x_1 = \lambda_1 v_1$, $x_2 \ne 0$, or $x_2 = 0$. In the first case,

$$\varphi_M(x) \le (x \cdot v_1)^2 / x^2 = (x_2 \cdot v_1)^2 / x^2 \le (x_2 \cdot v_1)^2 / x_2^2 \le A$$

(since $x^2 = x_2^2 + 2\lambda_1(x_2 \cdot v_1) \ge x_2^2$). In the second case,we can use the argument above (by splitting one summand) to assume that $x = \lambda_i v_i + \lambda_j v_j$, where $i, j \le K$. We have

$$\varphi_M(x) \le \min\{(x \cdot v_i)^2 / x^2, (x \cdot v_j)^2 / x^2\} =$$

$$\min\{\tfrac{1}{2}(\lambda_i / \lambda_j)(v_i \cdot v_j), \tfrac{1}{2}(\lambda_j / \lambda_i)(v_i \cdot v_j)\} \le \tfrac{1}{2}(v_i \cdot v_j).$$

Clearly,if $x = v_i + v_j$, $i, j \le K$, then

$$(x \cdot v_i)^2 / x^2 = \tfrac{1}{2}(v_i \cdot v_j),$$

and, if $x = v_j$, $j > K$, then for $i \le K$

$$(x \cdot v_i)^2 / x^2 = (v_j \cdot v_i)^2 / v_j^2.$$

Thus the estimate

$$\varphi_M(x) \le \max\{A, \tfrac{1}{2}B\}$$

is sharp.

Corollary 2.7.1. Let $M = \mathbf{E}$. Then

$$\max \varphi_E = 1.$$

PROOF. Applying the theorem and Proposition 2.5.7, we see that $\max \varphi_E$ is equal to $\max((\omega_i \cdot \omega_9)^2 / \omega_i^2 = (\omega_2 \cdot \omega_9)^2 / \omega_2^2 = 1$.

Corollary 2.7.2. Let $M = \mathbf{R}$. Then

$$\max \varphi_{\mathbf{R}} = 2.$$

PROOF. Let $B = \{\alpha_0,\dots,\alpha_9\}$ be a canonical root basis in R and $\bar{B} = \{\alpha_0,\dots,\alpha_{10}\}$ be the corresponding crystallographic root basis. In the notation of Corollary 2.6.2

$$C(\bar{B}) = \mathbb{R}_{\geq 0}\omega_0 + \dots + \mathbb{R}_{\geq 0}\omega_{17}.$$

There are two improper vertices in P(B). They are represented by the vectors $v_1 = \omega_1$ and $v_2 = \omega_8$. In the notation of the Theorem, we have

$$A = (\omega_{17}\bullet\omega_1)^2/\omega_{17}^2 = (\omega_2\bullet\omega_8)^2/\omega_2^2 = 2,$$

$$B = 1.$$

Corollary 2.7.3. Let $M = Q_{2,4,5}$. Then

$$\max \varphi_M = 2 .$$

PROOF. Use the Theorem and Corollary 2.6.3 and observe that

$$\max \varphi_M = \max \{(\omega_i\bullet\omega_8)^2/\omega_i^2\} = 2.$$

Let $\alpha\in\mathbf{E}_{-2}$, $\mathbf{E}_\alpha = (\mathbf{Z}\alpha)^\perp$. For every $x\in\mathbf{E}_{>0}$ we define

$$\varphi_\alpha(x) = \inf_{f\in(E_\alpha)_0'} \{(x\bullet f)^2/x^2\}.$$

Let $x\in\mathbf{E}_{>0}$, then

$$y = 2x+(x\bullet\alpha)\alpha\in \mathbf{E}_\alpha .$$

It follows from the proof of Corollary 2.6.1 that

$$\mathbf{E}_\alpha \cong Q_{2,4,5}.$$

By Corollary 2.7.3,

$$\varphi_{E_\alpha}(y) \leq 2,$$

i.e there exists $f\in\mathbf{E}_\alpha$ such that $(y\bullet f)^2 = (2x\bullet f)^2 \leq 2y^2 = 8x^2 + 4(x\bullet\alpha)^2$, hence

$$(x\bullet f)^2 \leq 2x^2 + (x\bullet\alpha)^2.$$

This shows that

$$\varphi_\alpha(x) \le (x\cdot f)^2/x^2 \le (x\cdot\alpha)^2/x^2 + 2.$$

In fact, one can prove a better result.

Theorem 2.7.2. Let $x\in\mathbf{E}_{>0}$ and $\alpha\in\mathbf{E}_{-2}$. Assume that

$$(x\cdot\alpha)^2 > 9x^2/\varphi_E(x) .$$

Then

$$\varphi_\alpha(x) < (x\cdot\alpha)^2/x^2.$$

PROOF. Without loss of generality, we may assume that $\alpha = \mathfrak{r}$, where $\{\alpha_0,\ldots,\alpha_9\}$ is the Reye root basis in $\mathbf{E}$ constructed in Proposition 2.6.4. Then $B = \{\alpha_0,\ldots,\alpha_8\}$ is a canonical root basis in $\mathbf{E}_\alpha$ and the corresponding fundamental chamber $C(B) = \mathbb{R}_{\ge 0}\omega_0' + \ldots + \mathbb{R}_{\ge 0}\omega_8'$. Let

$$y = 2x + (x\cdot\alpha)\alpha \in \mathbf{E}_\alpha.$$

Applying transformations from $W_B(\mathbf{E}) = W(\mathbf{E}_\alpha)$, we may assume that $y\in C(B)$. The expressions for the ω_i's allow us to write

$$y = a_0\alpha + a_1f_1 + a_2f_2 + a_3f_3 + a_4f_{10} + a_5f_{5,6} + a_6f_{6,7} + a_7f_{7,8} + a_8f_{8,9} + a_9f_{9,10},$$

where f_i and $f_{j,j+1}$ are isotropic vectors in $\mathbf{E}$. Also, we know that

$$\alpha\cdot f_i = 1,\ i=1,2,3,\ \alpha\cdot f_{10} = 3,\ \alpha\cdot f_{j,j+1} = 2,\ j=5,6,7,8,\ \alpha\cdot f_{9,10} = 0$$

$$f_i\cdot f_{9,10} = 1,\ i=1,2,3,\ f_{10}\cdot f_{9,10} = 2,\ f_{j,j+1}\cdot f_{9,10}=2,\ j=5,6,7,\ f_{8,9}\cdot f_{9,10} = 1.$$

Since $y\cdot\alpha = 0$, we have

$$2a_0 = a_1 + a_2 + a_3 + 3a_4 + 2a_5 + 2a_6 + 2a_7 + 2a_8.$$

Assume that $x\cdot\alpha > a_0$. Then,

$$4\varphi_\alpha(x) \le 4(x\cdot f_{9,10})^2/x\cdot x \le (y\cdot f_{9,10})^2/x^2 =$$

$$= (a_1+a_2+a_3+2a_4+2a_5+2a_6+2a_7+a_8)^2/x^2$$

$$= (2a_0-a_4-a_8)^2/x\cdot x \le 4a_0^2/x^2$$

$$< 4(x\cdot\alpha)^2/x^2,$$

i.e. $\varphi_\alpha(x) < (x\cdot\alpha)/x^2$.

Assume that $x\cdot\alpha \ge a_0$. Then

$$2x = (y-a_0\alpha) + (a_0-(x\bullet\alpha))\alpha,$$

and

$$2x^2 = ((y-a_0\alpha)\bullet x + (a_0-(x\bullet\alpha)(x\bullet\alpha) \geq (y-a_0\alpha)\bullet x \geq (\varphi_E(x)(x^2)^{1/2}\sum_{i\geq 1} a_i,$$
$$2x\bullet\alpha = (y-a_0\alpha)\bullet\alpha -2(a_0-(x\bullet\alpha)) \leq (y-a_0\alpha)\bullet\alpha \leq 3\sum_{i\geq 1} a_i.$$

Here we have used that $(\varphi_E(x)(x^2))^{1/2} \geq x\bullet f_i \geq 0$, $(\varphi_E(x)(x^2))^{1/2} \geq x\bullet f_{j,j+1} \geq 0$, and $\alpha\bullet f_i \leq 3$, $\alpha\bullet f_{j,j+1} \leq 3$.

Thus

$$2x\bullet\alpha \leq 3\sum_{i\geq 1} a_i \leq 6(x^2)^{\frac{1}{2}}/\varphi_E(x))^{\frac{1}{2}}, \text{ i.e. } (x\bullet\alpha)^2 \leq 9x^2/\varphi_E(x).$$

The theorem is proven.

§8. 2-congruence subgroups of finite Weyl groups

Let M be a free **Z**-module, ℓ be a prime number, $M_\ell = M/\ell M$ has a natural structure of a vector space over $\mathbb{F}_\ell$ and the canonical homomorphism of groups

$$r_\ell : GL(M) \to GL(M_\ell)$$

is well-defined. For every subgroup G of GL(M) , let

$$G(1) = G\cap Ker(r_\ell) = \{g\in G: g(x) \equiv x \bmod \ell M\},$$

and call it the **ℓ-congruence subgroup** of G. Clearly, it is a normal subgroup of G.

In this section we will recall the known results about the groups W(2) , where $W = W_B(M)$ is the Weyl group of a root basis of finite type in a lattice M

Lemma 2.8.1. Let M be a negative definite even lattice spanned by an irreducible root basis B. Then

$$O(M)(2) = \{\pm id_M\}.$$

Proof. Let $\sigma \in O(M)(2)$ and $\alpha \in B$. Then

$$\sigma(\alpha) = \alpha + 2x \text{ for some } x \in M.$$

Thus

$$4x \cdot x = \sigma(\alpha)\cdot\sigma(\alpha) - 2\sigma(\alpha)\cdot\alpha + \alpha^2 = -4 - 2\sigma(\alpha)\cdot\alpha.$$

Suppose $x \neq 0$; then $x^2 \leq -2$, and $\sigma(\alpha)\cdot\alpha \geq 2$. This implies that $\sigma(\alpha)$ and α are proportional (otherwise the sublattice spanned by α and $\sigma(\alpha)$ is not negative definite. Thus $\sigma(\alpha) = \pm\alpha$. If $\sigma(\alpha) = -\alpha$ and $\sigma(\beta) = \beta$, for some $\beta \in B$, then $\alpha\cdot\beta = \sigma(\alpha)\cdot\sigma(\beta) = -\alpha\cdot\beta$, i.e. $\alpha\cdot\beta = 0$. Therefore $B = B_1 \amalg B_2$, where $B_1 = \{\alpha \in B : \sigma(\alpha) = \alpha\}$, $B_2 = \{\alpha \in B : \sigma(\alpha) = -\alpha\}$, and $\alpha\cdot\beta = 0$ for any $\alpha \in B_1, \beta \in B_2$. This contradicts the irreducibility of B.

Proposition 2.8.1. Let $B = \{\alpha_0, \ldots, \alpha_r\}$ be an irreducible root basis of finite type in a lattice M, $W = W_B(M)$. Assume that M_B is a primitive sublattice of M. Then

$$W(2) = \{id_M\} \text{ if B is not of type } A_1, D_{2K}, E_7, E_8.$$
$$= \{\pm id_M\} \text{ otherwise.}$$

PROOF. This can be found in **[Bou 2]**, but since it is not stated explicitly there, we sketch a proof. Obviously, we may assume that $M = M_B$. By the previous lemma,

$$W(2) \subset \{\pm id_M\}.$$

It is known that W acts transitively on the set of unordered canonical root bases (corresponding to so-called Weyl chambers). Thus there exists an element $w_0 \in W$ such that $w_0(B) = -B$. Hence $w_0^2 = 1$. Since $(-w_0)(B) = B$, $-w_0 \in O(M)$ cannot belong to W, unless it is the identity. On the other hand, we know that $A(B) = O(M)/W$ is isomorphic to the group of symmetries of the Dynkin diagram of B.

If A(B) is trivial (this happens in the cases E_7 and E_8), $w_0 = -id_M \in W$ and $W = \{\pm id_M\}$. If B is of type A_n, then $W \cong A_{n+1}$, the symmetric group on n+1 letters. It is well known that this group contains a normal subgroup of order

2 only in the case $n = 1$. The cases D_n and E_6 are more complicated, and we have to refer to **[Bou 2]** Chapter V, §6, n°2, Corollary 3, to see that $w_0 = -id_M$, if B is of type D_{2k}, and $-id_M \notin W$, if B is of type D_{2k+1} or E_6.

Remark 2.8.1. The element $w_0 \in W$ such that $w_0(B)=-B$ has the following remarkable properties:

(i) $w_0^2 = 1$;

(ii) w_0 generate W(2) in the cases A_1, D_{2r}, E_7, E_8;

(iii) $l(w_0) = \# R(B) > l(w)$ for any $w \in W \setminus \{w_0\}$;

(iv) $l(ww_0) = l(w_0) - l(w)$ for any $w \in W$;

(v) $w_0 = (s_{\alpha_0} \circ \ldots \circ s_{\alpha_{n-1}})^{h/2}$, where $h = \# R(B)/\# B$ is the Coxeter number of B (which can be found in Tables of **[Bou 2]**);

(vi) $w_0 = g^{\dim \mathfrak{g}}$ for a certain element $g \in W$, where $\mathfrak{g}$ is the simple Lie algebra associated to the root system B ($\dim \mathfrak{g} = \# R(B) + \# B$).

Properties (i) and (ii) were proven before. The proof of properties (iii)-(v) can be found in **[Bou 2]**. Property (vi) is proven in **[St]**.

Let V be a vector space of dimension n over the field $\mathbb{F}_2$. Recall that a quadratic form $q: V \to \mathbb{F}_2$ is a map such that $(x,y) \mapsto q(x+y)-q(x)-q(y)$ defines a bilinear map b_q on $V \times V$. It is well known that there are at most two isomorphism classes of non-degenerate (in the sense of b_q) quadratic forms on V : in an appropriate coordinate system a non-degenerate quadratic form is one of the following forms (see **[Ar]**, Chapter III, §6):

$$x_1x_2 + \ldots + x_{2k-1}x_{2k} + x_{2k+1}^2 \quad \text{if } n = 2k+1;$$

$$\text{(I) } x_1x_2 + \ldots + x_{2k-1}x_{2k} \quad \text{or} \quad \text{(II) } x_1x_2 + \ldots + x_{2k-1}x_{2k} + x_{2k-1}^2 + x_{2k}^2 \quad \text{if } n = 2k.$$

The two types when $\dim V = n = 2k$ differ by the following property

$$\#\{x \in V : q(x) = 0\} = 2^{k-1}(2^k+1) \quad \text{in the first case}$$

$$= 2^{k-1}(2^k-1) \quad \text{in the second case}$$

We will call the quadratic forms of the first (resp. second) type

even (resp. **odd)** quadratic forms. The corresponding orthogonal groups will be denoted by

$$O^+(V,q) \text{ or } O^+(2k,\mathbb{F}_2), \text{ (resp. } O^-(V,q) \text{ or } O^-(2k,\mathbb{F}_2)).$$

if $V = \mathbb{F}_2^{2k}$ and q is as above of type (I) (resp. (II)).

Note that the bilinear form b_q is a a non-degenerate symplectic form on V (i.e. a bilinear form b: $V\times V \to \mathbb{F}_2$ with $b(x,x) \equiv 0$) and its isomorphism class is independent of q and in an appropriate coordinate system (corresponding to q as above) can be given by the formula

$$x_1y_{k+1}+\ldots+x_ky_{2k}+x_{k+1}y_1+\ldots+x_{2k}y_k.$$

The group of symplectic isometries of V is denoted by Sp(V) or $Sp(2k,\mathbb{F}_2)$, if $V = \mathbb{F}_2^n$ and b is as above. In particular, we obtain

$$O^{\pm}(V) \text{ is a subgroup of } Sp(V).$$

Proposition 2.8.2.

$$\#Sp(2k,\mathbb{F}_2) = 2^{k^2}(2^{2k}-1)(2^{2k-2}-1)\ldots(2^2-1),$$

$$[Sp(2k,\mathbb{F}_2):O^{\pm}(2k,\mathbb{F}_2)] = 2^k(2^{k-1}\mp 1).$$

PROOF. See **[Ar]**, Chapter III, §6.

Proposition 2.8.3. Let W be the Weyl group of an irreducible root basis B of finite type of rank n. Then

$$\begin{aligned} \#W &= n! \text{ if B is of type } A_n \\ &= 2^{n-1}n! \text{ if B is of type } D_n, \\ &= 2^7.3^7.5 \text{ if B is of type } E_6, \\ &= 2^{10}.3^4.5.7 \text{ if B is of type } E_7, \\ &= 2^{14}.3^5.5^2.7 \text{ if B is of type } E_8. \end{aligned}$$

PROOF. Of course, it can be found in **[Bou 2]**.

Proposition 2.8.4. Let W be the Weyl group of an irreducible root basis B of finite type of rank n. Then

$$\begin{aligned} W/W(2) &\cong \{1\} \text{ (type } A_1\text{)},\\ &\cong S_{n+1} \text{ (type } A_n,\ n>1),\\ &\cong S_n \ltimes (\mathbb{Z}/2)^{n-1} \text{ (type } D_n,\ n \text{ odd)},\\ &\cong S_n \ltimes (\mathbb{Z}/2)^{n-2} \text{ (type } D_n,\ n \text{ even)},\\ &\cong O^-(6,\mathbb{F}_2) \text{ (type } E_6\text{)},\\ &\cong Sp(6,\mathbb{F}_2) \text{ (type } E_7\text{)},\\ &\cong O^+(8,\mathbb{F}_2) \text{ (type } E_8\text{)}. \end{aligned}$$

PROOF. We may assume that $M = M_B$. Let $\bar{W} \cong W/W(2)$ be the image of W under the reduction homomorphism r_2: $W \to GL(\bar{M})$, where $\bar{M} = M/2M$. In the cases A and D, the assertion follows from the description of W given in **[Bou 2]**. The other cases are given there as Exercises 1-3, §4, Chapter VI. Let us solve them. The quadratic form $x \to x \cdot x$ on M defines a quadratic form on $\bar{M} \cong \mathbb{F}_2^n$ by

$$q(\bar{x}) = \tfrac{1}{2} x \cdot x \bmod 2,$$

where the bar denotes the image under the reduction homomorphisms $M \to \bar{M}$. Its associated bilinear form b_q is given by

$$b_q(\bar{x},\bar{y}) = x \cdot y \bmod 2.$$

Assume that B is of type E_6. Then M is an even lattice of rank 6 and discriminant 3, hence b_q is non-degenerate and q is a non-degenerate quadratic form on $\bar{M} \cong \mathbb{F}_2^{\,6}$. Clearly, $\bar{W} \subset O(\bar{M},q)$. Comparing their orders by Propositions 2.8.2 and 2.8.3, we find $\#\bar{W} = \#O^-(6,\mathbb{F}_2)$. Thus q is of odd type and $W \cong \bar{W} = O(M,q) \cong O(6,\mathbb{F}_2)$.

The case E_8 is similar to the case E_6. However, in this case, comparing the orders of $\bar{W}$ and $O(\bar{M},q)$, we find that q is of even type and $\bar{W} \cong O^+(8,\mathbb{F}_2)$.

In the case E_7, we note that $M \subset M'$, where $M' = E_8$. This defines an inclusion $W \hookrightarrow W' = W(E_8)$ and the induced inclusion $\bar{W} \hookrightarrow \bar{W}' = O(\bar{M}') \cong O^+(8,\mathbb{F}_2)$. The group $\bar{W}$ stabilizes the vector $x_0 \in \bar{M}'$ which spans $Rad(\bar{M})$. Thus $W \hookrightarrow O(\bar{M}')_{x_0}$. It is known (and is easy to verify) that the latter group is isomorphic to

$O^{\varepsilon}(6,\mathbb{F}_2)\ltimes(\mathbb{Z}/2)^6$ or $Sp(6,\mathbb{F}_2)$ depending on whether $x_0^2 = 0$. Comparing the orders, we find that

$$W = O(M')_{x_0} \cong Sp(6,\mathbb{F}_2).$$

S9. The factor group W/W(2).

The goal of this section is to generalize Proposition 2.8.4 to the cases of some root bases of not necessarily finite type. No general theorems are known. However, this can be done in some particular cases.

Theorem 2.9.1. Let B be a standard root basis in the lattice $M = Q_{p,q,r}$, $p\le q\le r$, of rank $n = p+q+r-2$ and $W = W_B(M)$. Let $d = |pqr-pq-pr-qr|$ be the discriminant of M. Assume that B is not of finite type (i.e. $p^{-1}+q^{-1}+r^{-1} \le 1$). Then $W/W(2)$ is isomorphic to one of the following groups:

(i) $Sp(n-1,\mathbb{F}_2)$ if $n,p,q,r \equiv 1 \bmod 2$, $n\equiv 0 \bmod 3$, $d \equiv 0 \bmod 2$, or exactly two of p,q,r, say p and q, are even, $p+q\equiv 2 \bmod 4$;

(ii) $Sp(n-2,\mathbb{F}_2)\ltimes(\mathbb{Z}/2)^{n-2}$ if $p,q,r \equiv 0 \bmod 2$, one of $p,q,r \equiv 0 \bmod 4$;

(iii) $O^{\varepsilon}(n,\mathbb{F}_2)$ if $d\equiv 1 \bmod 2$;

(iv) $O^{\varepsilon}(n-1,\mathbb{F}_2)\ltimes(\mathbb{Z}/2)^{n-1}$, if $n,p,q,r \equiv 1 \bmod 2$, $n\equiv 0 \bmod 3$, d even or exactly two of p,q,r, say p and q are even, $p+q\equiv 0 \bmod 4$;

(v) $O^{\varepsilon}(n-2,\mathbb{F}_2)\ltimes(\mathbb{Z}/2)^{2n-4}$, if $p,q,r \equiv 0 \bmod 2$, $p\equiv q\equiv r \bmod 4$.

The sign $\varepsilon = \pm 1$ is determined as follows. If $r \ge 5$, then $\varepsilon = -\varepsilon'$, where ε' is the sign corresponding to the lattice $Q_{p,q,r-4}$. In this way we can proceed until we come to the case $p,q,r \le 4$ where we take $\varepsilon = -1$.

PROOF. See **[Gri]**. Note that, in the case p=2, the result is due to A.Coble **[Co 1]**. The proof is based on the analysis of the homomorphism

$$W/W(2) \to O(\bar{M}/Rad(\bar{M})) \subset GL(n',\mathbb{F}_2), \; \bar{M} = M/2M.$$

Its image is an irreducible subgroup generated by transvections. By a result of J. McLaughlin **[McL]**, it is isomorphic to one of the groups $O^{\varepsilon}(n',\mathbb{F}_2)$, $Sp(n',\mathbb{F}_2)$ or S_r. Its kernel is isomorphic to a subgroup of the 2-elementary group $Hom(\bar{M}/Rad(\bar{M}),Rad(\bar{M})) \cong (\bar{M}/Rad(\bar{M}))^d$, where $d = \dim Rad(\bar{M})$. The rest of the proof consists of a clever analysis which allows one to decide which group is realized as the image and as the kernel.

To be a little more self-contained, let us prove this theorem in the following cases, where it will be applied: $(p,q,r) = (2,3,6)$, $(2,3,7)$, $(2,4,5)$, $(2,4,6)$. We denote by $\bar{W}$ the image of W in $GL(\bar{M})$. Obviously, $W/W(2) \cong \bar{W}$

Case (2,3,6) : $M = \tilde{E}_8$, $W/W(2) \cong O^+(8,\mathbb{F}_2)\ltimes(\mathbf{Z}/2\mathbf{Z})^8$, $\#O^+(8,\mathbb{F}_2) = 2^{13}.3^5.5^2.7$.

In this case $M' = M/RadM \cong E_8$ and we have an exact sequence

$$1 \to M' \xrightarrow{t} W \to W' \to 1 \quad , W' = W(E_8).$$

The first homomorphism t is defined by

$$t(m')(x) = x + (m'\bullet x')f \ , \ x' = x + Rad(M) \in M', \ Rad(M) = \mathbf{Z}f.$$

The homomorphism t is well-defined for the following reason. Choose a canonical root basis $B = \{\alpha_0,\dots,\alpha_8\}$ of M. Then , we may take

$$f = 3\alpha_0+2\alpha_1+4\alpha_2+6\alpha_3+5\alpha_4+4\alpha_5+3\alpha_6+2\alpha_7+\alpha_8,$$

and for every root α_i, $f-\alpha_i = \beta_i$ is also a root. Clearly, the images $\alpha_0',\dots,\alpha_7'$ of $\alpha_0,\dots,\alpha_7$ in M' form a free basis of M'. It is immediately checked that

$$t(\alpha_i') = s_{\alpha_i}\circ s_{\beta_i} \in W.$$

Note that $t(w(m')) = w\circ t(m)\circ w^{-1}$ for any $w \in W$. Thus $t(M')$ is a minimal normal subgroup containing $s_{\alpha_i}\circ s_{\beta_i}$.

By reduction, we easily get a split exact sequence

$$1 \to M'/2M' \to W \to W' \to 1,$$

which gives us the required isomorphism

$$W(\tilde{E}_8)/W(\tilde{E}_8)(2) \cong O^+(8,\mathbb{F}_2)\ltimes(\mathbf{Z}/2)^8.$$

Case (2,3,7): $M = E$, $W/W(2) \cong O^+(10,\mathbb{F}_2)$, $\#O^+(10,\mathbb{F}_2) = 2^{20}.3^5.5^2.7.17.31$.

Since M is unimodular, the quadratic form $q(x) = \frac{1}{2}x\bullet x \bmod 2$ on M is non-degenerate. Hence $W/W(2) \cong \bar{W} \subset O\varepsilon(10,\mathbb{F}_2)$. To find the type of the quadratic form we argue as follows (cf. **[B-P]**). Recall that $M = M_1 \perp M_2$, where $M_1 \cong E_8$,

$M_2 \cong U$. Let $\bar{M}_1$ and $\bar{M}_2$ be the images of these sublattices in $\bar{M}$. The restriction of q to $\bar{M}_1$ is a form of even type (Proposition 2.8.4). Thus it has $136 = 2^3(2^4+1)$ zeroes in M_1. Obviously it has 3 zeroes on $\bar{M}_2$. Thus q has 3.136 zeroes of type $\bar{x}_1+\bar{x}_2$, $\bar{x}_1 \in \bar{M}_1, \bar{x}_2 \in \bar{M}_2$, $q(\bar{x}_1) = q(\bar{x}_2) = 0$, and 2^8-136 zeroes of type $\bar{x}_1+\bar{x}_2$, $\bar{x}_1 \in M_1$, $\bar{x}_2 \in M_2$, $q(\bar{x}_1) = q(\bar{x}_2) = 1$. Altogether, we find $528 = 2^4(2^5+1)$ zeroes of q. Thus it must be an even quadratic form.

Thus $\bar{W} \subset O^+(10,\mathbb{F}_2)$. To prove the equality, we use the well known fact that $O^+(10,\mathbb{F}_2)$ is generated by $496 = 2^{10}-528 = 2^4(2^5-1)$ transvections, the transformations $\bar{x} \to \bar{x}+(\bar{x}\cdot\bar{a})\bar{a}$, where $q(\bar{a}) = 1$ (see **[Ar]**). Obviously, each reflection $s_\alpha \in O(M)$, $\alpha \in M_{-2}$, induces a transvection. On the other hand, we find the following 496 noncongruent mod 2M roots in M (we use the model of **E** as a sublattice of $H^{1,10}$):

45 of type e_i-e_j,

120 of type $e_0-e_i-e_j-e_k$,

210 of type $2e_0-(e_1+\ldots+e_{10}-e_i-e_j-e_k)$,

120 of type $3e_0-2e_i-(e_1+\ldots+e_{10}-e_i-e_j)$,

1 of type $4e_0-3e_1-e_2-\ldots-e_{10}$.

This proves that $\bar{W} = O^+(10,\mathbb{F}_2)$.

Case (2,4,5) : $M \cong \mathbf{E}_\alpha$, where $\alpha \in \mathbf{E}_{-2}$, $W/W(2) \cong Sp(8,\mathbb{F}_2)$ of order $2^{16}3^5 5.7.17$. In this case $Rad(\bar{M}) = \mathbb{F}_2\bar{x}_0$, where $x_0 = \alpha_0+\alpha_1+\alpha_3$ with respect to a canonical root basis on M. We have a canonical homomorphism $W \to Sp(8,\mathbb{F}_2)$, corresponding to the representation $W \to GL(M/Rad(M))$. This representation is irreducible. In fact, if L is a proper subspace of $M' = M/Rad(M)$ invariant with respect to W, then we may assume that at least one reflection s_{α_i} induces a non-trivial transformation $\bar{s}_{\alpha_i}$ of L. Then $s_{\alpha_i}(x)-x = (x\cdot\alpha)\alpha_i$ for $x \in L$ implies that $\bar{\alpha}_i \in L$. For every $\alpha = \alpha_j$ such that $\alpha_i\cdot\alpha_j = 1$, we have $s_\alpha(\alpha_i)-\alpha_i = \alpha$. This shows that $\bar{\alpha} \in L$. Since we can join any two simple roots α_j's by a chain of simple edges in the Dynkin diagram, we see that all $\bar{\alpha}_j$'s belong to L. This is certainly impossible, because they span $\bar{M}$. Now, the image $\bar{W}'$ of $\bar{W}$ in $Gl(\bar{M}')$ is an irreducible subgroup generated by transvections. By a theorem of McLaughlin **[McL]**, $\bar{W}' \cong O^{\varepsilon}(8,\mathbb{F}_2)$, or $Sp(8,\mathbb{F}_2)$, or S_r, $r = 9$ or 10. Since M contains a sublattice isomorphic to E_8 (spanned by $\alpha_0,\alpha_2,\dots,\alpha_8$), $\bar{W}' \supset O^+(8,\mathbb{F}_2)$. This easily excludes the case S_r (comparing the orders). On the other hand, $Q_{2,4,5}$ is obviously included in the lattice $Q_{2,5,5}$ of discriminant $5 \not\equiv 0 \bmod 2$. This shows that $\bar{W} \subset O^+(10,\mathbb{F}_2)_{\bar{x}_0}$ (compare the proof of Theorem 2.8.1, case E_7). Since

$$\tfrac{1}{2}x_0 \cdot x_0 = (\alpha_0+\alpha_1+\alpha_3)\cdot(\alpha_0+\alpha_1+\alpha_3) = -3,$$

$\bar{x}_0$ is not an isotropic vector in $Q_{2,5,5}$. Hence $O^+(10,\mathbb{F}_2)_{\bar{x}_0} \cong Sp(8,\mathbb{F}_2)$ and the canonical homomorphism $\bar{W}' \rightarrow \bar{W}'$ must be isomorphism. Since, obviously, $\bar{W} \neq O^+(8,\mathbb{F}_2)$, we get $\bar{W} \cong Sp(8,\mathbb{F}_2)$.

Case (2,4,6): $M \cong \mathbf{R}$, $\bar{W} \cong Sp(8,\mathbb{F}_2) \ltimes (\mathbf{Z}/2)^8$, $\#\bar{W} = 2^{24}\cdot 3^5\cdot 5^2\cdot 7\cdot 17$.

In this case $Rad(\bar{M})$ is two-dimensional and is spanned by $\bar{x}_0$ and $\bar{x}_1$, where

$$x_0 = \alpha_0+\alpha_1+\alpha_3 \quad , \quad x_1 = \alpha_0+\alpha_1+\alpha_3+\alpha_5+\alpha_7+\alpha_9.$$

We have a canonical homomorphism $\varphi\colon \bar{W} \rightarrow GL(\bar{M}/Rad(\bar{M})) \cong GL(8,\mathbb{F}_2)$, whose image is contained in a $Sp(8,\mathbb{F}_2)$. Since M contains $Q_{2,4,5}$, by the previous case, we see that $\varphi(\bar{W}) = Sp(8,\mathbb{F}_2)$. Contrarily to this case, $Ker(\varphi) \neq \{1\}$. Indeed, let M_1 be the sublattice of M isomorphic to E_7 which is spanned by $\alpha_0,\alpha_2,\dots,\alpha_7$. Let w_0 be the corresponding generator of $W(M_1)(2)$. Consider the element

$$G = w_0 \circ s_{\alpha_9} \in W.$$

We have

$$G(\alpha_0) = -\alpha_0,\ G(\alpha_2) = -\alpha_2,\ \dots\ ,G(\alpha_7) = -\alpha_7,\ g(\alpha_9) = -\alpha_9\ ,$$

Since $f_2 = 2\alpha_0+\alpha_1+2\alpha_2+3\alpha_3+4\alpha_4+3\alpha_5+2\alpha_6+\alpha_7 \in M_1^{\perp}$, $w_0(f_2) = f_2$. This shows that

$$\begin{aligned} G(\alpha_1) &= G(f-2\alpha_0-2\alpha_2-3\alpha_3-4\alpha_4-3\alpha_5-2\alpha_6-\alpha_7) \\ &= f_2+2\alpha_0+2\alpha_2+3\alpha_3+4\alpha_4+3\alpha_5+2\alpha_6+\alpha_7 \end{aligned}$$

$$= \alpha_1+4\alpha_0+4\alpha_2+6\alpha_3+8\alpha_4+6\alpha_5+4\alpha_6+2\alpha_7,$$

Since $f_1 = 3\alpha_0+2\alpha_2+4\alpha_3+6\alpha_4+5\alpha_5+4\alpha_6+3\alpha_7+2\alpha_8+\alpha_9\epsilon(M'+\mathbf{Z}\alpha_9)^\perp$, we get similarly

$$G(\alpha_8) = \tfrac{1}{2}G(f_1-3\alpha_0-2\alpha_2-4\alpha_3-6\alpha_4-5\alpha-4\alpha_6-3\alpha_7-\alpha_9)$$

$$= \alpha_8+3\alpha_0+2\alpha_2+4\alpha_3+6\alpha_4+5\alpha_5+4\alpha_6+3\alpha_7+\alpha_9$$

Thus we see that

$$G(\alpha_i) \equiv \alpha_i \bmod 2M,\ i \neq 8,$$

$$G(\alpha_8) \equiv \alpha_8+\alpha_0+\alpha_5+\alpha_7+\alpha_9 = \alpha_8+x_1 \bmod 2M,$$

i.e. G defines a nontrivial element of $\mathrm{Ker}(\varphi)$.

For any $g\epsilon\mathrm{Ker}(\varphi)$ anf for any $\bar{x}\epsilon\bar{M}$,$g(\bar{x})-\bar{x}\epsilon\mathrm{Rad}(\bar{M})$. This defines a linear transformation $T(g) : \bar{M}/\mathrm{Rad}(\bar{M}) \to \mathrm{Rad}(\bar{M})$, $\bar{x} \to g(\bar{x})-\bar{x}$. It is easy to verify that this also defines an injective homomorphism

$$\mathrm{Ker}(\varphi) \to \mathrm{Hom}(\bar{M}/\mathrm{Rad}(\bar{M}),\mathrm{Rad}(\bar{M})) = (\bar{M}/\mathrm{Rad}(\bar{M}))^{2*} \cong (\mathbb{F}_2)^{16}.$$

The argument from the proof of the previous case shows that $M/\mathrm{Rad}(M)$ is an irreducible W-module. This implies that only two possibilities may occur:

$$\mathrm{Ker}(\varphi) = \bar{M}/\mathrm{Rad}(\bar{M})^* \cong (\mathbf{Z}/2)^8,$$

$$\mathrm{Ker}(\varphi) = (\bar{M}/\mathrm{Rad}(\bar{M}))^{2^*} \cong (\mathbf{Z}/2)^{16}.$$

To exclude the second possibility, we embed M into $M' = Q_{2,4,7}$ of discriminant 6. Arguing as before, we see that $\mathrm{Ker}(\varphi') \subset (\bar{M}'/\mathrm{Rad}(\bar{M}'))^* \cong \mathbb{F}_2^{10}$, where φ' is the homomorphism $\bar{W} \to GL(\bar{M}'/\mathrm{Rad}(\bar{M}'))$ similar to φ. Since $\mathrm{Ker}(\varphi) \subset \mathrm{Ker}(\varphi')$, we obtain what we need.

§10. The structure of W(2).

Our next goal is to study the 2-level congruence subgroup W(2) in the cases $W = W_B(M)$, where B is a canonical root basis in a lattice $Q_{p,q,r}$ with $(p,q,r) = (2,3,6), (2,3,7), (2,4,4), (2,4,5)$ and $(2,4,6)$. Without reminder, we use freely the notations of §5 and §6.

First we consider the case where $M = E_r$, $r \geq 9$. We choose its model as a sublattice of $H^{1,r}$, the orthogonal complement of the vector

$$k_r = -3e_0+e_1+\dots+e_r.$$

We will call an exceptional vector $e \in \mathcal{E}_r$ **strongly exceptional** if $e \in W_r e_r$. The set of strongly exceptional vectors in $H^{1,r}$ will be denoted by $\mathcal{E}'_r$.

Lemma 2.10.1. Let $e = m_0e_0-m_1e_1-\dots-m_re_r \in \mathcal{E}'_r$. Then
(i) if $m_0 = e\cdot e_0 = 0$,then $e = e_i$ for some $i = 1,\dots,r$;
(ii) if $m_0 \neq 0$, then $m_0,\dots,m_r \geq 0$.

PROOF. (i) is obvious, since $e_r \in \mathcal{E}_r$. To prove (ii) we apply the reflections s_{α_i}, $i\neq 0$, which permute e_i and e_{i-1} leaving the other e_j's unchanged, to assume that $m_1 \geq \dots \geq m_r$. Or, equivalently, $e\cdot\alpha_i \geq 0$, $i\neq 0$. Let $e = w(e_r)$. Write $w = s_{\alpha_i}\circ w'$, where $l(w') = l(w)-1$. Since $e_r \in C(B)$, $\alpha_i\cdot w'(e_r) \geq 0$ (Proposition 2.2.4). If $w'(e_r)\cdot e_0 = 0$, then $w'(e_r) = e_j$ for some $j \neq 0$. Hence

$$e = s_{\alpha_i}(e_j) = e_j+(e_j\cdot\alpha_i)\alpha_i,$$

and in this case the assertion is easily verified for each $i = 0,\dots,r$. Then, by induction on $l(w)$, we may assume that $w'(e_r)\cdot e_j \geq 0$ for any j. We have

$$e\cdot e_j = s_{\alpha_i}w'(e_r)\cdot e_j = w'(e_r)\cdot s_{\alpha_i}(e_j) = w'(e_r)\cdot(e_j+(e_j\cdot\alpha_i)\alpha_i)$$
$$= w'(e_r)\cdot e_j + (e_j\cdot\alpha_i)(w'(e_r)\cdot\alpha_i).$$

This implies that $e\cdot e_j \geq 0$ if $i = 0$. However, if $i \neq 0$,

$$e = s_{\alpha_i}w'(e_r) = s_{\alpha_i}(m_0e_0-m_1e_1-\dots-m_ie_i-m_{i+1}e_{i-1}-\dots-m_re_r)$$
$$= m_0e_0-m_1e_1-\dots-m_{i+1}e_i-m_ie_{i+1}-\dots-m_re_r$$

also satisfies the assertion of the lemma.

Lemma 2.10.2. Let $1 \leq i < j < k \leq r$ and $A(i,j,k) = s_\alpha$, where $\alpha = e_0 - e_i - e_j - e_k \in R(B)$. Then, for any $x = m_0 e_0 - m_1 e_1 - \ldots - m_r e_r$

$$A(i,j,k)(x) = m_0' e_0 - m_1' e_1 - \ldots - m_r' e_r,$$

where

$$m_0' = 2m_0 - m_i - m_j - m_k \ , \ m_i' = m_0 - m_j - m_k,$$

$$m_j' = m_0 - m_i - m_k, \ m_k' = m_0 - m_i - m_j,$$

$$m_q' = m_q \ , \text{ if } q \neq i,j,k.$$

PROOF. Straightforward.

Remark 2.10.1. Recall from §5 that $\mathcal{E}_r \neq \mathcal{E}_r'$ for $r \geq 10$. Also

$$\mathcal{E}'_{10} = \{x \in \mathcal{E}'_{10} : x - k_{10} \text{ is primitive}\}.$$

It follows from Noether's inequality (see **[D-O]**) that

$$x = me_0 - m_1 e_1 - \ldots - m_r e_r \in \mathcal{E}_r, \quad m_1 \geq \ldots \geq m_r,$$

is strongly exceptional if and only if

$$m_1, \ldots, m_r \geq 0 \text{ if } m \neq 0 \text{ and } m \geq m_1 + m_2 \text{ if } r > 2.$$

For example,

$$x = 9e_0 - 5e_1 - 5e_2 - 2e_3 - \ldots - 2e_{10} \notin \mathcal{E}'_{10}.$$

Indeed, $A(1,2,3)(x) = 6e_0 - 2e_1 - 2e_2 + e_3 - 2e_4 - \ldots - 2e_{10}$ does not satisfy Lemma 2.10.1.

By Proposition 2.5.1

$$E_r = E_8 \perp (\mathbf{Z}f + \alpha_9) \perp (\mathbf{Z}\alpha_8 + \mathbf{Z}\alpha_{10} + \mathbf{Z}\alpha_{11} + \ldots + \mathbf{Z}\alpha_{r-1}) \cong E_8 \oplus U \oplus A_{r-10},$$

We define the **Bertini involution** σ_0 of M_r with respect to B_r by

$$\sigma_0 = -id_{E_8} \oplus id_{E_8^\perp} = w_0 \ ,$$

where $W_{E_8}(2) = (w_0)$.

Lemma 2.10.3. σ_0 acts on the basis $\{\alpha_0,\dots,\alpha_r\}$ as follows

$$\sigma_0(\alpha_i)=-\alpha_i,\quad i=0,\dots,7,$$

$$\sigma_0(\alpha_8)=-\alpha_8+2f=\alpha_8+2\alpha_0+4\alpha_1+8\alpha_2+12\alpha_3+10\alpha_4+8\alpha_5+6\alpha_6+4\alpha_7,$$

$$\sigma_0(\alpha_j)=\alpha_j,\quad j>8$$

PROOF. The only thing to check is the formula for $\sigma_0(\alpha_8)$. Note that $\sigma_0(f)=f$, hence $\sigma_0(\alpha_8) = \sigma_0(f+\alpha_8-f) = 2f-\alpha_8$.

Corollary 2.10.1. σ_0 acts in $H^{1,r}$ by the formula

$$\sigma_0(m_0e_0-m_1e_1-\dots-m_re_r) = m_0'e_0-m_1'e_1-\dots-m_r'e_r,$$

where

$$m_0' = 17m_0-6m_1-\dots-6m_8,$$

$$m_i' = 6m_0-2(m_1+\dots+m_8)-m_i,\quad i = 1,\dots,8$$

$$m_j' = m_j,\quad j>8.$$

In particular, if $x = m_0e_0-m_1-\dots-m_re_r\in\mathcal{E}'_r$, then

$$m_0' = 6+6(m_9+\dots+m_r)-m_0$$

$$m_i' = 2+2(m_9+\dots+m_r)-m_i,\quad i = 1,\dots,8,$$

$$m_j' = m_j,\quad j>8.$$

PROOF. Use that $e_i\in E_8^{\perp}$, if $i>8$, thus $\sigma_0(e_i)_0=e_i$, $i>8$. Now,

$$\sigma_0(e_8) = \sigma_0(e_9+\alpha_8) = e_9+\sigma_0(\alpha_8) = e_9-\alpha_8+2f =$$

$$= 6e_0-2e_1-\dots-2e_7-3e_8,$$

$$\sigma_0(e_7) = \sigma_0(e_8+\alpha_7) = \sigma_0(e_8)+\sigma_0(\alpha_7) =$$

$$= \sigma_0(e_8)-\alpha_7 = 6e_0-2e_1-\dots-2e_6-3e_7-2e_8$$

$$\sigma_0(e_i) = 6e_0-2e_1-\dots-2e_7-2e_8-e_i\quad,\quad i=8,\dots,1.$$

$$\sigma_0(e_0) = \sigma_0(\alpha_0+e_1+e_2+e_3) = -\alpha_0+\sigma_0(e_1)+\sigma_0(e_2)+\sigma_0(e_3)=$$

$$= 17e_0-6(e_1+\dots+e_8).$$

Let K_r = $\{x= m_0e_0-m_1-...-m_re_r\in\mathcal{E}'_r: m_8,m_9,...,m_r \le 1\}$

Lemma 2.10.4. Let $x= m_0e_0-m_1-...-m_{10}e_{10}\in K_{10}$. Then up to a permutation of m_i's, $i\neq 0$, $(m_0,m_1,...,m_{10})$ is equal to one of the following vectors:

(0,-1,...,0),(1,1,1,0,...,0), (2,1,1,1,1,1,0,...,0), (3,2,1,1,1,1,1,1,0,0,0),

(4,2,2,2,1,1,1,1,1,0,0), (5,2,2,2,2,2,2,1,1,0,0), (5,3,2,2,2,1,1,1,1,1,0),

(6,3,3,2,2,2,2,1,1,1,0), 6,3,3,3,2,1,1,1,1,1,1), (6,4,2,2,2,2,1,1,1,1,1),

(7,3,3,3,3,2,2,2,1,1,0), (7,4,3,3,2,2,2,1,1,1,1), (7,3,3,3,3,3,1,1,1,1,1),

(8,4,4,3,3,2,2,2,1,1,1), (8,4,3,3,3,3,3,1,1,1,1), (8,3,3,3,3,3,3,3,1,1,0),

(9,5,3,3,3,3,3,3,1,1,1), (9,4,4,4,3,3,3,2,1,1,1),

PROOF. Let $x= m_0e_0-m_1-...-m_re_r\in K_{10}$. By Corollary 2.10.1,

$$\sigma_0(x) = m_0'e_0-m_1'e_1-...-m_{10}'e_{10},$$

where $m_0' = 6+6(m_9+m_{10})-m_0 \le 18-m_0$. Applying Lemma 2.10.1, we obtain that $m_0< 18$. If $m_0> 9$, then $m_0' \le 8$, $m_9' = m_{10}' =1$. It is not hard to list all the vectors $x = m_0'e_0-m_1'-...-m_{10}'e_{10}\in K_{10}$ with this property (by computer, or by hand). By inspection, we find that $x = \sigma_0(x')\notin K_{10}$. Thus we get that $m_0 \le 9$. It remains to list all of them, which is easy.

Lemma 2.10.5. Suppose that r = 9 or 10 .Then

$$\mathcal{E}'_r = W_rK_r = W_r(2)'K_r,$$

where $W_r(2)'$ is the minimal normal subgroup of W_r containing σ_0.

PROOF. Let $w\in W_r$ and $x\in K_r$. We prove by induction on $l(w)$ that $w(x)\in W_r(2)'K_r$. Write $w = s_\alpha w'$, where $l(w') = l(w)-1$, and assume that $w'(x)=w_0(y)$ for some $y\in K_r$, $w_0\in W_r(2)'$. Then $s_\alpha w'(x) = s_\alpha w_0(y) = w_0's_\alpha(y)$ for some $w_0'\in W_r(2)'$. Thus the assertion would follow if we prove

(*) $s_\alpha(y) \in W_r(2)'K_r$ for any $\alpha\in B$ and $y\in K_r$.

Let $y = m_0e_0-m_1e_1-...-m_re_r\in K_r$, i.e. $m_8,...,m_r\le 1$. Clearly $s_\alpha(y)\in K_r$ if $\alpha\neq\alpha_7$, and $s_{\alpha_7}(y) = m_0e_0-m_1e_1-...-m_8e_7-m_7e_8-m_9e_9-...-m_re_r$. Thus it suffices to show

(*)' Let $y \in K_r' = \{m_0e_0 - m_1e_1 - \dots - m_re_r \in \mathcal{E}_r : m_7, m_9, \dots, m_r \le 1\}$, then there exists an element $w \in W_r(2)'$ such that $w(y) \in K_r$.

Using Lemma 2.10.4, we can list the set K'_r for $r = 9$ and 10. Table 1 gives the corresponding element w which is written as a product of the elements $A(i,j,k)B(i_1,\dots,i_r)A(i,j,k)$, where $B(i_1,\dots,i_r) = s\sigma_0 s^{-1}$, $s \in W_{\{\alpha_1,\dots,\alpha_{r-1}\}} \subset W_r$ sends e_j to e_{i_j}, $j=9,\dots,r$ and e_j to e_j, $j \le 8$. Of course, it is enough to do it for one vector in each equivalence class with respect to a permutation of the vectors e_i, $i \ne 0,7,8$.

Theorem 2.10.1. Assume that $8 \le r \le 10$. Then

$$W_r(2) = \overline{\langle \sigma_0 \rangle}$$

PROOF. Let $W_r(2)'$ be the group on the right side. Embed E_r into $H^{1,r}$ by means of B_r. We use the notation of Lemma 2.10.5 .Let $e_r \in K_r$. For every $w \in W_r(2)$ we can find an element $w' \in W_r(2)'$ such that $w(e_r) = w'(x)$ for some $x \in K_r$. Thus $w'^{-1} \circ w(e_r) = x$. Since $w'^{-1} \circ w \in W_r(2)$, $x \equiv e_r \mod 2E_r$. However,

$$x = m_0e_0 - \dots - m_re_r \in K_r.$$

This implies that $m_r = 1$, $m_{r-1} = \dots = m_8 = 0$, $m_i \equiv 0 \mod 2$, $i < 8$. It follows from Lemma 2.10.4 that $x = e_r$. Thus $w' \circ w^{-1}(e_r) = e_r$. Next, we recall Proposition 2.5.4 to observe that the linear function $e_r^* \in E_r^*$ defined by

$$e_r^*(x) = x \cdot e_r = \omega_r \cdot x$$

takes the value 0 on α_i, $i \ne r-1$ and 1 on α_{r-1}. If we identify $(E_r)_{\mathbb{R}}$ with $(E_r)^*_{\mathbb{R}}$, then $e_r^* \in C(B)$ and its stabilizer in W_r is the Weyl group $W_{\{\alpha_0,\dots,\alpha_{r-2}\}} \cong W_{r-1}$. This shows that

$$w'^{-1} \circ w \in W_{r-1} \cap W_r(2) = W_{r-1}(2)$$

(because E_{r-1} is a primitive sublattice of E_r). Now, we can finish the proof easily.

Let $r = 9$. Since $W_8(2) = \overline{\langle \sigma_0 \rangle}$, $w'^{-1}w \in W_8(2) \subset W_9(2)$, hence $w \in W_9(2)'$. This proves the assertion for $r = 9$.

Let $r = 10$. Since, $W_9(2)' = W_9(2)$, $w'^{-1}w \in W_9(2) \subset W_{10}(2)'$, hence $w \in W_{10}(2)'$. This proves the assertion for $r = 10$.

Remark 2.10.2. We conjecture that the previous theorem is true also for any $r \geq 10$. It would follow from the validity of (*)' in the proof of Lemma 2.10.5, which can be checked by computer for relatively small r.

Remark 2.10.3. In the case (2,3,6), i.e. $E_9 = \tilde{E}_8$, we know from the proof of Theorem 2.9.1 (case (2,3,6)) that W(2) contains the subgroup $\mathbf{Z}^8 \cong 2E_9$ and the quotient $W(2)/\mathbf{Z}^8 \cong \mathbf{Z}/2 \cong (w_0) = W(E_8)(2)$.

Next we will deal with the lattices $R_r = Q_{2,4,r-4}$ $(r \geq 7)$. Let $B_r = \{\alpha_0,\ldots,\alpha_{r-1}\}$ be a canonical root basis in R_r. We embed R_r into $L_r = \langle 2\rangle \oplus \langle -1\rangle^r$ by using Proposition 2.6.2. Thus B_r is given in terms of the standard basis of L_r by the formulas stated there. As above, we assume that each R_i is embedded into R_{i-1} as a sublattice spanned by $\{\alpha_0,\ldots,\alpha_{i-1}\} = B_i$. Let $W_r = W_{B_r}(R_r)$. If $r \geq 9$ we consider the lattice E_{r-1} as a sublattice of R_r spanned by $B_r \backslash \{\alpha_1\}$. Let

$$f_1 = 3\alpha_0+2\alpha_2+4\alpha_3+6\alpha_4+5\alpha_5+4\alpha_6+3\alpha_7+2\alpha_8+\alpha_9$$

be the generator of $Rad(E_9)$,

$$f_2 = 2\alpha_0+\alpha_1+2\alpha_2+3\alpha_3+4\alpha_4+3\alpha_5+2\alpha_6+\alpha_7$$

be the generator of $Rad(R_8)$.

As we know from Proposition 2.6.3

$$R_r = R_7 \perp (\mathbf{Z}f_2+\mathbf{Z}\alpha_8) \perp (\mathbf{Z}(f_2-\alpha_9)+\mathbf{Z}\alpha_{10}+\ldots+\mathbf{Z}\alpha_{r-1}) \cong E_7 \oplus U \oplus A_{r-8},$$

$$R_r = E_8 \perp (\mathbf{Z}f_1+\mathbf{Z}(f_2-\alpha_9)) \perp (\mathbf{Z}\alpha_{10}+\ldots+\mathbf{Z}\alpha_{r-1}).$$

Define

$$K = -id_{N_7} \oplus id_{N_7^\perp} = w_0$$

(resp.

$$B = -\mathrm{id}_{E_8} \oplus \mathrm{id}_{E_8^{\perp}} = w_0', \text{ if } r \geq 10),$$

where w_0 (resp. w_0') is the generator of $W(E_7)(2)$ (resp. $W(E_8)(2)$).

We call K (resp. B) the **Kantor involution** (resp. **Bertini involution**) of N_r.

Lemma 2.10.6. K and B act on the basis B_r as follows:

$$K(\alpha_i) = -\alpha_i,\ i < 7,\quad K(\alpha_i) = \alpha_i\ ,\ i > 7\ ,$$

$$K(\alpha_7) = \alpha_7+2\alpha_0+2\alpha_1+4\alpha_2+6\alpha_3+4\alpha_4+2\alpha_5,$$

$$B(\alpha_i) = -\alpha_i,\ i = 0,2,\ldots,8\ ,\ B(\alpha_i) = \alpha_i,\ i\neq 0,\ i > 9,$$

$$B(\alpha_1) = \alpha_1+10\alpha_0+8\alpha_2+14\alpha_3+20\alpha_4+16\alpha_5+12\alpha_6+8\alpha_7+4\alpha_8,$$

$$B(\alpha_9) = \alpha_9+6\alpha_0+4\alpha_2+8\alpha_3+12\alpha_4+10\alpha_5+8\alpha_6+6\alpha_7+4\alpha_8.$$

PROOF. By definition, $K(\alpha_i) = -\alpha_i$, $i<7$. Since $\alpha_j \in R_7^{\perp}$ for $j > 7$, $K(\alpha_j)=\alpha_j$ for such j. Since $f_2 \in R_7^{\perp}$, $f_2-\alpha_7 \in R_7$, we have

$$K(\alpha_7) = K(f_2+\alpha_7-f_2)=f_2+f_2-\alpha_7 =$$

$$= 2f_2-\alpha_7=\alpha_7+2\alpha_0+2\alpha_1+4\alpha_2+6\alpha_3+4\alpha_4+2\alpha_5.$$

By definition, $B(\alpha_i) = -\alpha_i$, $i=0,\ldots,8$. Since $\alpha_j \in E_8^{\perp}$ for $j \neq 0$, $j > 9$, $B(\alpha_j) = \alpha_j$ for such j. Now, since $f_1 \in E_8^{\perp}$, $f_1-\alpha_9 \in E_8$,

$$B(\alpha_9) = B(f_1+\alpha_9-f_1) = f_1+f_1-\alpha_9$$

$$= 6\alpha_0+4\alpha_2+8\alpha_3+12\alpha_4+10\alpha_5+8\alpha_6+6\alpha_7+4\alpha_8+\alpha_9.$$

Similarly, since $f_2-\alpha_9 \in E_8^{\perp}$, $f_2-\alpha_1 \in E_8$, we have

$$B(\alpha_1) = B((f_2-\alpha_9)+(\alpha_1-f_2)+\alpha_9)) = f_2-\alpha_9+f_2-\alpha_1+B(\alpha_9)$$

$$=4\alpha_0+\alpha_1+4\alpha_2+6\alpha_3+8\alpha_4+6\alpha_5+4\alpha_6+2\alpha_7+$$

$$+6\alpha_0+4\alpha_2+8\alpha_3+12\alpha_4+10\alpha_5+8\alpha_6+6\alpha_7+4\alpha_8$$

$$= \alpha_1+10\alpha_0+8\alpha_2+14\alpha_3+20\alpha_4+16\alpha_5+12\alpha_6+8\alpha_7+4\alpha_8.$$

Corollary 2.10.2. Let $x = m_0e_0-m_1e_1-\ldots-m_re_r \in H^{1,r}$. Then

(i) $K(x) = m_0'e_1'-m_1'e_1-\ldots-m_r'e_r,$

where

$$m_0' = 15m_0 - 4(m_1+\dots+m_7), \quad m_i' = 8m_0 - 2(m_1+\dots+m_7) - m_i, \quad i=1,\dots,7,$$

$$m_j' = m_j, \; j > 7.$$

(ii) $B(x) = m_0'e_1' - m_1'e_1 - \dots - m_r'e_r$,

where

$$m_0' = 33m_0 - 16m_1 - 6m_2 - \dots - 6m_9, \quad m_1' = 32m_0 - 15m_1 - 6(m_2+\dots+m_9),$$

$$m_i' = 12m_0 - 6m_1 - 2(m_2+\dots+m_9) - m_i, \; i = 2,\dots,9, \; m_i' = m_i, \; i \geq 10.$$

PROOF. Since $e_i \in R_7^\perp$, $i \geq 8$, $K(e_i) = e_i$, $i \geq 8$. Now,

$$K(e_7) = K(e_8+\alpha_7) = e_8 - \alpha_7 + 2f_2 = 2e_8 - e_7 + 4e_0 - 2e_1 - \dots - 2e_8$$

$$= 4e_0 - 2e_1 - \dots - 2e_6 - 3e_7,$$

$$K(e_i) = K(e_{i-1}+\alpha_i) = -\alpha_i + K(e_{i-1}) = 4e_0 - 2(e_1+\dots+e_7) - e_i, \; i = 6,\dots,1.$$

$$K(e_0) = K(\alpha_0+e_1+e_2+e_3+e_4) = -\alpha_0 + K(e_1)+K(e_2)+K(e_3)+K(e_4) =$$

$$= 15e_0 - 8(e_1+\dots+e_7).$$

Since $e_i \in E_8^\perp$, $i \geq 10$, $B(e_i) = e_i$, $i \geq 10$. Hence

$$B(e_9) = B(e_{10}+\alpha_9) = e_{10} + B(\alpha_9) = e_{10} + 2f_1 - \alpha_9$$

$$= 6e_0 - 6e_1 - 2(e_2+\dots+e_9) - e_9,$$

$$B(e_i) = B(e_{i-1}+\alpha_i) = -\alpha_i + B(e_{i-1})$$

$$= 6e_0 - 6e_1 - 2(e_2+\dots+e_9) - e_i, \quad i=8,\dots,2,$$

$$B(e_1) = B(e_2+\alpha_1) = B(e_2)+B(\alpha_1) = B(e_2) - \alpha_1 + 2f_1 + 2f_2 - 2\alpha_9 =$$

$$= 6e_0 - 6e_1 - 2(e_2+\dots+e_9) - e_2 - (e_1 - e_2)$$

$$= 2(3e_0 - 3e_1 - e_2 - \dots - e_{10}) + 2(2e_0 - e_1 - \dots - e_8) - 2(e_9 - e_{10})$$

$$+ 16e_0 - 15e_1 - 6(e_2+\dots+e_9),$$

$$B(e_0) = B(\alpha_0+e_1+e_2+e_3+e_4) = -\alpha_0 + B(e_1)+B(e_2)+B(e_3)+B(e_4) =$$

$$= 33e_0 - 32e_1 - 12(e_2 - \dots - e_9).$$

Lemma 2.10.7. Let $\alpha = e_0-e_i-e_j-e_\kappa-e_p \in R(B_r)$. Denote s_α by $A(i,j,\kappa,p)$. Then

$$A(i,j,\kappa,p)(m_0e_0-m_1e_1-\ldots-m_re_r) = m_0'e_0-m_1'e_1-\ldots-m_r'e_r,$$

where

$$m_0' = 3m_0-m_i-m_j-m_\kappa-m_p, \quad m_i'= 2m_0-m_j-m_\kappa-m_p,$$

$$m_j'= 2m_0-m_j-m_\kappa-m_i, \quad m_\kappa'= 2m_0-m_j-m_i-m_p,$$

$$m_p'= 2m_0-m_j-m_\kappa-m_i, \quad m_q' = m_q\ , q \neq i,j,\kappa,p.$$

PROOF. Obvious.

Let $e = e_0-2e_1$, $\mathfrak{M}_r = W'_r e$.

Lemma 2.10.8. Let $x = m_0e_0-m_1e_1-\ldots-m_re_r \in \mathfrak{M}_r$. Then

(i) $x\cdot\kappa = 4m_0-m_1-\ldots-m_r = 2$,

(ii) $x\cdot x = 2m_0^2-m_1^2-\ldots-m_r^2 = -2$,

(iii) $m_0 > 0$, $m_0 \in 2\mathbf{Z}+1$, $m_i \geq 0$, $m_i \in 2\mathbf{Z}$, $i \geq 1$,

(iv) x is primitive.

PROOF. (i) and (ii) are obvious (since they hold for $x = e$).

(iii) Since $x\cdot e \in 2\mathbf{Z}$ for any $x \in L_r = {}'H^{1,r}$, $e\cdot\alpha_i \in 2\mathbf{Z}$, $\forall \alpha_i \in B_r$. Hence $x \equiv e \mod 2L_r$, for any $x \in \mathfrak{M}_r$. The assertion about the positivity of the m_i's is proven as in Lemma 2.10.1.

Lemma 2.10.9. Let $x = m_0e_0-m_1e_1-\ldots-m_re_r \in \mathfrak{M}_r$. Then

(i) $K(x) = m_0'e_0-m_1'e_1-\ldots-m_r'e_r,$

where

$$m_0' = 4(m_8+\ldots+m_r)+8-m_0,$$
$$m_i' = 4+2(m_8+\ldots+m_r)-m_i \quad ,i=1,\ldots,7,$$
$$m_i' = m_i\ , \ i>7.$$

(ii) $B(x) = m_0'e_0-m_1'e_1-\ldots-m_r'e_r,$

where

$$m_0' = 9m_0-10m_1+6(m_{10}+...+m_r)+12,$$

$$m_1' = 8m_0-9m_1+6(m_{10}+...+m_r)+12,$$

$$m_i' = 4m_0-4m_1+2(m_{10}+...+m_r)+4-m_i,$$

$$m_j' = m_j,\ j \geq 10.$$

PROOF. This follows immediately from the previous lemmas.

Let

$$S_r = \{\ x = m_0e_0-m_1e_1-...-m_re_r \in \mathcal{E}_r:\ m_0 \leq m_i+1 \text{ for some } i \in \{1,2,3,4\},$$
$$m_j \leq 2 \text{ for } j \in \{1,2,3,4\}\setminus\{i\}\}.$$

Lemma 2.10.10. Let $x = m_0e_0-m_1e_1-...-m_{10}e_{10} \in S_{10}$. Then, up to a permutation of the m_i's, $i \neq 0$, $(m_0,m_1,...,m_{10})$ is equal to one of the following vectors:

(1,2,0,0,0,0,0,0,0,0,0), (3,2,2,2,2,2,0,0,0,0,0), (5,4,4,2,2,2,2,2,0,0,0),

(7,6,6,2,2,2,2,2,2,2,0), (7,6,4,4,4,2,2,2,2,0,0), (9,8,6,4,4,4,2,2,2,2,0),

(11,10,8,4,4,4,4,2,2,2,2), (11,10,6,6,6,4,2,2,2,2,2), (13,12,8,6,6,4,4,4,2,2,2).

PROOF. Applying $w \in W_{B_{10}\setminus\{\alpha_0\}}$, which permutes the e_i's, $i \geq 1$, we may assume that $m_0 \leq m_1+1$, $m_i \leq 2$ for $i = 2,3,4$. Let w be as above which sends $\{8,9,10\}$ to $\{2,3,4\}$. Then

$$w \circ K \circ w^{-1}(x) = m_0'e_0- m_1e_1-m_2e_2-m_3e_3-m_4'-...-m_{10}',$$

where $m_0' = 4(m_1+m_2+m_3+4)-m_0 \leq 32-m_0$ (Lemma 2.10.9). Since $m_0>0$, we get $m_0 < 32$. If $m_0>16$, then

$$w \circ B \circ w^{-1}\ (x) = m_0'e_0- m_1e_1-m_2e_2-m_3e_3-m_4'-...-m_{10}',$$

where $m_0' = 9m_0-10m_1+6m_4+12 \leq 17$, $m_1' = m_0' -(m_0-m_1) \geq m_0'-1$ (Lemma 2.10.9). Now, we can list (by computer, or by hand) the set S'_{10} of vectors $(m_0,m_1,...,m_{10})$ with $m_0 \leq 17$, $m_1 \geq m_0+1$, satisfying properties (i)-(iv) of Lemma 2.10.8 and check which of them belong to $\mathcal{E}_{10}$ (trying to reduce them to e by using Lemma 2.10.7). We find that only those which are listed in the statement of the lemma belong to S_{10}. On the other hand, if $m_0>17$, then

$x = w \circ B \circ w^{-1}(y)$, $w \in W_{B_{10} \setminus \{\alpha_0\}}$, where $y \in S_{10}'$. By inspection, we find that none of them belong to S_{10}.

Lemma 2.10.11. Let G_r denote the minimal normal subgroup of W'_r containing the Kantor and Bertini involutions K and B (resp. K , if $r = 9$). Let

$$S_r = \{x = m_0e_0 - \ldots - m_re_r \in \mathcal{E} \ : \ m_0 \leq m_i - 1 \text{ for some } i \in \{1,2,3,4\},$$

$$m_j \leq 2, \ j \in \{1,2,3,4\} \setminus \{i\}\}.$$

Assume that $r = 9$ or 10. Then

$$\mathfrak{M}_r = W_r' S_r = G_r S_r.$$

PROOF. This is similar to the proof of Lemma 2.10.5. It suffices to check that $s_i(S_r) \subset G_rS_r$ for every simple reflection $s_i = s_{\alpha_i}$. It is immediately observed that $s_i(x) \in S_r$, if $i \neq 0,4$. Thus we have to show that for every $x \in S_r$, $s_0(x)$ or $s_4(x)$ can be reduced to an element from S_r by conjugates of Kantor and Bertini involutions. Table 2 (resp. Table 3) shows how to do it in the case $s = s_4$ (resp. $s = s_0$). We use the notation $K(i,j,k)$ (resp.$B(i,j)$) for the conjugates wKw^{-1} (resp. wBw^{-1}), where w is a permutation of the e_i's which sends e_i,e_j,e_k to e_8,e_9,e_{10} (resp. sending e_i,e_j to e_1,e_{10}) respectively and leaving the other e_i's unchanged.

Theorem 2.10.2. Assume that $9 \leq r \leq 10$. Then

$$W'_r(2) = \overline{<B,K>} .$$

PROOF. Let $w \in W'_r(2)$. By Lemma 2.10.11, $w(e) = w'(e')$ for some $e' \in S_r$ and $w' \in W_r(2)'$, the normal subgroup of $W_r(2)$ generated by B and K. Since

$$\tfrac{1}{2}e' = w'^{-1} \circ w(\tfrac{1}{2}e) \equiv \tfrac{1}{2}e \equiv \tfrac{1}{2}e_0 - e_1 \mod 2L_r,$$

we easily find that $e' = e$. Thus $w'^{-1} \circ w \in W_r'(2)_e$. Since $e \bullet \alpha_i = 0$, $i \neq 1$, $W_r'(2)_e$ can be identified with $W(2)$, where $W = W_{B \setminus \{\alpha_1\}} \cong W(E_9)$. By Theorem 2.10.1, the latter group is a minimal normal subgroup of W generated by B. This proves the assertion.

Remark 2.10.4. It follows from the proof that the assertion of the theorem holds for any $r \geq 9$, if Theorem 2.10.1 holds for every $r \geq 9$.

Let $r = 10$, i.e. $R_r = \mathbf{R}$. Consider the sublattice $\mathbf{R}_{\{\alpha_0,\alpha_2,\dots,\alpha_7\}} \cong E_7$ and let w_0'' be the generator of $W_{\{\alpha_0,\alpha_2,\dots,\alpha_7\}}(2)$. We set

$$G = w_0''\circ s_{\alpha_9} \in W'_{10} = W_B(\mathbf{R}),$$

and call it the **Geiser involution** of R.

Lemma 2.10.12. G acts on the basis B_{10} as follows:

$$G(\alpha_i) = -\alpha_i, \; i \neq 1,8,$$

$$G(\alpha_1) = 2f_2-\alpha_1=\alpha_1+2\alpha_0+4\alpha_2+6\alpha_3+8\alpha_4+6\alpha_5+4\alpha_6+2\alpha_7,$$

$$G(\alpha_8) = f_1-\alpha_8=\alpha_8+3\alpha_0+2\alpha_1+4\alpha_2+6\alpha_3+5\alpha_4+4\alpha_5+3\alpha_6+\alpha_9.$$

PROOF. By definition,

$$G(\alpha_i) = -\alpha_i, \; i=0,2,\dots,7,9.$$

Now, since $f_2 \in (\mathbf{R}_{\{\alpha_0,\alpha_2,\dots,\alpha_7\}})^{\perp}$,

$$G(\alpha_1) = G(-f_1+\alpha_1+f_1)=2f_1-\alpha_1=\alpha_1+2\alpha_0+4\alpha_2+6\alpha_3+8\alpha_4+6\alpha_5+4\alpha_6+2\alpha_7.$$

Since $f_1 \in (\mathbf{R}_{\{\alpha_0,\alpha_2,\dots,\alpha_9\}})^{\perp}$, we have

$$G(\alpha_8) = \tfrac{1}{2}G(f_1+2\alpha_8-f_1) = \tfrac{1}{2}(f_1-2\alpha_8+f_1) = f_1-\alpha_8 =$$

$$= \alpha_8+3\alpha_0+2\alpha_1+4\alpha_2+6\alpha_3+5\alpha_4+4\alpha_5+3\alpha_6+\alpha_9.$$

Corollary 2.10.3. Let $\tilde{W}(2)$ be the minimal normal subgroup of $W = W_B(\mathbf{R})$ containing the Bertini, Kantor and Geiser involutions. Then

$$W/\tilde{W}(2) \cong Sp(8,\mathbb{F}_2).$$

PROOF. It follows from the lemma that G acts on $\mathbf{R}/2\mathbf{R}$ by the formula $\bar{\alpha}_i \to \bar{\alpha}_i$, $i\neq 8$, $\bar{\alpha}_8 \to \bar{\alpha}_8+\bar{\alpha}_0+\bar{\alpha}_4+\bar{\alpha}_6+\bar{\alpha}_9$, where $\bar{\alpha}_i = \alpha_i + 2\mathbf{R}$. Thus under the canonical homomorphism $W \to W/W(2) \to GL(\mathbf{R}/2\mathbf{R})$ its image coincides with the transformation G defined in the proof of Theorem 2.8.2 (case $Q_{2,4,6}$). Now the assertion follows immediately from this proof.

Corollary 2.10.4. Let $W_r' = W_{B_r}(R_r)$. Then

$$\bar{W}_{10}(2) \cap W'_9 = W'_9(2).$$

PROOF. Obviously $W_9'(2) \subset \bar{W}_{10}(2) \cap W_9'$. On the other hand,

$$W_9'/W_9'(2) \cong W/\bar{W}(2) \cong Sp(8, \mathbb{F}_2)$$

Bibliographical notes to Chapter II.

The material of the first 4 sections is rather well-known. We gave the references in the text. We take here an approach from **[Lo 2]**, where the notion of a root basis is introduced.

The Enriques lattice **E** from §5 and the corresponding canonical root basis with Dynkin diagram of type $E_n$was studied first in algebraic geometry in the works of M.Kantor **[Kan]**, A.Coble **[Co 3]**, and P. Du Val **[DuV 2]** in connection with birational transformations of the projective plane. We refer to **[D-O]** for an exposition of this theory. Later on, it reappeared in the theory of surface singularities (see**[Arn, Lo 1]**). The model of **E** based on the notion of isotropic sequences appears first in the work of **[Cos 1]** and **[B-P]**.

The Reye lattice from §6 appears again in the works of A. Coble and P. Du Val in connection with birational transformation of the three-dimensional projective space. It has been studied in detail by the authors in connection with their work on automorphisms of Enriques surfaces **[C-D 2]**.

The function φ_M from §7 appears implicitly in **[Cos 1]** where first estimates for its values are given. Theorem 2.7.1 generalizes an earlier unpublished result of E.Looijenga.

The structures of the group W(2) and the quotient group W/W(2) in the case of finite Weyl groups is well known **[Bou 2]**. The generaliation of the corresponding results to the infinite case was first undertaken by A.Coble **[Co 1]**.His results about the structure of W/W(2) were generalized and updated by B.Griess **[Gri]**. The proofs of Theorems 2.10.1 and 2.10.2 about the structure of W(2) given by Coble **[Co 2]** is incorrect. The first rigorous proof of Theorem 2.10.1 is due to E.Looijenga (unpublished). We gave a modified version of his proof and generalized its idea to the case of the Reye lattice considered in Theorem 2.10.2.

	x	s(x)	APPLY	OBTAIN	SEE
1	3 0 0 0 1 1 1 2 1 1 1	3 0 0 0 1 1 1 1 2 1 1	B(1,2)	3 0 0 2 1 1 1 1 0 1 1	
2	3 0 0 1 1 1 1 2 1 0 1	3 0 0 1 1 1 1 1 2 0 1	B(2,9)	3 2 0 1 1 1 1 1 0 0 1	
3	3 0 1 1 1 1 1 2 1 0 0	3 0 1 1 1 1 1 1 2 0 0	B(9,10)	3 2 1 1 1 1 1 1 0 0 0	
4	4 0 1 1 1 1 1 3 1 1 1	4 0 1 1 1 1 1 1 3 1 1	A(6,7,8)B(1,6)A(6,7,8)	4 0 1 1 1 1 1 3 1 1 1	
5	4 1 1 1 1 1 1 3 0 1 1	4 1 1 1 1 1 1 0 3 1 1	A(5,6,7)B(6,7)A(5,6,7)	4 1 1 1 1 1 3 0 1 1 1	
6	4 0 0 1 1 2 2 2 1 1 1	4 0 0 1 1 2 2 1 2 1 1	B(1,2)	2 0 0 1 1 0 0 1 0 1 1	
7	4 0 1 1 1 2 2 2 1 0 1	4 0 1 1 1 2 2 1 2 0 1	B(1,9)	2 0 1 1 1 0 0 1 0 1 1	
8	4 1 1 1 1 2 2 2 1 0 0	4 1 1 1 1 2 2 1 2 0 0	B(1,10)	2 1 1 1 1 0 0 1 0 0 0	
9	5 0 1 1 3 2 2 2 1 1 1	5 0 1 1 3 2 2 1 2 1 1	A(4,5,6)B(1,5)A(4,5,6)	3 0 1 1 1 1 2 0 0 1 1	
10	5 1 1 1 3 2 2 2 1 0 1	5 1 1 1 3 2 2 1 2 0 1	A(4,5,6)B(1,9)A(4,5,6)	3 0 1 1 1 1 2 0 0 1 1	
11	5 0 1 1 2 2 2 3 1 1 1	5 0 1 1 2 2 2 1 3 1 1	A(5,6,8)K(1,6)A(5,6,8)	3 0 1 1 1 2 0 0 1 1 1	
12	5 1 1 1 2 2 2 3 1 0 1	5 1 1 1 2 2 2 1 3 0 1	A(5,6,8)K(6,9)A(5,6,8)	3 1 1 1 1 2 0 0 1 0 1	
13	5 0 2 2 2 2 2 2 0 1 1	5 0 2 2 2 2 2 0 2 1 1	B(1,7)	1 0 0 0 0 0 0 0 0 1 1	
14	5 1 2 2 2 2 2 2 0 1 1	5 1 2 2 2 2 2 0 2 1 1	B(7,9)	1 1 0 0 0 0 0 0 0 0 1	
15	6 4 1 1 2 2 2 2 1 1 1	6 4 1 1 2 2 2 1 2 1 1	A(1,5,6)B(5,6)A(1,5,6)	4 2 1 1 0 2 2 1 0 1 1	
16	6 1 1 2 2 2 2 4 1 1 1	6 1 1 2 2 2 2 1 4 1 1	A(3,6,8)B(3,6)A(3,6,8)	4 1 1 2 0 0 2 1 2 1 1	
17	6 1 1 1 3 3 3 2 1 1 1	6 1 1 1 2 2 2 1 2 1 1	A(4,5,6)B(4,5)A(4,5,6)	3 1 0 0 1 1 2 1 0 1 1	
18	6 1 1 1 2 3 3 3 3 2 1 1	6 1 1 1 2 3 3 1 3 1 1	A(5,6,8)B(6,8)A(5,6,8)	4 1 1 1 0 3 1 1 1 1 1	
19	7 1 2 2 3 3 4 2 1 1 1	7 1 2 2 3 3 4 1 2 1 1	A(4,5,6)B(4,5)A(4,5,6)	3 1 0 0 1 1 2 1 0 1 1	
20	7 1 2 2 2 3 4 3 1 1 1	7 1 2 2 2 3 4 1 3 1 1	A(4,5,6)A(2,3,8)B(3,4)A(2,3,8)A(4,5,6)	5 1 2 0 2 3 2 1 1 1 1	
21	7 1 2 2 2 3 3 4 1 1 1	7 1 2 2 2 3 3 1 4 1 1	B(2,3)A(5,6,8)B(5,6)A(5,6,8)	3 1 0 0 2 1 1 1 0 1 1	
22	7 3 3 3 3 2 2 2 0 1 1	7 3 3 3 3 2 2 0 2 1 1	A(2,3,4)A(1,5,6)B(5,6)A(1,5,6)A(2,3,4)	3 1 1 1 1 2 0 0 0 1 1	
23	7 2 2 2 3 3 3 3 0 1 1	7 2 2 2 3 3 3 0 3 1 1	A(4,5,6)A(2,3,8)B(2,7)A(2,3,8)A(4,5,6)	3 0 0 2 1 1 1 0 1 1 1	
24	8 4 4 3 3 2 2 2 1 1 1	8 4 4 3 3 2 2 1 2 1 1	A(2,3,4)A(1,5,6)B(5,6)A(1,5,6)A(2,3,4)	6 0 0 1 1 0 0 1 0 1 1	
25	8 4 4 2 2 2 3 3 1 1 1	8 4 4 2 2 2 3 1 3 1 1	A(1,2,6)A(4,5,8)B(5,6)A(4,5,8)A(1,2,6)	4 2 2 0 2 0 1 1 1 1 1	
26	8 2 2 2 3 3 4 4 1 1 1	8 2 2 2 3 3 4 1 4 1 1	A(4,5,6)A(3,6,8)B(3,8)A(3,6,8)A(4,5,6)	4 0 0 2 1 1 2 1 2 1 1	6
27	8 1 3 3 3 3 4 3 1 1 1	8 1 3 3 3 3 4 1 3 1 1	A(4,5,6)A(2,3,8)B(3,8)A(2,3,8)A(4,5,6)	3 1 2 0 1 1 0 1 0 1 1	
28	8 1 3 3 3 3 3 4 1 1 1	8 1 3 3 3 3 3 1 4 1 1	A(4,5,6)A(1,2,3)B(6,7)A(1,2,3)A(4,5,6)	4 1 1 1 1 3 1 0 1 1 1	
29	8 3 3 3 3 3 3 3 0 1 1	8 3 3 3 3 3 3 0 3 1 1	B(7,10)	4 1 1 1 1 1 0 1 1 3 1	5
30	9 3 3 3 4 4 4 2 1 1 1	9 3 3 3 4 4 4 1 2 1 1	A(4,5,6)A(1,2,3)(B(2,3)A(1,2,3)A(4,5,6)	5 3 1 1 2 2 2 1 0 1 1	
31	9 2 3 3 4 4 4 3 1 1 1	9 2 3 3 4 4 4 1 3 1 1	A(4,5,6)A(2,3,8)B(2,3)A(2,3,8)A(4,5,6)	5 0 1 1 2 2 2 1 3 1 1	11
32	9 2 3 3 3 4 4 4 1 1 1	9 2 3 3 3 4 4 1 4 1 1	A(4,5,6)A(2,3,8)B(2,3)A(2,3,8)A(4,5,6)	5 0 1 1 3 2 2 1 2 1 1	9

Table 1

	X	S(X)	APPLY	OBTAIN	SEE
1	5 2 2 2 4 2 4 2 0 0 0	5 2 2 2 2 4 4 2 0 0 0	K(8,9,10)	3 2 2 2 2 0 0 2 0 0 0	
2	5 2 2 2 4 0 4 2 2 0 0	5 2 2 2 0 4 4 2 2 0 0	K(4,9,10)	3 2 2 2 0 0 0 2 2 0 0	
3	5 2 2 4 2 4 2 2 0 0 0	5 2 2 4 4 2 2 2 0 0 0	K(8,9,10)	3 2 2 0 0 2 2 2 0 0 0	
4	7 2 2 2 6 2 4 4 4 0 0	7 2 2 2 2 6 4 4 4 0 0	A(4,5,6,7)K(8,9,10)K(4,9,10)A(4,5,6,7)	3 2 2 2 2 2 0 0 0 0 0	
5	7 2 2 2 6 0 4 4 4 2 0	7 2 2 2 0 6 4 4 4 2 0	K(4,8,10)A(5,6,7,8)K(4,6,10)A((5,6,7,8)	3 2 2 2 0 2 0 0 0 2 0	
6	7 2 2 2 6 4 4 4 2 0 0	7 2 2 2 4 6 4 4 2 0 0	A(4,5,6,7)K(6,9,10)A(4,5,6,7)	5 2 2 2 4 2 4 0 2 0 0	
7	7 2 2 6 2 4 4 4 2 0 0	7 2 2 6 4 2 4 4 2 0 0	A(3,4,6,7)K(7,9,10)A(3,4,6,7))	5 2 2 2 4 2 4 0 2 0 0	
8	7 2 2 2 6 2 6 2 2 2 0	7 2 2 2 2 6 6 2 2 2 0	A(4,5,6,7)K(4,7,10)K(5,7,10)A(4,5,6,7)	3 2 2 2 2 2 0 0 0 0 0	
9	7 2 2 6 2 6 2 2 2 2 0	7 2 2 6 6 2 2 2 2 2 0	A(3,4,5,6)K(4,6,10)K(5,6,10)A(3,4,5,6)	7 2 2 6 2 6 2 2 2 2 0	
10	7 2 2 2 6 0 6 2 2 2 0	7 2 2 2 0 6 6 2 2 2 2	A(5,6,7,8)A(4,5,6,9)K(6,7,8)K(4,7,8)A(4,5,6,9)A(5,6,7,8)	11 2 2 2 4 10 6 6 6 2 2	20
11	9 2 2 2 8 2 6 4 4 4 0	9 2 2 2 2 8 6 4 4 4 0	A(5,6,7,8)K(7,8,10)A(5,6,7,8)	7 2 2 2 2 4 6 4 4 0 0	4
12	9 2 2 2 8 0 6 4 4 4 2	9 2 2 2 0 8 6 4 4 4 2	A(5,6,7,8)K(4,8,9)K(4,7,8)A(5,6,7,8)	5 2 2 2 0 4 2 4 0 0 2	
13	9 2 2 8 2 4 4 4 6 2 0	9 2 2 8 4 2 4 4 6 2 0	A(4,6,7,8)K(5,6,10)A(4,6,7,8)	7 2 2 0 4 6 4 4 2 2 0	6
14	9 2 2 2 8 4 6 4 4 2 0	9 2 2 2 4 8 6 4 4 2 0	A(5,6,7,8)K(7,8,10)A(5,6,7,8)	7 2 2 2 2 4 6 4 4 0 0	8
15	9 2 2 2 8 6 4 4 4 2 0	9 2 2 2 6 8 4 4 4 2 0	A(4,5,6,7)K(7,8,10)K(6,7,10)A(4,5,6,7)	5 2 2 2 2 4 4 0 0 2 0	1
16	9 2 2 8 2 4 6 4 4 2 0	9 2 2 8 4 2 6 4 4 2 0	A(3,5,7,8)K(5,7,10)A(3,5,7,8)	7 2 2 4 0 2 4 4 6 2 0	6
17	9 2 2 8 2 6 4 4 4 2 0	9 2 2 8 6 2 4 4 4 2 0	A(5,6,7,8)A(3,4,6,8)K(6,8,10)A(3,4,6,8)A(5,6,7,8)	7 2 2 6 4 2 4 0 4 2 0	7
18	11 2 2 2 10 2 8 4 4 4 2	11 2 2 2 2 10 8 4 4 4 2	A(5,6,7,8)A(5,6,9,10)K(8,9,10)A(5,6,9)A(5,6,7)	9 2 2 2 2 6 8 4 0 4 4	11
19	11 2 2 2 10 4 8 4 4 4 2	11 2 2 2 4 10 8 4 4 4 2	A(5,6,8)A(4,5,6,9)K(7,8,9)A(4,5,6,9)A(5,6,8)	5 2 2 2 4 2 0 0 0 4 2	
20	11 2 2 2 10 8 4 4 4 4 2	11 2 2 2 8 10 4 4 4 4 2	A(4,5,6,7)A(4,5,8,9)K(7,8,9)A(4,5,8,9)A(4,5,6,7)	9 2 2 2 8 6 4 0 4 4 2	
21	11 2 2 10 2 8 4 4 4 4 2	11 2 2 10 8 2 4 4 4 4 2	A(3,4,6,9)A(3,4,7,8)K(6,7,8)A(3,4,7,8)A(3,4,6,9)	9 2 2 6 8 4 2 4 4 0 2	17
22	11 2 2 10 2 4 8 4 4 4 2	11 2 2 10 4 2 8 4 4 4 2	A(3,4,6,8)A(3,5,6,7)K(4,5,8)A(3,5,6,7)A(3,4,6,8)	9 2 2 6 4 2 8 4 4 0 2	17
23	11 2 2 2 10 2 6 6 6 4 2	11 2 2 2 2 10 6 6 6 4 2	A(5,6,7,8)K(7,8,9)K(6,7,8)A(5,6,7,8)	7 2 2 2 2 6 6 2 2 0 2	4
24	11 2 2 2 10 4 6 6 6 2 2	11 2 2 2 4 10 6 6 6 2 2	A(5,6,7,8)K(4,5,6)K(4,6,7)A(5,6,7,8)	7 2 2 2 0 2 2 6 6 2 2	4
25	11 2 2 2 10 6 6 6 4 2 2	11 2 2 2 6 10 6 6 4 2 2	A(4,5,6,7)K(6,7,8)K(4,6,7)A(4,5,6,7)	7 2 2 2 6 6 2 2 0 2 2	
26	11 2 2 10 2 6 6 6 4 2 2	11 2 2 10 6 2 6 6 4 2 2	A(3,4,6,7)K(5,6,7)K(4,6,7)A(3,4,6,7)	7 2 2 6 6 2 2 2 2 0 2	9
27	11 2 2 10 2 4 6 6 6 2 2	11 2 2 10 4 2 6 6 6 2 2	A(3,6,7,9)K(4,6,9)K(6,7,9)A(3,6,7,9)	7 2 2 6 0 2 2 6 2 2 2	
28	13 2 2 2 12 4 6 6 8 4 4	13 2 2 2 4 12 6 6 8 4 4	A(4,5,6,8)A(5,7,8,9)K(4,9,10)K(4,8,9)A(5,7,8,9)A(4,5,6,8)	5 2 2 2 0 4 2 2 4 0 0	1
29	13 2 2 2 12 6 8 6 4 4 4	13 2 2 2 6 12 8 6 4 4 4	A(5,6,7,8)A(4,5,6,9)K(8,9,10)K(6,8,90A(4,5,6,9)A(5,6,7,8)	5 2 2 2 2 4 4 2 0 0 0	1
30	13 2 2 2 12 8 6 6 4 4 4	13 2 2 2 8 12 6 6 4 4 4	A(4,5,6,7)K(6,7,8)A(5,8,9,10)K(6,7,8)A(5,8,9,10)A(4,5,6,7)	5 2 2 2 4 4 2 2 0 0 0	

Table 2

x	s(x)	APPLY	OBTAIN	see
1 3 2 2 0 0 0 2 2 2 0 0	5 4 4 2 2 0 2 2 2 0 0	K(5,9,10)	3 0 0 2 2 0 2 2 2 0 0	
2 3 2 0 0 0 2 2 2 2 0 0	7 6 4 4 4 2 2 2 2 0 0	K(5,9,10)A(1,2,3,4)K(4,9,10)A(1,2,3,4)	3 2 0 0 0 2 2 2 2 0 0	
3 5 2 2 0 4 4 2 2 2 0 0	7 4 4 2 6 4 2 2 2 0 0	K(5,9,10)A(1,2,4,5)K(5,9,10)A(1,2,4,5)	3 0 0 2 2 0 2 2 2 0 0	
4 5 2 0 0 4 4 2 2 2 2 0	9 6 4 4 8 4 2 2 2 2 0	A(1,2,3,4)K(2,3,10)A(1,2,3,4)	7 6 4 4 4 0 2 2 2 2 0	2
5 5 0 0 0 4 4 2 2 2 2 2	11 6 6 6 10 2 2 2 2 2	A(3,4,6,10)K(4,5,6)K(5,6,10)A(3,4,6,10)A(1,2,3,4)K(1,2,5)K(1,2,3)A(1,2,3,4)	7 2 2 6 2 0 2 2 2 2 6	
6 7 6 2 2 2 4 4 4 2 0 0	9 8 4 4 4 4 4 4 2 0 0	K(1,8,9)K(8,9,10)	10 0 0 0 0 0 0 2 0 0	
7 7 6 2 2 0 4 4 4 2 2 0	11 10 6 6 4 4 4 4 2 2 0	K(5,9,10)A(1,2,3,4)A(1,5,6,7)K(4,5,10)A(1,5,6,7)A(1,2,3,4)	7 2 2 2 4 0 4 4 6 2 0	
8 7 6 2 0 0 4 4 4 2 2 2	13 12 8 6 6 4 4 4 2 2 2	A(1,2,3,5)A(1,2,4,6)K(5,6,7)K(2,,5,6)A(1,2,4,6)A(1,2,3,5)	5 4 4 2 2 0 0 0 2 2 2	1
9 9 8 2 2 2 6 4 4 4 2 0	13 12 6 6 6 6 4 4 4 2 0	B(1,10)	9 8 2 2 2 2 4 4 4 6 0	
10 9 8 2 2 0 6 4 4 4 2 2	15 14 8 8 6 6 4 4 4 2 2	A(1,2,3,5)A(1,4,6,7)K(5,6,7)K(5,7,8)A(1,4,6,7)A(1,2,3,5)	7 6 4 4 2 2 4 0 0 2 2	3
11 11 10 2 2 2 8 4 4 4 4 2	17 16 8 8 8 8 4 4 4 4 2	A(1,2,3,6)A(1,2,4,5)A(1,3,7,8)K(7,8,9)K(2,7,8)A(1,3,7,8)A(1,2,4,5)A(1,2,3,6)	9 8 4 4 4 4 0 4 4 0 0	5
12 11 10 2 2 2 6 6 6 4 2 2	17 16 8 8 8 6 6 6 4 2 2	A(1,2,3,6)A(1,4,5,7)A(1,2,3,8)K(5,7,8)A(1,2,3,8)A(1,4,5,7)A(1,2,3,6)K(8,9,10)	9 0 4 4 4 6 6 6 0 2 2	14
13 13 12 2 2 2 6 6 8 4 4 4	21 20 10 10 10 10 6 6 8 4 4 4	A(1,2,3,7)A(1,4,5,6)K(1,5,6)K(5,6,7)A(1,4,5,6)A(1,2,3,7)	5 4 2 2 2 2 2 4 0 0 0	
14 9 8 4 4 4 6 6 6 0 2 2	11 6 2 2 2 6 6 6 0 2 2	K(1,2,8)A(2,3,4,10)K(1,2,8)A(2,3,4,10)K(1,8,9)	3 0 0 0 0 2 2 2 0 2 2	

Table 3

Chapter III

THE GEOMETRY OF THE ENRIQUES LATTICE

§1. Divisors of canonical type.

Let

$$D = \sum_{i \in I} m_i R_i$$

be an effective divisor on a nonsingular projective surface X with irreducible components R_i. We say that D is of **canonical type** if

$$K_X \cdot R_i = D \cdot R_i = 0 \text{ for all } i.$$

Such a D is called **indecomposable** if D is connected and g.c.d.(m_i) = 1.

If D is an irreducible divisor of canonical type, then D is of the following possible three types:

(i) a smooth elliptic curve,

(ii)) a cuspidal rational curve,

(iii) a nodal rational curve.

This follows easily from the formula for the arithmetic genus of an irreducible curve.

If D is a reducible divisor of canonical type and R_i is its irreducible component, then

$$R_i^2 = -2 , \ R_i \cong \mathbb{P}^1.$$

In particular, the set $B(D) = \{[R_i]\}_{i \in I}$ is a root basis in the lattice $M = \mathrm{Num}(F)$. Since $D \cdot R_i = 0$ for all i, the sublattice $M_{B(D)}$ of Num(F) spanned by the classes $[R_i]$ lies in the orthogonal complement of [D]. Since $D^2 = 0$, by the Hodge Index theorem, $M_{B(D)}$ is semi-negative definite. Thus B(D) is a root basis of affine type. If D is reducible and indecomposable, then B(D) is an irreducible root system of affine type. Moreover the class [D] belongs to the radical of $M_{B(D)}$.

Applying the classification of irreducible root systems of affine type from Chapter 2, §3, we obtain:

Proposition 3.1.1. Let $D = \Sigma m_i R_i$ be an indecomposable divisor of canonical type. Then D is one of the following divisors:

$\tilde{A}_0(I_0)$: smooth elliptic curve

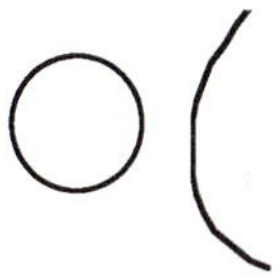

$\tilde{A}_0^*(I_1)$: rational curve with a node

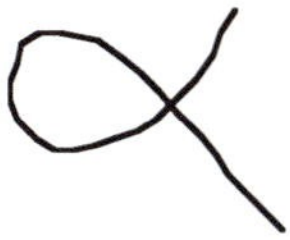

$\tilde{A}_0^{**}(II)$: rational curve with a cusp

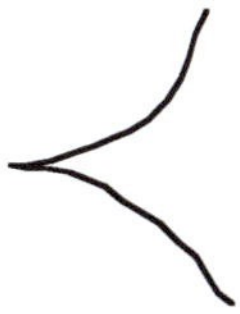

$\tilde{A}_1(I_2)$: $D = R_1+R_2$, R_1 transversally intersects R_2 at two points

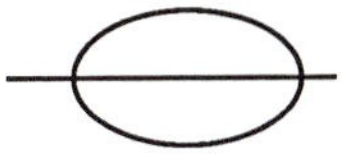

$\tilde{A}_1^*(III)$: $D = R_1+R_2$, R_1 is tangent to R_2 at one point, $R_1 \cdot R_2 = 2$

$\tilde{A}_2(I_3)$: $D = R_1+R_2+R_3$, $R_i \cdot R_j = 1, i \neq j$, $R_1 \cap R_2 \cap R_3 = \emptyset$

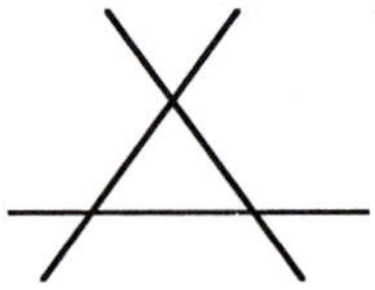

$\tilde{A}_2^*(IV)$: $D = R_1+R_2+R_3$, $R_i \cdot R_j = 1, i \neq j$, $R_1 \cap R_2 \cap R_3 \neq \emptyset$

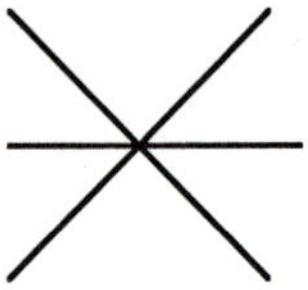

$\tilde{A}_{n-1}$ (I_n), $n \geq 4$: $D = R_1+\ldots+R_n$, $R_1 \cdot R_2 = R_2 \cdot R_3 = \ldots = R_{n-1} \cdot R_n = R_n \cdot R_1 = 1$, $R_i \cdot R_j = 0$ otherwise

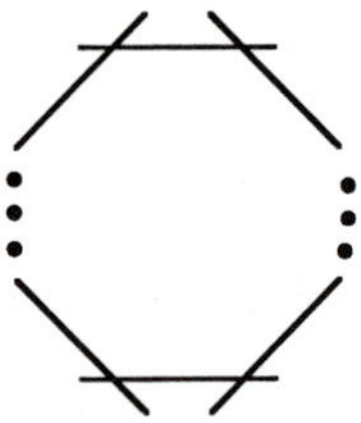

$\tilde{D}_{n+4}(I_n^*), n \geq 0$: $D = R_0+R_1+R_2+R_3+2R_4+\ldots+2R_{4+n}$, $R_1 \bullet R_4 = R_2 \bullet R_4 = R_4 \bullet R_5 = \ldots = R_{3+n} \bullet R_{4+n} = \; = R_{4+n} \bullet R_2 = R_{4+n} \bullet R_3$, $R_i \bullet R_j = 0$ otherwise

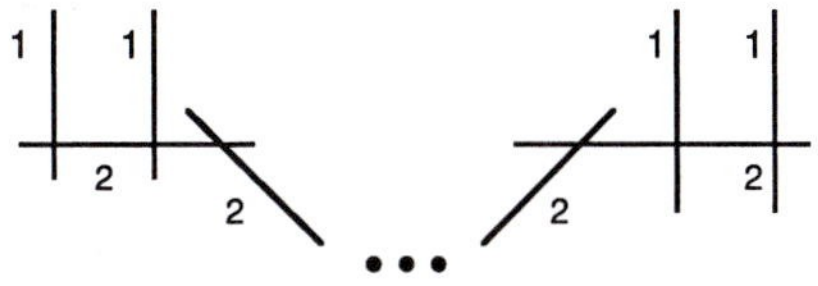

$\tilde{E}_6(IV^*)$: $D = R_0+2R_1+R_2+2R_3+3R_4+2R_5+R_6$, where $R_0 \bullet R_1 = R_1 \bullet R_2 = R_1 \bullet R_4 = R_2 \bullet R_3 = R_3 \bullet R_4 = R_4 \bullet R_5 = R_5 \bullet R_6 = 1$ and $R_i \bullet R_j = 0$ otherwise

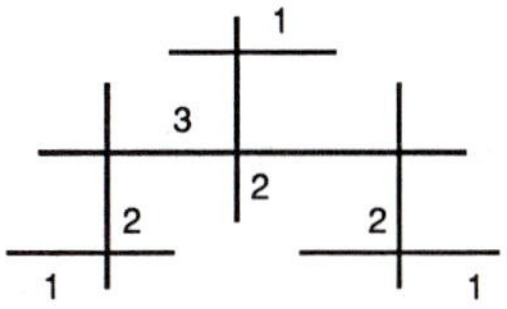

$\tilde{E}_7(III^*)$: $D = 2R_0+R_1+2R_2+3R_3+4R_4+3R_5+2R_6+R_7$,where $R_0 \bullet R_4 = R_1 \bullet R_2 = R_2 \bullet R_3 = \ldots = R_6 \bullet R_7 = 1$ and $R_i \bullet R_j = 0$ otherwise

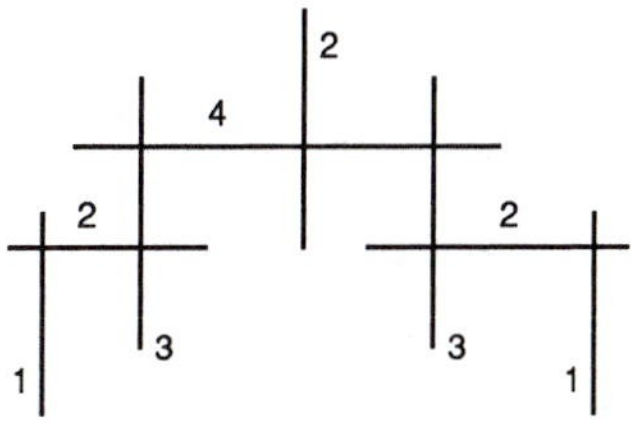

$\tilde{E}_8(II^*)$: $D = 2R_0+2R_1+4R_2+6R_3+5R_4+4R_5+3R_6+2R_7+R_8$,where $R_0 \bullet R_3 = R_1 \bullet R_2 = R_2 \bullet R_3 = \ldots = R_7 \bullet R_8 = 1$ and $R_i \bullet R_j = 0$ otherwise.

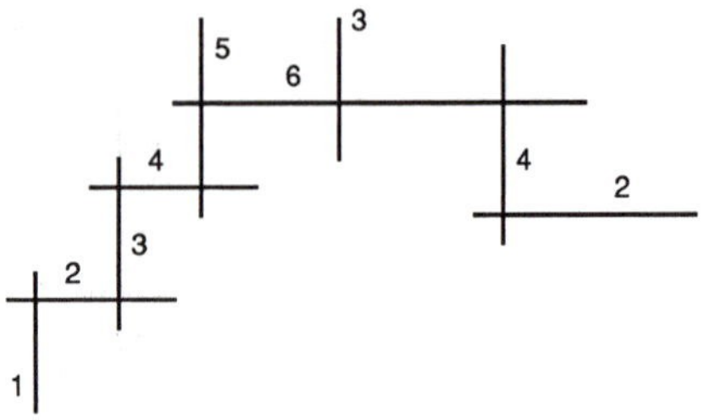

Here we use the notation indicating the type of the root basis B_D (except for the first three cases) and put in the brackets the notation introduced first by K.Kodaira. We will say D is a **divisor of canonical type** $\tilde{A}_n$ and so on if D is an indecomposable divisor of canonical type equal to a divisor of type $\tilde{A}_n$ and so on as above.

We will see later in Chapter V that all the above types do occur.

Lemma 3.1.1. Let $D_i = \Sigma m_i R_i$ be an indecomposable divisor of canonical type. Then D satisfies the following properties:

(i) $H^i(D,\mathcal{O}_D) = \kappa$ $(i = 0,1)$;

(ii) $\omega_D \cong \mathcal{O}_D$;

(iii) for any effective divisor D' with $D' \bullet R_i = 0$, one has $D' = nD + D''$, where D" is an effective divisor disjoint from D, $n > 0$;

(iv) $D^2 = D \bullet K_X = 0$, $D \bullet D' \geq 0$ for every effective divisor D' on F.

PROOF. (i) Obviously, it suffices to consider the case where D is reducible. It follows from the previous classification that $D= R_i + D'$, where D' does not contain R_i and coincides with the fundamental cycle Z of a rational double point (Chapter 0,§3). Thus $H^0(D',\mathcal{O}_{D'}) = \kappa$, $H^1(D',\mathcal{O}_{D'}) = 0$. The assertion follows easily from the exact sequence

$$0 \to \mathcal{O}_{R_i} \otimes \mathcal{O}_F(-D') \to \mathcal{O}_D \to \mathcal{O}_{D'} \to 0$$

and the observation that $\deg_R(\mathcal{O}_R \otimes \mathcal{O}_F(-D')) = -2$, $R \cong \mathbb{P}^1$.

(ii) Tensoring the above exact sequence by $\mathcal{O}_X(K_X+D)$, we get another one:

$$0 \to \omega_{R_i} \to \omega_D \to \mathcal{O}_{D'} \otimes \mathcal{O}_X(K_X+D) \to 0.$$

By restriction to D', a non-zero section $s \in H^0(D,\omega_D) \cong H^1(D,\mathcal{O}_D) \cong k$, defines a non-zero section of $\mathcal{L} = \mathcal{O}_{D'} \otimes \mathcal{O}_X(K_X+D) \cong \mathcal{O}_{D'} \otimes \omega_D$. Since $\deg(\mathcal{L} \otimes \mathcal{O}_{R_j}) = 0$ for all components R_j of D', $\mathcal{L} \cong \mathcal{O}_{D'}$ (Chapter 0, Proposition 0.2.2). This shows that s generates ω_D at every point of D'. Since $\deg(\omega_D \otimes \mathcal{O}_{R_j}) = 0$, s generates $\omega_D \otimes \mathcal{O}_{R_j}$ or is identically zero. Since R_i intersects D', the latter case does not occur. Thus s generates ω_D everywhere.

(iii) Let $\mathcal{L} = \mathcal{O}_X(D')$. Replacing $\mathcal{O}_X(K_X+D)$ by $\mathcal{L}$ in the previous argument and taking for s the section of 1 defined by D', we obtain that $\mathcal{L} \otimes \mathcal{O}_D \cong \mathcal{O}_D$. From this and (i) the assertion follows immediately.

(iv) Obvious.

Proposition 3.1.2. Let D be an indecomposable divisor of canonical type on an Enriques surface F. Then $|D|$ or $|2D|$ does not have base points and is of dimension 1.

PROOF. Since $2K_F \sim 0$ and $\omega_D \cong \mathcal{O}_D$, we have

$$\mathcal{O}_D(D) \cong \mathcal{O}_D(K_F),\ \mathcal{O}_D(2D) \cong \mathcal{O}_D.$$

Assume $\dim |D| = 0$. Then, by Riemann-Roch, $H^1(F,\mathcal{O}_F(D)) = 0$ and the cohomology sequence for the exact sequence

$$0 \to \mathcal{O}_F(D) \to \mathcal{O}_F(2D) \to \mathcal{O}_D(2D) \to 0$$

implies $\dim |2D| = 1$. Here, we use that $H^0(D,\mathcal{O}_D) \cong k$. Let $|2D| = |M| + Z$ be the decomposition of $|2D|$ into its moving part $|M|$ and the fixed part Z. Then

$$0 = 2D^2 = M \cdot D + Z \cdot D\ .$$

Since $D \cdot M \geq 0$, $D \cdot Z \geq 0$ (Lemma 3.1.1 (iv)), we get $M \cdot D = 0$, hence (Lemma 3.1.1(iii)), $|M| = |2D|$. Since $D^2 = 0$, $|2D|$ is free of base points.

Assume $\dim |D| \neq 0$. Then

$$h^0(\mathcal{O}_D(D)) = h^0(\mathcal{O}_D(K_F)) \leq h^0(\mathcal{O}_D(2K_F)) = h^0(\mathcal{O}_D)) = 1.$$

The exact cohomology sequence for

$$0 \to \mathcal{O}_F \to \mathcal{O}_F(D) \to \mathcal{O}_D(D) \to 0$$

gives dim $|D| = 1$. One concludes as before that $|D|$ has no base points.

We will see in Chapter 5, §4 that $\dim|D| = 0$ if the class $[D]$ of D is primitive in $\mathrm{Num}(F)$.

A linear system of dimension 1 (a pencil) on a surface X whose general member is an irreducible divisor of arithmetic genus 1 will be called **a genus 1 pencil**. If its general member is smooth (resp. non-smooth) we will call it an elliptic (resp. quasi-elliptic) pencil. Note that the latter case happens only if $\mathrm{char}(k) = 2,3$ (see Chapter 5,§1). The previous proposition says that for every indecomposable curve of canonical type $|D|$ or $|2D|$ is a genus 1 pencil.

Proposition 3.1.3. Let $|P|$ be a genus 1 pencil on an Enriques surface F. Then there exists an indecomposable divisor of canonical type E such that $P \sim 2E$. If F is classical (resp. non-classical), then there exist two such E's which differ by K_F in $\mathrm{Pic}(F)$ (resp. E is defined uniquely). Moreover, the divisor D is of type $\tilde{A}_n$ unless $p = 2$ and F is not a μ_2-surface.

PROOF. See Chapter 5 , §7.

The divisor E such that $|2E|$ is a genus 1 pencil is said to be a **half-fibre** of $|2E|$.

Corollary 3.1.1. Let $|P|$ be a genus 1 pencil on F. Then $P \cdot D$ is even for every divisor D on F.

Proposition 3.1.4. Let D be a divisor on an Enriques surface F. Assume that $|D|$ is not empty and has no fixed components. Then one of the following holds:

(i) $D^2 > 0$ and there exists an irreducible curve C in $|L|$;

(ii) $D^2 = 0$ and there exists a genus 1 pencil $|P|$ such that $|D| = |kP|$ for some $k \geq 1$.

PROOF. If $D^2 = 0$, then $|D|$ must be composed of a pencil, that is, $|D| = |kP|$ for pencil $|P|$. By the genus formula, $|P|$ is a genus 1 pencil.

If $D^2 > 0$, Bertini's theorem tells us that the general member of $|D|$ is reducible only if $|D|$ is composed of a pencil which is necessary linear (since $(\mathrm{Pic}_{F/K})^{\circ}_{red} = \{0\}$). Then $|D| = |nH|$, and by Riemann-Roch,

$$n = \dim |D| \geq \tfrac{1}{2}n^2H^2.$$

This implies that $n = 1$, $H^2 = 2$, hence the general member of $|L|$ is irreducible.

Corollary 3.1.2. Let $|D|$ be a non-empty linear system with $D^2 > 0$ on F. Then the moving part $|M|$ of $|D|$ is irreducible. In particular, $|M|$ is a genus 1 pencil only if $D^2 = 2$.

PROOF. Let $|D| = |M| + Z$ be the decomposition of $|D|$ into its moving part and its fixed part. Assume that $|M|$ is reducible. Then, by Propositions 3.1.3 and 3.1.4, $|M| = |2\kappa E|$ for a genus 1 pencil $|2E|$ and $\kappa \geq 2$. Since $D^2 \neq 0$, there exists a component R of Z such that $E \cdot R \neq 0$. By Riemann-Roch,

$$\kappa = \dim |D| \geq \tfrac{1}{2}(2\kappa E+R)^2 = 2\kappa(E \cdot R) + \tfrac{1}{2}R^2 \geq 2\kappa-1.$$

Hence $\kappa = 1$, a contradiction.

Proposition 3.1.5. Let C be an irreducible curve on an Enriques surface F with $C^2 > 0$. Assume $K_F \neq 0$. Then $|C+K_F|$ has a basis component if and only if $|C| = |2E+R+K_F|$, where $|2E|$ is a genus 1 pencil and R is a smooth rational curve with $R \cdot E = 1$.

PROOF. It follows from the proof of Step 6 in Theorem 1.4.1 of Chapter 1 that $|C+K_F| = |M| + Z$, where $|M|$ does not have fixed components and $Z = 0$ or $Z^2 = -2$, $M^2 = 0$. The latter case implies $|C| = |2E+R+K_F|$ as above ($|M| = |2E|$, $Z = R$). The converse is obvious.

Proposition 3.1.6. Let D be an effective divisor on an Enriques surface F with $D^2 > 0$ and $D \cdot C \geq 0$ for any curve C. Then $|D|$ has no fixed components unless $|D| = |2E+R|$ for a genus 1 pencil $|2E|$ and a smooth rational curve R with $R \cdot E = 1$.

PROOF. Let $|D| = |M| + Z$ be the decomposition of $|D|$ into its moving part and the fixed part. Assume $M^2 > 0$. Then

$$\tfrac{1}{2}M^2 = \dim |M| = \dim |D| \geq \tfrac{1}{2} D^2.$$

Since $D^2 = D\bullet M+D\bullet Z = M^2+M\bullet Z+D\bullet Z$, we conclude that $M\bullet Z = D\bullet Z = Z^2 = 0$. Hence $Z = 0$ by the Hodge Index Theorem.

Assume $M^2 = 0$. Then $|M| = |2E|$ is a genus 1 pencil. One easily checks that $Z = R$ is a smooth rational curve and $E\bullet R = 1$.

Corollary 3.1.3. Let D be an effective divisor on F with $D^2 > 0$ and $D\bullet C \geq 0$ for any curve C. Then $H^1(F,\mathcal{O}_F(-D)) = 0$.

PROOF. If $|D|$ has no fixed components, then $|D|$ contains an irreducible curve and the assertion follows from Theorem 1.4.1. Otherwise, $|D| = |2E+R|$ as in the above proposition. The exact sequence of sheaves

$$0 \to \mathcal{O}_F(-2E-R) \to \mathcal{O}_F(-2E) \to \mathcal{O}_R(-2E) \to 0$$

together with Serre duality and Riemann-Roch show that

$$\begin{aligned} h^1(\mathcal{O}_F(-D)) &= h^1(\mathcal{O}_F(-2E)-h^1(\mathcal{O}_R(-2E))+h^2(\mathcal{O}_F(-D))-h^2(\mathcal{O}_F(-2E) \\ &= h^1(\mathcal{O}_F(K_F+2E)-h^0(\mathcal{O}_R)+h^0(\mathcal{O}_F(K_F+D))-h^0(\mathcal{O}_F(K_F+2E) \\ &= -2+h^0(\mathcal{O}_F(K_F+D)). \end{aligned}$$

It follows from the proof of Proposition 3.1.6 that R is the fixed component of $|D|$. Hence $h^0(\mathcal{O}_F(K_F+D)) = h^0(\mathcal{O}_F(K_F+2E)$. If $K_F \sim 0$, this number is 2 and we are done. If $K_F \nsim 0$, by Proposition 3.1.3, $|K_F+2E| = |E_1+E_2|$, where E_1 and E_2 are disjoint indecomposable divisors of canonical type. This easily implies that $h^0(\mathcal{O}_F(K_F+2E)) = 1$ in this case and the assumption that $|D|$ has fixed components is contradictory.

Corollary 3.1.4. Let $|2E|$ be a genus 1 pencil and R be a smooth rational curve such that $E\bullet R = 1$. Assume $K_F \nsim 0$. Then $|2E+R+K_F|$ has no fixed components.

PROOF. Assume that $|M| = |2E+R+K_F|$ has a fixed component. Then $|M| = |2E'+R'|$, where $|2E'|$ is a genus 1 pencil and R' is a smooth rational curve with $E'\bullet R' = 1$. Then $E\bullet(2E+R) = E\bullet(2E'+R') = 1$ implies $E\bullet E' = 0$, hence, $|2E| = |2E'|$, $R+K_F \sim R'$. However, by Riemann-Roch, $|R+K_F| = \emptyset$.

S2. The nodal chamber.

Let X be a nonsingular projective algebraic surface. Recall from Chapter 0 that a **nodal curve** on X is a smooth rational curve R with $R\bullet K_X = 0$ (equivalently, $R^2 = -2$).

We denote by $\mathfrak{R}(X)$ the set of nodal curves on X. Obviously,

$$R_1 \equiv R_2 \Rightarrow R_1 = R_2, \text{ for any } R_1, R_2 \in \mathfrak{R}(X).$$

This allows us to identify the class $[R]$ of a nodal curve R in Num(X) with R itself. This identification of $\mathfrak{R}(X)$ with a subset of Num(X) always will be assumed.

Since $R^2 = -2$ for any $R\in\mathfrak{R}(X)$ and since R intersects nonnegatively every other nodal curve, the subset $\mathfrak{R}(X)$ is a root basis in Num(X). The Weyl group of this basis is said to be the **Weyl group of the surface** X and is denoted by W_X. This is a subgroup of O(Num(X)). Let

$$C_X = \{x\in \mathrm{Num}(X)_{\mathbb{R}}: x\bullet R \geq 0 \text{ for any } R\in\mathfrak{R}(X)\}.$$

This is a fundamental chamber for $\mathfrak{R}(X)$. We will call it the **nodal chamber** of X. Finally, we denote by V_X^+ the connected component of the cone

$$V_X = \{x\in \mathrm{Num}(X)_{\mathbb{R}}: x^2 \geq 0\}.$$

which contains the class of an ample divisor and set

$$\mathrm{Num}(X)^+ = \mathrm{Num}(X)\cap V_X^+,$$

$$C_X^+ = C_X \cap \mathrm{Num}(X)^+.$$

It is clear that $\mathrm{Num}(X)^+$ consists of the classes of effective divisors D with

$D^2 \geq 0$. It follows from Chapter 2, §4 that $C_X{}^+$ is a fundamental domain in $V_X{}^+$ with respect to the group W_X.

Theorem 3.2.1. Let D be an effective divisor on an Enriques surface F with $D^2 \geq 0$. Then

$$D \sim D' + \Sigma m_i R_i,\ m_i \geq 0,$$

where R_i are nodal curves, $w([D]) = [D']$ and one of the following cases occurs:
(i) D' is an irreducible curve with $D^2 > 0$;
(ii) D' is a divisor of canonical type;
(iii) D'=2E+R, where |2E| is a genus 1 pencil and R is a nodal curve with $R \cdot E = 1$.
Moreover, the class of D' in Num(F) is uniquely determined by the class of D.

PROOF. By Corollary 2.2.2. of Chapter 2, we can write D or $D+K_F$ as above for some D' with $[D'] \in \mathrm{Num}(X)^+$. Let us show that $\mathrm{Num}(X)^+$ consists of the classes of divisors D' satisfying one of the three properties from above. Let D' be a divisor such that $[D'] \in \mathrm{Num}(X)^+$. If $D'^2 > 0$, by Proposition 3.1.6, |D'| contains a divisor of type (i) or (iii). If $D'^2 = 0$, then $|D'| \neq \emptyset$ and we may assume that D' is effective and connected. Let $D = \Sigma m_i C_i$ be its decomposition into irreducible components. Since $D' \cdot C_i \geq 0$ for every C_i with $C_i^2 \geq 0$ and $D' \cdot C_i \geq 0$ for every $C_i \in \mathcal{R}(X)$, we obtain $D'^2 = D' \cdot C_i = 0$ for every irreducible component C_i. This immediately implies that D' is of canonical type.

Finally, we observe that if D' satisfies one of the properties (i) - (iii), then $|D'+K_F|$ contains a divisor satisfying one of the these properties as well.

Corollary 3.2.1. Every Enriques surface contains a genus 1 pencil.

PROOF. Since rk(Num(F)) = 10 (Theorem 1.2.3 of Chapter 1) and Num(F) is indefinite, Num(F) contains some isotropic vectors (**[Ser 4]**). It remains to apply Theorem 3.2.1 and Proposition 3.1.2.

Corollary 3.2.2.(Reducibility lemma). Every effective divisor D on F is linearly equivalent to a sum of irreducible curves of arithmetic genus 0 or 1.

PROOF. Obviously, we may assume that D is an irreducible curve with $D^2 > 0$. Fixing a canonical root basis in Num(F), we find an element $w \in W(\mathrm{Num}(F))$ such that w([D]) belongs to the fundamental chamber. By Remark 2.5.3 of Chapter 2, we can write w([D]) as a non-negative linear combination of isotropic vectors f_i for which $f_i \cdot f_j > 0$, $i \neq j$. Applying w^{-1} to w([D]), we obtain

$$D \equiv \Sigma n_i E_i$$

where $E_i^2 = 0$ and $E_i \cdot E_j > 0$, $i \neq j$. Since $D \cdot E_i > 0$, E_i is an effective divisor. By the previous theorem, the assertion is true for every E_i, and hence for D or $D+K_F$. This shows that we may assume that $K_F \neq 0$, i.e. F is a classical Enriques surface. In this case the argument is due to Enriques himself. By induction on the degree of D with respect to some very ample divisor, it suffices to show that |D| contains a reducible divisor. Let $D' \in |D+K_F|$. If |D'| does not contain an irreducible divisor, then we write $D' \sim C + \Sigma m_i R_i$ as in the theorem. This implies that |D| contains a reducible divisor $C' + \Sigma m_i R_i$ for some $C' \in |C+K_F|$. Thus, we may assume that both D and D' are irreducible divisors. The linear system |2D| contains two algebraic subvarieties V and V' formed by the divisors D_1+D_2 and $D_1'+D_2'$ respectively, where $D_i \in |D|$ and $D_i' \in |D'|$. Since dim |D| = dim |D'| = n = $\frac{1}{2}D^2$ (Corollary 1.4.1), dim V = dim V' = 2n. Since |2D| does not have fixed components, it contains an irreducible divisor and its dimension is equal to $\frac{1}{2}(2D)^2 = 4n$. Thus, $V \cap V' \neq \emptyset$ and there exists a divisor $C \in |2D|$ which can be written as

$$C = D_1 + D_2 = D_1' + D_2',$$

where $D_1, D_2 \in |D|$, $D_1', D_2' \in |D'|$. Since $|D| \cap |D'| = \emptyset$, this happens only if D_1 or D_2 is reducible.

Corollary 3.2.3. Pic(F) is generated by the classes of curves of arithmetic genus 1 and 0.

Remark 3.2.1. Note that the proof of the latter corollary does not use the argument of Enriques. It follows immediately from Corollary 2.5.6 and the fact that K_F is equal to the differences of the classes of curves of arithmetic genus 1.

Also compare the previous result with the obvious fact that for every rational surface X, Pic(X) is generated by the classes of smooth rational curves.

An Enriques surface is said to be **nodal** if $\mathcal{R}(F) \neq \emptyset$, **unnodal** otherwise.

Corollary 3.2.4. Let F be an unnodal Enriques surface. Then every effective divisor on F is linearly equivalent to a sum of irreducible curves of arithmetic genus 1.

We will denote by

$$\Phi: \mathrm{Num}(F)^+ \rightarrow \mathbf{Z}_{\geq 0}$$

the function defined by

$$\Phi(x) = \tfrac{1}{2}\inf\{[P]\cdot x, \ |P| \text{ is a genus 1 pencil}\}.$$

If no confusion arises we will write

$$\Phi(D) = \Phi([D])$$

for every divisor D with $[D] \in \mathrm{Num}(F)^+$.

§3. Canonical r-sequences and $U_{[r]}$-markings.

Recall that an **isotropic r-sequence** in Num(F) is an ordered set of r isotropic vectors f_i such that $f_i \cdot f_j = 1$ if $i \neq j$ and f_1 is primitive (Chapter 5, §5). An isotropic r-sequence $(f_1, \dots, f_r)$ will be called **non-degenerate canonical** if every $f_i \in C_F$, i.e. if f_i is equal to the class of a divisor E_i of canonical type. Since all f_i's are primitive vectors, each E_i is an indecomposable divisor of canonical type and $|2E_i|$ is a genus 1 pencil.

Conversely, if $(|P_1|,\ldots,|P_r|)$ is an ordered set of genus 1 pencils with $P_i\cdot P_j = 4$ for $i\neq j$, it defines a unique non-degenerate canonical isotropic r-sequence $(f_1,\ldots,f_r)$ such that $|P_i| = |2E_i|$, $|E_i| = f_i$.

The unordered isotropic set corresponding to an isotropic r-sequence will be called an **isotropic r-set**. A non-degenerate canonical isotropic r-set has a similar meaning.

Note that $r\leq 10$ for every isotropic r-sequence. An isotropic 10-sequence is called **maximal**.

Lemma 3.3.1. Let $(e_1,\ldots e_r)$ be an isotropic r-sequence in Num(F). There exists a unique non-degenerate canonical isotropic c-sequence $(f_1,\ldots,f_c)$, $c\leq r$, and some nodal curves $R_{j,K}$ such that for some $w\in W_F$ one has after reindexing

$$w(e_1) = f_1,\ w(e_2) = f_1+R_{1,1},\ \ldots\ ,w(e_K) = f_1+R_{1,1}+\ldots+R_{1,K-1},$$

$$w(e_{K+1}) = f_2+R_{2,1},\ \ldots\ ,w(e_r) = f_c+\ldots+R_{c,i(c)}.$$

The number c is equal to the number of distinct W_F-orbits of the vectors $e_1,\ldots,e_r$. The curves $R_{j,1},\ldots,R_{j,i(j)}$ form a root basis of type $A_{i(j)}$.

PROOF. Let $e = e_1+\ldots+e_r$ and choose $e\in W_F$ such that $w(e) = h\in C_F$. Then

$$h^2 = r(r-1),\ \Phi(h) = r-1.$$

By Lemma 2.7.1, we can find isotropic vectors $f_1,\ldots,f_c$ from C_F such that

$$\Phi(h) = h\cdot f_i,\ i = 1,\ldots,c.$$

Since $\Phi(h) = h\cdot w(e_i) = r-1$ for every $i = 1,\ldots,r$ and $h\cdot f > r-1$ for every isotropic vector different from the $w(e_i)$'s, each f_i must be equal to $w(e_j)$ for some j. Up to reindexation, we may assume that $e_1,\ldots,e_K$ belong to the W_F-orbit of $f_1 = w(e_1)$. By Corollary 2.2.2,

$$w(e_2) = f_1 + \sum_{i\in I} m_iR_i\ ,$$

for some nodal curves R_i and some $m_i \geq 0$. Since

$$(w(e_2)-f_1)\cdot f_1 = w(e_2)\cdot f_1 = 1,$$

there exists $i\in I$ for which $f_1\cdot R_i = 1$ and $m_i = 1$. The vector f_1+R_i is isotropic and

$$h\cdot(f_1+R_i) = h\cdot(w(e_2)-\sum_{j\neq i} m_jR_j \le h\cdot w(e_2) \le r-1.$$

As above, $f_1+R_i = f_1+(f_1\cdot R_i)R_i$ must be one of the $w(e_i)$'s, $i = 1,\ldots,k$. After reindexing we may assume that $w(e_2) = f_1+R_i$ and set $R_{1,1} = R_i$. Similarly, we write

$$w(e_3) = f_1 + \sum_{i\in I} m_i'R_i = = f_1 + R_j + \sum_{j\neq i} m_i'R_i,$$

where $f_1\cdot R_j = 1$, $f_1\cdot R_i = 0$ for $i\neq j$. Since

$$1 = w(e_3)\cdot w(e_2) = w(e_3)\cdot(f_1+R_{1,1}) = 2 + (R_{1,1} + \sum_{j\neq i} m_i'R_i),$$

$R_{1,1} = R_i$ for some $i\in I$. Since $f_1\cdot R_{1,1} = 1$ and $f_1\cdot R_i = 0$ for $i\neq j$, $R_{1,1} = R_j$. Moreover, $w(e_3)\cdot R_{1,1} = 0$ implies that

$$R_{1,1}\cdot \sum_{j\neq i} m_i'R_i = 1.$$

This shows that there exists R_i, $i \neq j$, such that $R_{1,1}\cdot R_i = 1$, $m_i = 1$. We set $R_{1,2} = R_i$ and again as above, after reindexation, may assume that

$$w(e_3) = f_1 + R_{1,1} + R_{1,2} .$$

A repeated use of this argument proves the assertion.

An isotropic r-sequence $(f_1,\ldots,f_r)$ equal to the sequence $(w(e_1),\ldots,w(e_r))$ as above will be called a **canonical isotropic r-sequence**. It is called **c-degenerate** if $c < r$. Similarly, we define canonical (degenerate) isotropic r-sets. It is clear that an r-degenerate canonical isotropic r-sequence is non-degenerate.

Corollary 3.3.1. Every non-degenerate canonical r-sequence $(f_1,\ldots,f_r)$ can be extended to a canonical maximal isotropic sequence provided $r \neq 9$.

PROOF. By Corollary 2.5.6, $(f_1,\ldots,f_r)$ can be extended to a maximal isotropic sequence $(e_1,\ldots,e_{10})$ where $e_i = f_i$, $i = 1,\ldots,r$. Applying Lemma 3.3.1, we find $w\in W_F$ such that $\{w(e_1),\ldots,w(e_{10})\}$ is a canonical isotropic 10-set which contains a non-degenerate canonical isotropic c-set $\{f_1',\ldots,f_c'\}$. It follows from the proof of this lemma that $\{f_1,\ldots,f_r\}$ is a subset of $\{f_1',\ldots,f_c'\}$. We know that the W_F-orbits of the vectors f_i are distinct and $w(f_i)\in C_F$ implies $w(f_i) = f_i$. This proves that $c \geq r$ and we are done.

Recall from Chapter 2, §5 that the lattice $U_{[r]}$ is the lattice $\mathbb{Z}v_1+\ldots+\mathbb{Z}v_r$ where $v_i \cdot v_j = 1-\delta_{ij}$. An embedding (primitive if $r = 1$) of the lattice $U_{[r]}$ into $\mathrm{Num}(F)$ is called a **$U_{[r]}$-marking of F**. Two $U_{[r]}$-markings $j:U_{[r]} \hookrightarrow \mathrm{Num}(F)$ and $j':U_{[r]} \hookrightarrow \mathrm{Num}(F)$ are said to be **isomorphic** if there exist an isometry $\sigma \in O(U_{[r]})$ and an element $w\in W_F$ such that $j' = w\circ j\circ\sigma$. In other words, two $U_{[r]}$-markings are isomorphic if they belong to the same orbit with respect to the natural action of $O(U_{[r]})\times W_F$ on the set of $U_{[r]}$-markings.

Proposition 3.3.1. There are natural bijective correspondences between the following sets:

(i) isomorphism classes of $U_{[r]}$-markings;

(ii) canonical isotropic r-sets;

(iii) elements $h\in C_F$ with $h^2 = r(r-1)$ and $\Phi(h) = r-1$;

(iv) W_F-orbits of primitive divisor classes D with $D^2 = r(r-1)$ and $\Phi(D) = r-1$.

PROOF. This follows from the proof of Lemma 3.3.1, Theorem 3.2.1 and Corollary 3.3.1.

A $U_{[r]}$-marking defined by a canonical c-degenerate isotropic sequence will be called a **canonical c-degenerate $U_{[r]}$-marking**. It is said to be non-degenerate if $c = r$. A $U_{[2]}$-marking will be called a **U-marking**. It is called **degenerate** if it is canonical 1-degenerate. It is called **non-degenerate** if it is canonical non-degenerate. Thus, a non-degenerate U-marking is determined uniquely by a pair of genus 1 pencils $|2E_1|$ and $|2E_2|$ with $E_1 \cdot E_2 = 1$. We will call such a pair a **non-degenerate U-pair**. A degenerate U-marking is determined by a pair $(|2E|,R)$, where $|2E|$ is a genus 1 pencil and R is a nodal curve with $E\cdot R = 1$. Such a pair will be called a **degenerate U-pair**.

Theorem 3.3.1. For every $r \le 10$, F admits a canonical c-degenerate $U_{[r]}$-marking, $c \ge 1$. Moreover, if F is unnodal, $c = r$.

PROOF. The existence of some $U_{[r]}$-marking follows from the existence of maximal isotropic sequences in the Enriques lattice (Proposition 2.5.6). Now the assertion follows from Lemma 3.3.1.

We define the **non-degeneracy invariant** d(F) of an Enriques surface by

$$d(F) = \max\{r: \text{F admits a canonical non-degenerate } U_{[r]}\text{-marking}\}.$$

Clearly d(F) = 10 if F is unnodal. In the next two sections we will show that $d(F) \ge 3$ unless char(k) = 2 and F is of very special type. In Part 2 we will give a construction due to S.Kondō **[Kon]** of an Enriques surface over $\mathbb{C}$ with a finite crystallographic nodal root basis defined by the following diagram:

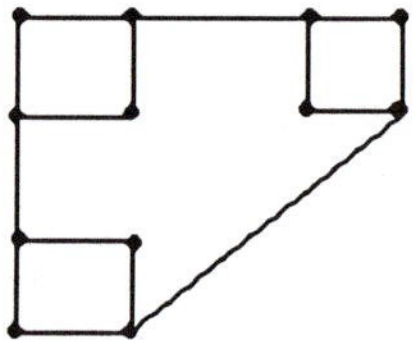

It is easy to list all genus 1 pencils on this surface. They correspond to the maximal root subbases of affine type. One checks that the non-degeneracy index d(F) of this surface is equal to 4. No surfaces with d(F) = 3 are known to us.

§4. U-markings.

An Enriques surface is said to be **$\tilde{E}_8$-special** if it contains

(i) a genus 1 pencil |2E|, where E is a divisor of canonical type of type $\tilde{E}_8$,

(ii) a nodal curve R such that $R \cdot E = 1$.

It is not known to us whether such a surface exists. It follows from

Chapter 5 that it does not exist if char(k) $\neq$ 2 (Proposition 5.1.8) or char(k) = 2 and F is a μ_2-surface (Theorem 5.7.5).

Proposition 3.4.1. An $\tilde{E}_8$-special Enriques surface does not admit a non-degenerate U-marking.

PROOF. Let E and R be as in the definition of an $\tilde{E}_8$-special surface. The irreducible components of E together with the nodal curve R form a root basis of type E_{10}. Since this is a crystallographic root basis, it coincides with the nodal root basis $\mathfrak{R}(F)$ (Proposition 2.4.3 of Chapter 2). Since there is only one orbit of isotropic primitive vectors in E_{10}, there is only one genus 1 pencil on F. This proves the assertion.

Lemma 3.4.1. Every reducible member of a genus 1 pencil on an Enriques surface F consists of at most 9 components.

PROOF. Since rk(Num(F)) = 10 and Num(F) is hyperbolic, every root basis of affine type in Num(F) consists of at most 9 components. Since the irreducible components of a divisor of canonical type form a root basis of affine type, we are done.

Theorem 3.4.1. Let F be an Enriques surface. Then F admits a non-degenerate U-marking, unless F is $\tilde{E}_8$-special. More generally, every canonical $U_{[1]}$-marking can be extended to a non-degenerate U-marking.

PROOF. Let f_1 be a primitive isotropic vector in C_F defining a canonical $U_{[1]}$-marking and E_1 be the corresponding divisor of canonical type. By Proposition 3.3.1, we can complete f_1 to a maximal canonical sequence . Clearly, the theorem holds unless the sequence is a degenerate 1-sequence. That is, we may assume that there exist 9 nodal curve $R_1,\ldots,R_9$ which together with E_1 form the intersection graph:

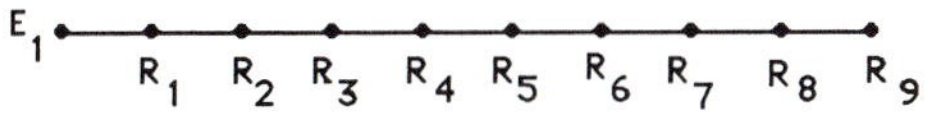

Since $R_2,\ldots,R_9$ are orthogonal to E_1 and form a connected intersection graph, they are irreducible components of one of the fibres E of the genus 1 pencil represented by f_1. It follows from the classification of indecomposable divisors of canonical type (Proposition 3.1.1) and Lemma 3.4.1 that E is of type $\tilde{A}_8$ or $\tilde{E}_8$, and $E \bullet E_1 = 1$ or 2 (Proposition 3.1.3). Let R_{10} denote the ninth component of E.

<u>Case 1</u>: E is of type $\tilde{A}_8$, $E \bullet E_1 = 2$.

Since $R_1 \bullet E = 2R_1 \bullet f_1$, R_1 intersects some other component R_i of E with multiplicity 1.

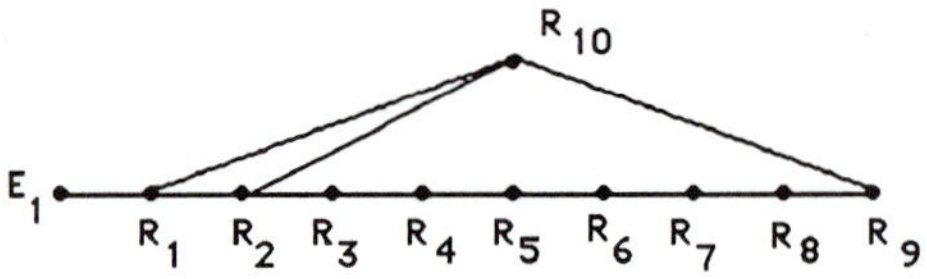

The divisor $E' = R_1+R_2+R_i$ is an indecomposable divisor of canonical type (of type $\tilde{A}_2$) and $E \bullet E' = 1$. Thus $([E_1],[E_2])$ is a non-degenerate 2-sequence.

<u>Case 2</u>: E is of type $\tilde{A}_8$, $E \bullet E_1 = 1$.

In this case $R_1 \bullet R_2 = 1$, $R_0 \bullet R_i = 0$, $i \neq 2$. We will learn from Chapter 5 ,§5 that F admits a nontrivial automorphism of order 3 which leaves E invariant and translates R_1 to another nodal curve R_1' which is disjoint from R_1 (it is a translation by a section of order 3 in the jacobian genus 1 fibration).

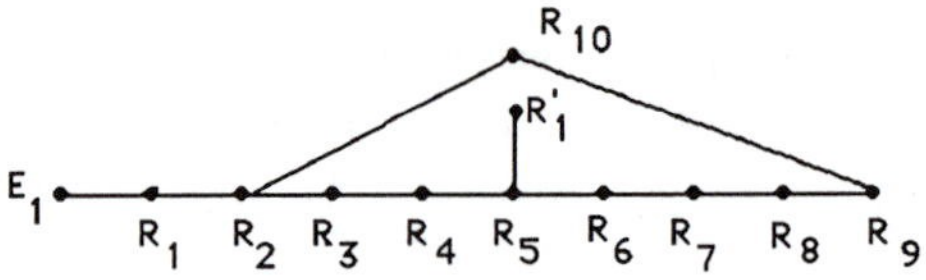

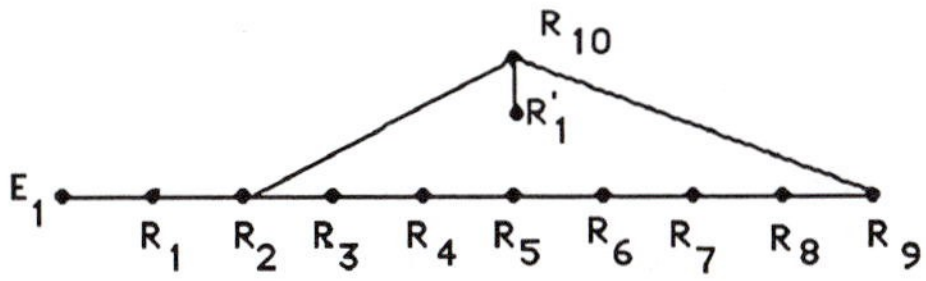

Let R_i be the irreducible component of E which intersects R_1'. Let

$$E' = R_1+R_1'+R_{10}+R_{i+1}+2(R_2+\ldots+R_i) \text{ if } i\leq 8$$

$$E' = R_1+R_1'+R_3+R_8+2(R_2+R_9+R_{10}) \text{ if } i = 9,$$

$$E' = R_1+R_1'+R_3+R_9+2(R_2+R_{10}) \text{ if } i = 10.$$

Then E' is a divisor of canonical type $\tilde{D}_n$ satisfying $E \cdot E' = 1$.

<u>Case 3</u>: E is of type $\tilde{E}_8$, $E \cdot E_1 = 2$.

Again, $R_1 \cdot E = 2$ and the structure of E implies that R_2 is one of the two components of multiplicity 2 of E. It is easy to see that R_{10} must intersect R_4 .

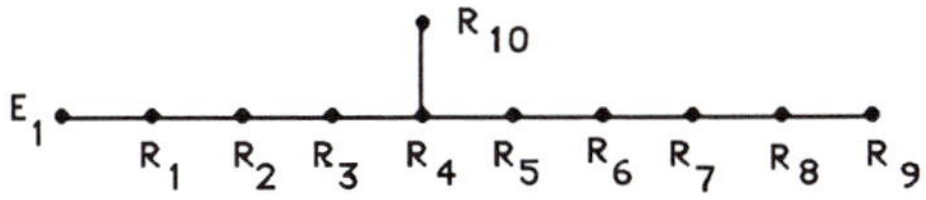

Let

$$E' = 2R_{10}+R_1+2R_2+3R_3+4R_4+3R_5+2R_6+R_7.$$

Then E' is an indecomposable divisor of canonical type $\tilde{E}_7$ and $E \cdot E' = 1$.

<u>Case 4</u>. Case 3: E is of type $\tilde{E}_8$, $E \cdot E_1 = 1$.

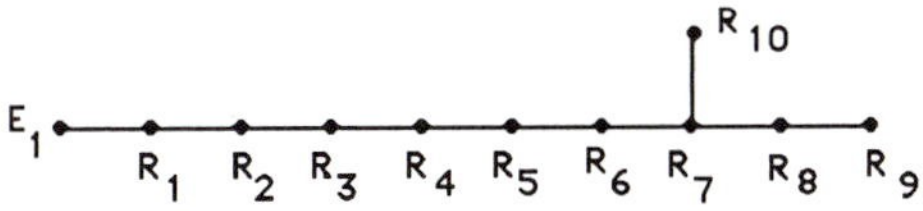

Since $R_1 \cdot E' = 1$, F is of $\tilde{\tilde{E}}_8$-special type.This proves the theorem.

Corollary 3.4.1. Assume that F is not $\tilde{E}_8$-special. Then it admits a non-degenerate or a 2-degenerate $U_{[3]}$-marking.

PROOF. Start from a nondegenerate U-marking and complete it to a maximal canonical isotropic sequence. Then choose a $U_{[3]}$-marking.

§5. $U_{[3]}$-markings

An Enriques surface F is said to be **$\tilde{D}_8$-special** if it contains

(i) a divisor D of canonical type $\tilde{D}_8$,

(ii) a nodal curve R such that $R \bullet D = 1$.

An Enriques surface F is said to be **$\tilde{A}_1+\tilde{E}_7$-special** if it contains

(i) a divisor D_1 of canonical type $\tilde{A}_1$,

(ii) a divisor D_2 of canonical type $\tilde{E}_7$,

(iii) a nodal curve R such that $R \bullet D_2 = 1$, $R \bullet D_1 = 1$ or 2.

An Enriques surface is said to be **extra special** if it is either $\tilde{E}_8$-special, or $\tilde{D}_8$-special or $\tilde{A}_1+\tilde{E}_7$-special.

It is not known to us whether extra special surfaces exist. We will see in Chapter 5 that they do not exist unless char(k) = 2 and F is not a μ_2-surface (Proposition 5.1.8 and Theorem 5.7.5)

Proposition 3.5.1. A $\tilde{D}_8$-special surface does not admit a non-degenerate $U_{[3]}$-marking.

PROOF. Let B be the root base in Num(F) formed by the irreducible components R_i of D and the curve R. Its Dynkin diagram looks like:

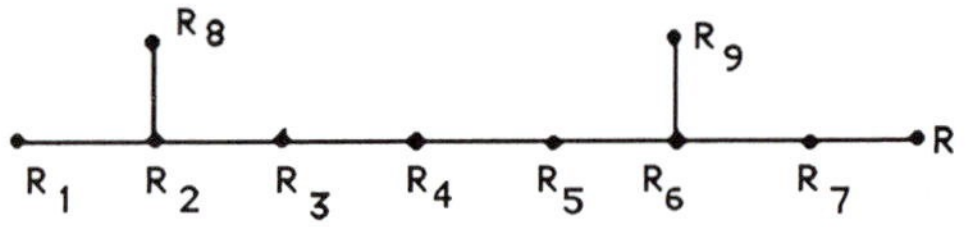

It contains 3 subdiagrams of affine type corresponding to the root bases $\{R_1,\dots,R_9\}$, $\{R,R_1,\dots,R_7,R_9\}$ and $\{R,R_2,\dots,R_9\}$. All of them are of maximal rank. By Proposition 2.4.2, B is a crystallographic base. Also we noted in Chapter 2,§4 that the number of W_B-orbits of primitive isotropic vectors is equal to the number of improper vertices of the fundamental polyhedron which is 3. Thus we have only 3 different genus 1 pencils and it suffices to show that they do not define a $U_{[3]}$-marking. We have the following three divisors of canonical type:

$$D_1 = R_1+R_7+R_8+R_9+2(R_2+\dots+R_6),$$

$$D_2 = R_1+2R_2+3R_3+4R_4+5R_5+6R_6+3R_9+4R_7+2R,$$

$$D_3 = R_8+ \ 2R_2+3R_3+4R_4+5R_5+6R_6+3R_9+4R_7+2R,$$

where

$$D_1\cdot D_2 = D_1\cdot D_3 = D_2\cdot D_3 = 2.$$

Since $D_1\cdot R = 1$, the class $[D_1]$ of D_1 in Num(F) is primitive. The classes $[D_2]$ and $[D_3]$ of the other two divisors can be both non-primitive because $D_2\cdot D_3 = 2$. This shows that the primitive isotropic vectors corresponding to D_1,D_2 and D_3 do not form an isotropic 3-set. Thus F does not admit a non-degenerate $U_{[3]}$-marking.

Proposition 3.5.2. A $\tilde{A}_1+\tilde{E}_7$-special surface does not admit a non-degenerate $U_{[3]}$-marking.

PROOF. Similar to the proof of the previous proposition. Let B be the root basis in Num(F) formed by the curve R and the irreducible components R_i of D_1 and D_2. Its Dynkin diagram looks like:

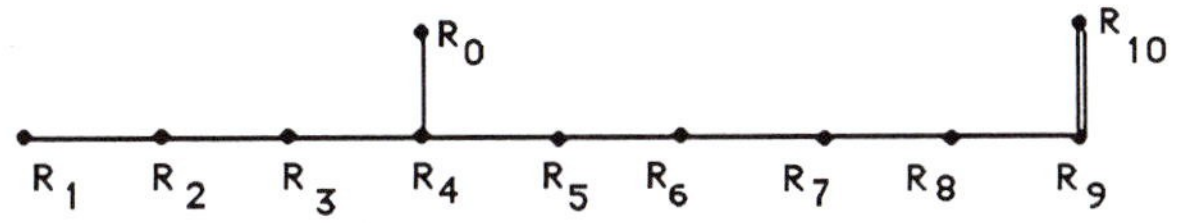

or like

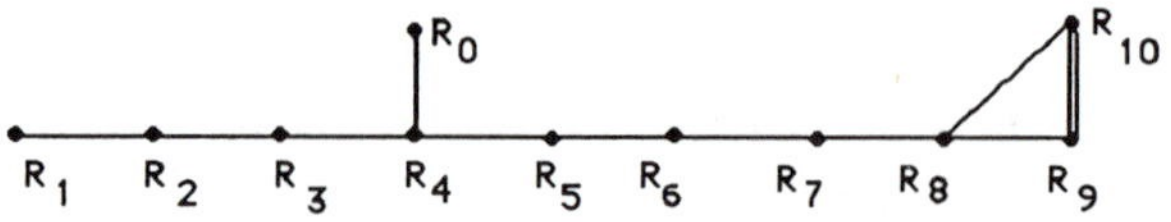

where R_{10} = R. It contains 3 (or 4) subdiagrams of affine type corresponding to the root basis $\{R_0,\dots,R_7\}$ of type $\tilde{E}_7$, $\{R_0,R_2,\dots,R_9\}$ (or $\{R_0,R_2,\dots,R_8,R_{10}\}$) of type $\tilde{E}_8$ and $\{R_9,R_{10}\}$ of type $\tilde{A}_1$. Each of them is contained in a subdiagram of affine type of maximal rank. By Proposition 2.4.2, B is a crystallographic basis. Also we noted in Chapter 2, §4 that the number of W_B-orbits of primitive isotropic vectors is equal to the number of improper vertices of the fundamental polyhedron which is 2 (or 3). Thus we have only 2 (or 3) different genus 1 pencils and it suffices to show that they do not define a $U_{[3]}$-marking. We leave this to the reader.

Theorem 3.5.1. Assume F is not extra special. Then F admits a non-degenerate $U_{[3]}$-marking.

PROOF. By Theorem 3.4.1, F admits a non-degenerate U-marking. Let $|2E_1|,|2E_2|$ be the corresponding pair of genus 1 pencils. By Corollary 3.4.1, we can extend it to a maximal canonical isotropic c-degenerate sequence. If $c \geq 3$, we are done. Assume c = 2. By Lemma 3.4.1, there exist nodal 8 curves $R_1,\dots,R_8$ which together with E_1 and E_2 form the intersection graph:

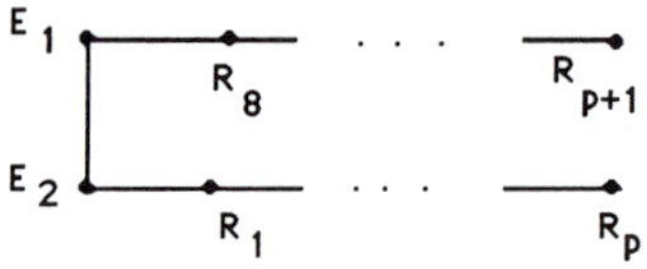

The curves $R_1,\dots,R_{8-\delta}$ (resp. $R_2,\dots R_8$) are irreducible components of a member of $|2E_1|$ (resp. $|2E_2|$), where δ = 0 if p = 8 and 1 otherwise. By Lemma 3.4.1, they can be completed to a divisor of canonical type F_1 (resp. F_2) by adding at most two components. The proof consists of considering all possible cases. We omit the cases, where it is obvious that $F_1 \cdot F_2 > 4$.

Case 1: p = 8.

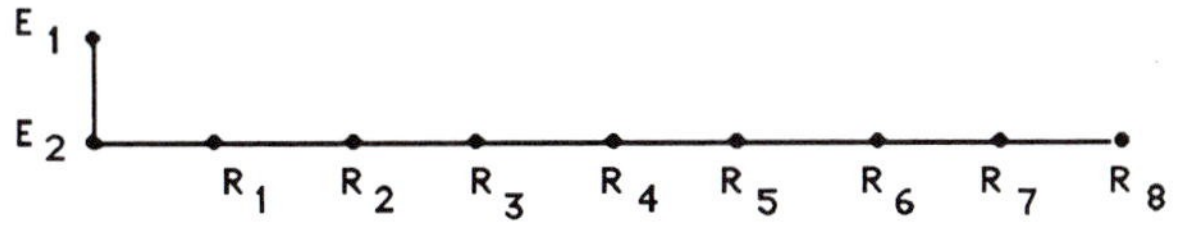

1.1: F_1 is of type $\tilde{A}_8$:

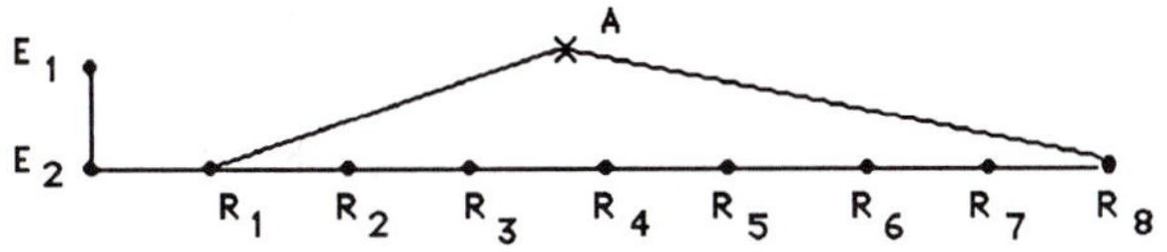

In this case either $F_1 \equiv E_1$, or $F_1 \equiv 2E_1$.

1.1.1: $F_1 \equiv E_1$.

We have $E_2 \cdot F_1 = 1$, hence $E_2 \cdot A = 0$. This implies that the component A belongs to F_2. Thus F_2 is obtained by adding 2 components, A and some B.

1.1.1.1: F_2 is of type $\tilde{A}_8$:

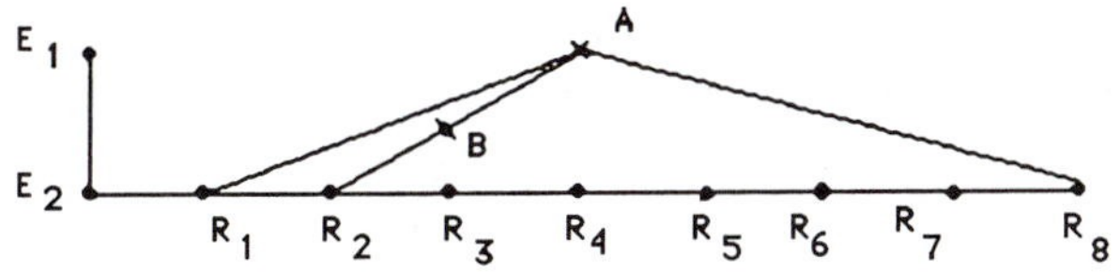

We observe that the curves R_1,R_2,B and A form a divisor D_1 of canonical type of type $\tilde{A}_3$. The curve R_3 intersects this divisor with multipliciy 1. Hence $|2D_1|$ varies in a genus 1 pencil. Clearly $E_2 \cdot D_1 = 1$. Thus the classes of D_1 and E_2 form a non-degenerate pair. We replace E_1 by D_1 to assume that $E_1 = D_1$. Since the curves R_4, R_5, R_6 and R_7 are disjoint from D_1, they are the components of another divisor of canonical type D_2 such that $|D_2|$ or $|2D_2|$ is equal to $|2D_1|$. By Lemma 3.4.1, the number of components in D_2 is at most 6. Thus $|2D_1| = |2E_1|$ does not contain a divisor of canonical type with 8 irreducible components as it should be in Case 1. This reduces this case to one of the next cases.

1.1.1.2: F_2 is of type $\tilde{E}_8$:

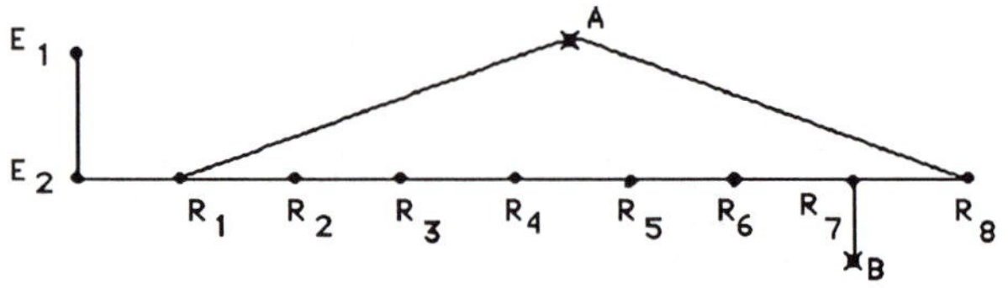

or

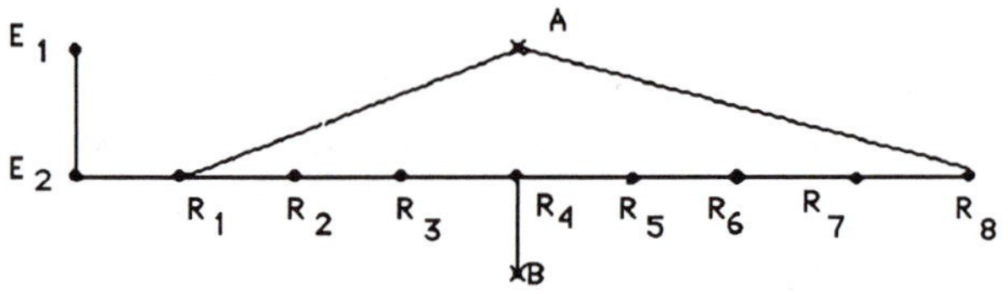

In this case , the curve R_1 intersects the divisor F_2 formed by the curves $R_2, \ldots, R_8$, A, B with multiplicity 3. Since R_1 may intersect only B which is of multiplicity 3, we obtain $F_1 \cdot F_2 = 3$. This is absurd.

1.1.2: $F_1 \equiv 2E_1$.

We know that A is not a component of F_2.

1.1.2.1: F_2 is of type $\tilde{A}_8$:

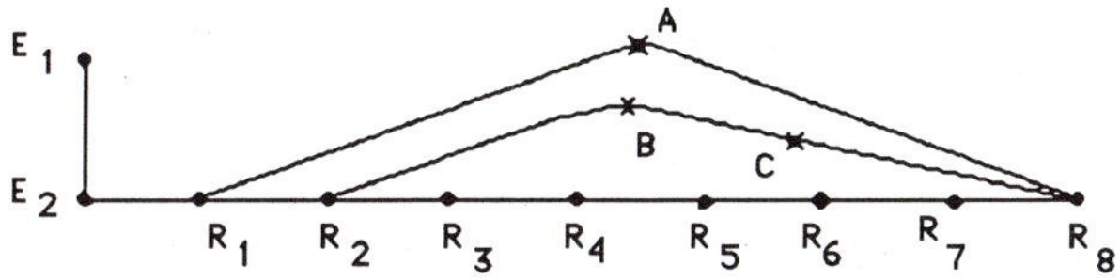

Since $F_1 \equiv 2E_1$, $B \cdot F_1 = C \cdot F_1 = 2$.

If $F_2 \equiv E_2$, $B \cdot R_1 = C \cdot R_1 = 0$, hence $B \cdot A = C \cdot A = 1$. The divisor $D = A+B+C$ is of canonial type and satisfies $D \cdot E_1 = D \cdot E_2 = 1$.

If $F_2 \equiv 2E_2$, $A \cdot F_2 = R_1 \cdot F_2 = 2$. This shows that $R_1 \cdot B = A \cdot C = 1, R_1 \cdot C = A \cdot B = 0$ or $R_1 \cdot B = A \cdot C = 0, R_1 \cdot C = A \cdot B = 1$. Assume the first case occurs. The divisor $D = R_1+R_2+B$ is of canonical type and $D \cdot E_1 = D \cdot E_2 = 1$. In the second case, the divisor $D = R_1+\ldots+R_8+C$ has the same properties.

1.1.2.2: The indecomposable part of F_2 containing $R_2,\ldots,R_8$ is of type $\tilde{A}_7$:

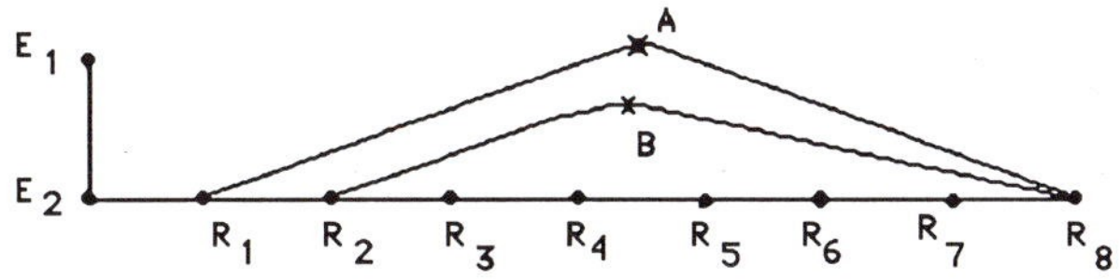

Similarly to 1.1.2.1 and is left to the reader.

1.1.2.3: F_2 is of type $\tilde{D}_8$:

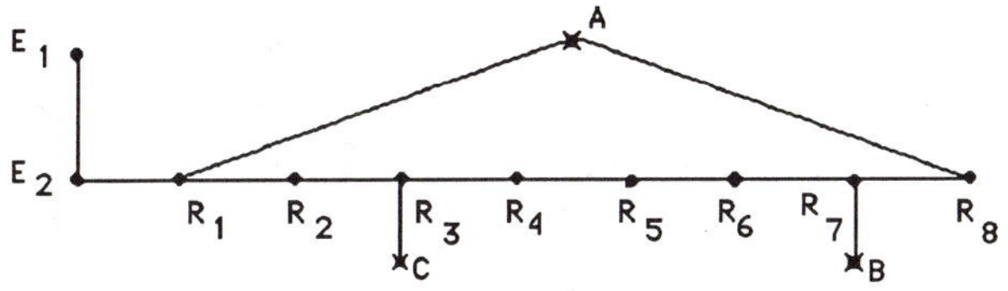

Since F_2 is of type $\tilde{D}_8$, R_1 must intersect C or B (**Here we use the assumption of the theorem**). Assume R_1 intersects C The divisor $D = R_1+R_2+R_3+C$ is of canonical type. Since $F_1 \cdot F_2 = 2$ and $E_2 \cdot F_1 = 2$, $F_2 \equiv E_2$. Thus $E_1 \cdot F_2 = 1$, hence E_1 intersects C with multiplicity 1 and $E_2 \cdot D = 1$. Since $E_2 \cdot D = 1$,

$\{D,E_1,E_2\}$ defines a non-degenerate isotropic 3-sequence. Similarly, we deal with the case $R_1 \bullet B = 1$.

1.1.2.4 :The indecomposable part F_2' of F_2 containing $R_2,\dots,R_8$ is of type $\tilde{E}_7$:

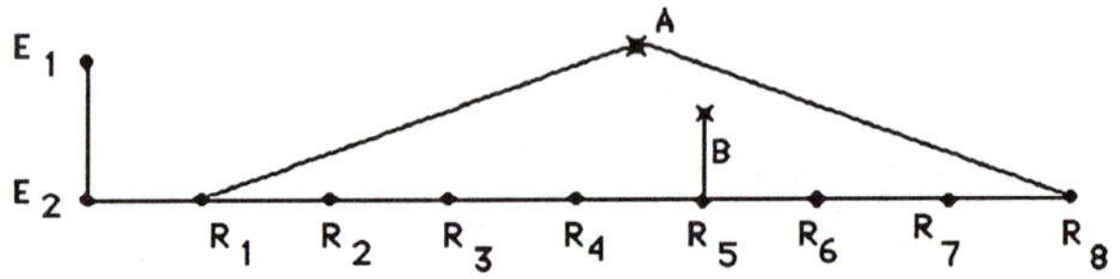

Since B is a component of F_2' of multiplicity 2, $R_1 \bullet B = 0$. This shows that $R_1 \bullet F_2 = 1$, hence $F_2' \equiv E_2$. However, B is not a component of F_1, hence $E_1 \bullet B \neq 0$. But then $E_1 \bullet F_2' = E_1 \bullet E_2 \geq 2$. This case is impossible.

1.2: F_1 is of type $\tilde{E}_8$:

1.2.1:

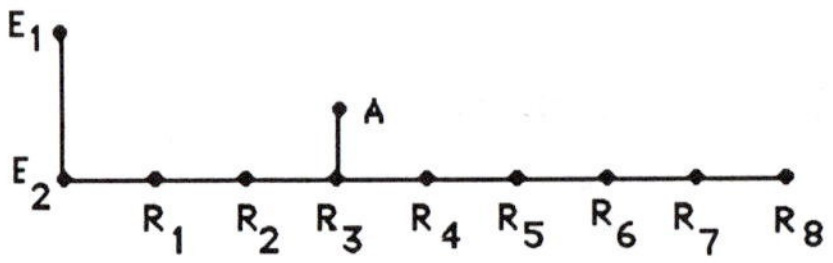

Since R_1 is a component of F_1 of multiplicity 2, $E_2 \bullet A = 0$. Thus A is a component of F_2. There must be another component B of F_2 different from A and $R_2,\dots,R_8$. It is easy to see that B must intersect R_7 which makes F_2 to be of type $\tilde{D}_8$ or intersects A and makes F_2 to be of type $\tilde{E}_8$. In the first case $R_1 \bullet F_2 = 1$, which is excluded by the **assumption of the theorem**.In the second case R_1 enters in F_2 with multiplcity 3. Since $R_1 \bullet E_2 = 1$, this is impossible.

1.2.2:

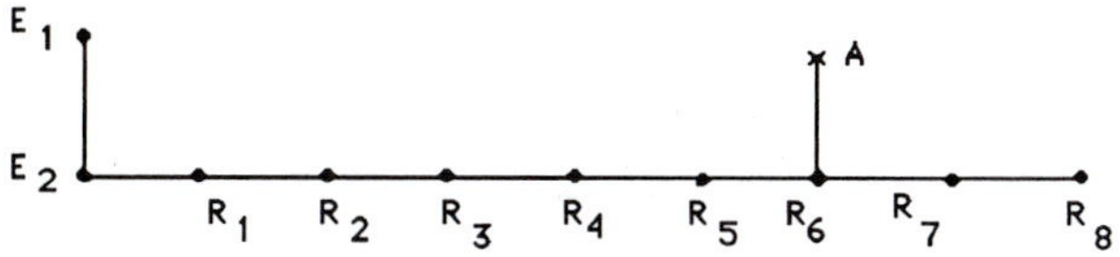

Again, $E_2 \bullet A = 0$, $E_2 \bullet F_1 = 1$, $E_1 \equiv F_1$ and A must be a component of F_2.This easily implies that F_2 is of type $\tilde{E}_8$ and its new component B intersects R_2. However, the divisor $R_1+B+A+R_7+2(R_2+...+R_6)$ is of type $\tilde{D}_8$ and intersects R_8 with multiplicity 1. This contradicts the assumption of the theorem.

Case 2: p = 7.

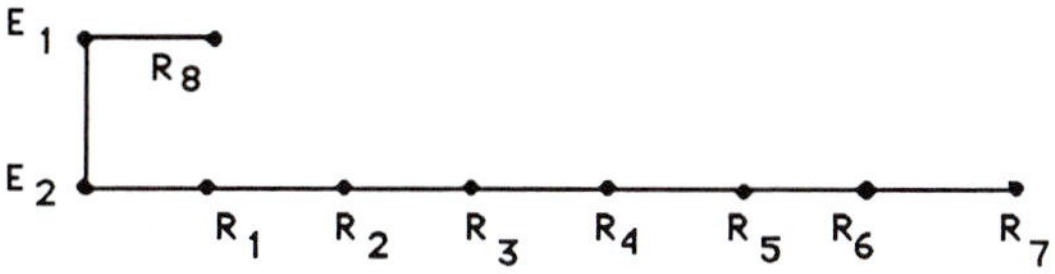

2.1:The indecomposable part F_1' of F_1 containing $R_2,...,R_8$ is of type $\tilde{E}_7$:

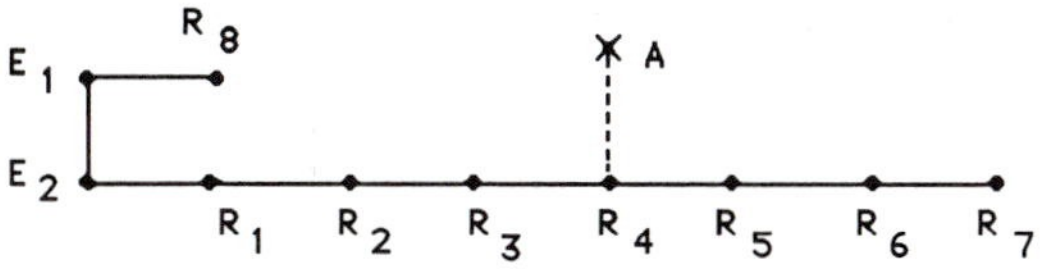

Since $E_1 \bullet R_8 = 1$, $R_8 \bullet F_1' = 1$ or 2. Clearly, R_8 must intersect A. Since A is a component of multiplicity 2 of F_2', $R_8 \bullet A = 1$. Thus $R_8 \bullet F_1' = 2$ and $F_1' \equiv 2E_1$. Also, we obtain $E_2 \bullet F_1' = 2$. However, this is impossible because E_2 may intersect only A which is of mutiplicity 2.

2.2. The indecomposable part F_1' of F_1 containing $R_2, \ldots, R_8$ is of type $\tilde{A}_7$:

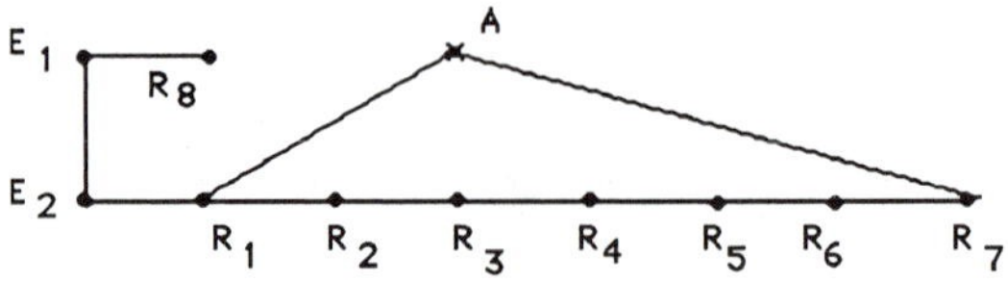

Since $E_1 \cdot R_8 = 1$, $R_8 \cdot F_1' = 1$ or 2. Clearly, R_8 must intersect A and $R_8 \cdot A = 1$ or 2.

2.2.1: $R_8 \cdot A = 2$:

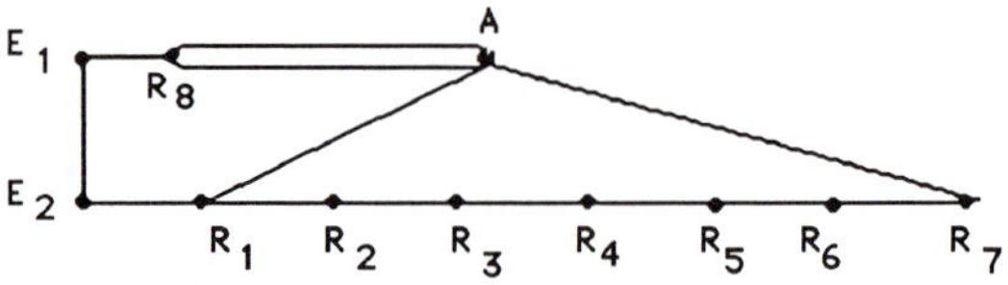

In this case $F_1' \equiv 2E_1$ and $E_2 \cdot A = 1$. Thus the divisor $D = R_8 + A$ is of canonical type such that $D \cdot E_1 = D_2 = 1$.

2.2.2. $R \cdot A = 1$.

In this case $F_1' \equiv E_1$ and A is a component of F_2.

2.2.2.1: F_2 is of type $\tilde{A}_8$ and is obtained by adding one more component B like this:

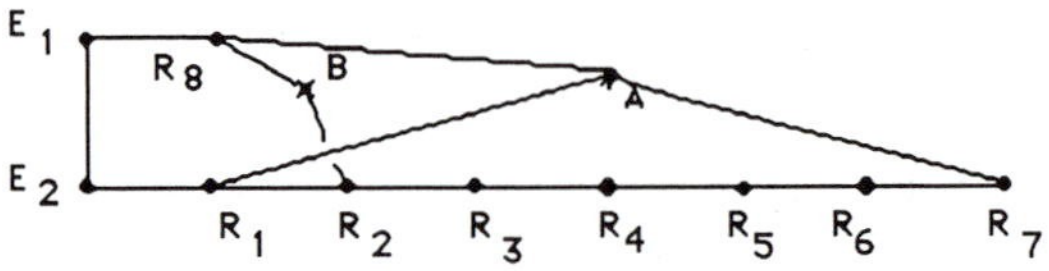

The divisor $D = R_8 + B + R_2 + R_1 + A$ is of canonical type $\tilde{A}_4$ satisfying $D \cdot E_2 = 1$. Replacing E_1 by D we reduce this case to one of the later cases with $p<7$ (compare with 1.1.1.1).

2.2.2.2. F_2 is of type $\tilde{E}_8$:

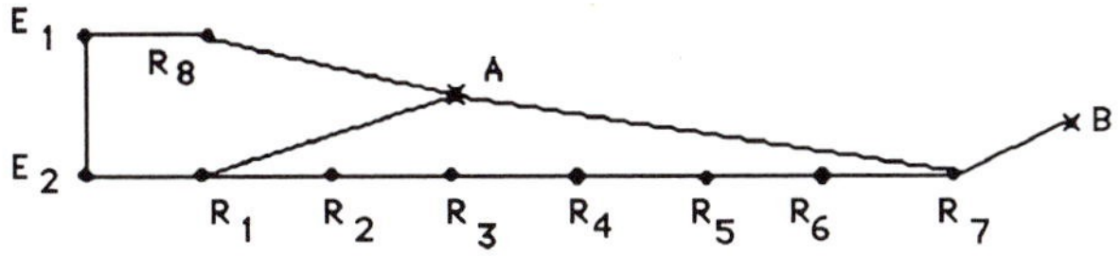

In this case $F_1' \cdot F_2 \geq 5$.

2.3: F_1 is of type $\tilde{A}_8$:

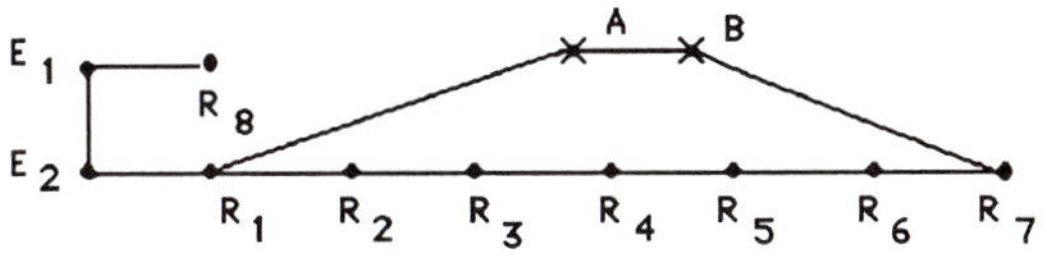

Again, $R_8 \cdot (A+B) = 1$ or 2.

2.3.1. $R_8 \cdot A = 1$, $R_8 \cdot B = 0$ or $R_8 \cdot A = 0$, $R_8 \cdot B = 1$.

In this case, $F_1 \equiv E_1$, A and B are components of F_2. We have already 9 components in F_2: $R_2, \ldots, R_8, A, B$, which do not form a divisor of canonical type. Thus F_2 consists of at least 10 components which contradicts Lemma 3.4.1.

2.3.2. $R_8 \cdot B = 2$, $R_8 \cdot A = 0$

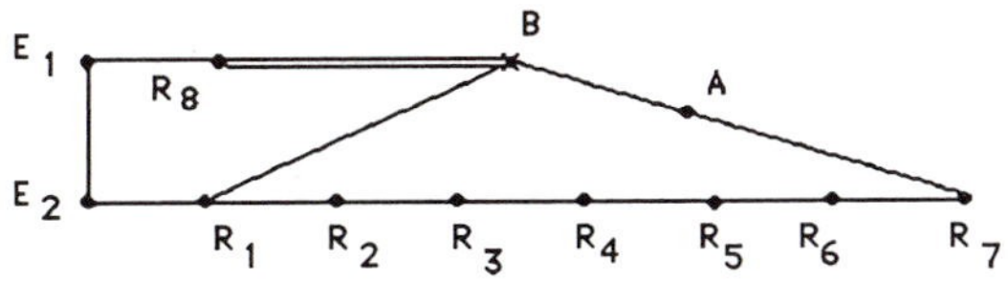

In this case, $F_1 \equiv 2E_1$ and E_2 intersects either A or B with multiplicity 1.

Assume $E_2 \cdot B = 1, E_2 \cdot A = 0$. Then $D = R_8 + A$ is of canonical type and

$D \cdot E_1 = D \cdot E_2 = 1$.

Assume $E_2 \cdot B = 0$, $E_2 \cdot A = 1$. Then $B+R_8$ is an indecomposable part of F_2. Let F_2' be the indecomposable part of F_2 containing the components $R_2, \ldots, R_7$.

2.3.2.1. F_2' is of type $\tilde{A}_6$:

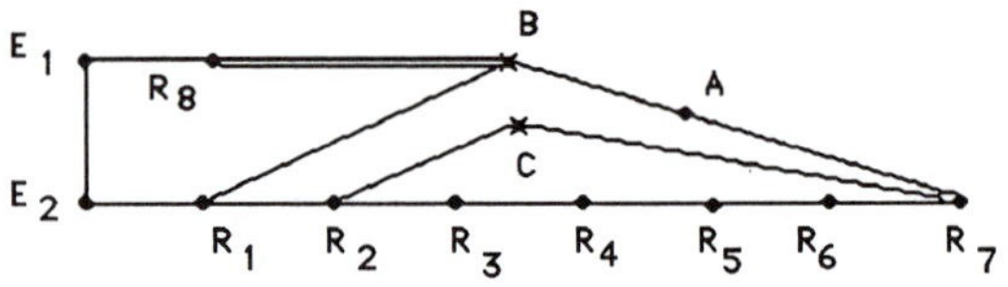

If $B+R_8 \equiv F_2'$, we have $C \cdot R_1 = C \cdot A = 0$. The divisor

$$D = R_1+R_2+C+R_7+A+B$$

is of canonical type $\tilde{A}_5$ and $D \cdot E_1 = 1$. We replace E_2 by D and use the argument from 1.1.1.1 to reduce this case to one of the later cases with $p < 7$.

If $F_2' = 2(B+R_8)$, $C \cdot R_1 = C \cdot A = 1$.

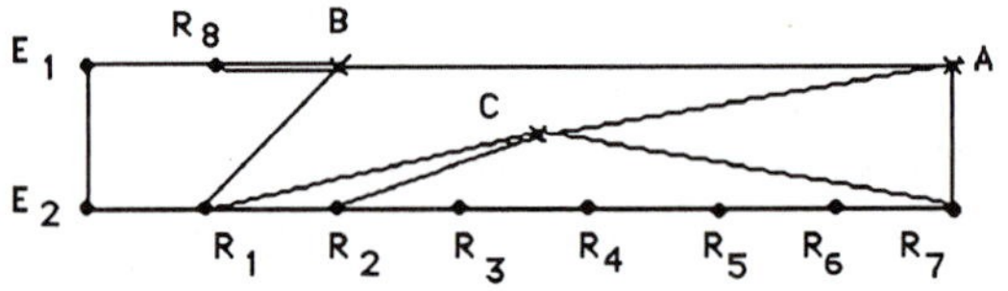

Let $D = R_1+R_2+C$. This is a divisor of canonical type $\tilde{A}_2$ with $D \cdot E_2 = 1$. Replacing E_1 by D, we see that $p < 7$ (cf. 1.1.1.1).

2.3.2.2. F_2' is of type $\tilde{A}_7$:

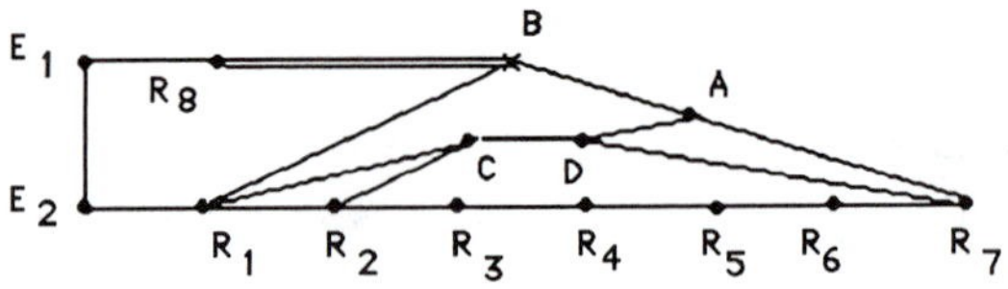

Since $F_1 \equiv 2E_1$, the new components C and D must intersect F_1 with multiplicity 2. We assume that $R_1 \cdot C = A \cdot D = 1$. The other case $R_1 \cdot D = A \cdot C = 1$ is considered similarly.

Let $D_1 = R_1+R_2+C$, $D_2 = D+A+R_7$. Then $D_1 \cdot D_2 = E_2 \cdot D_1 = E_2 \cdot D_2 = 1$. Since D_1 and D_2 are of canonical type, we are done.

2.3.2.3: F_2' is of type $\tilde{E}_7$:

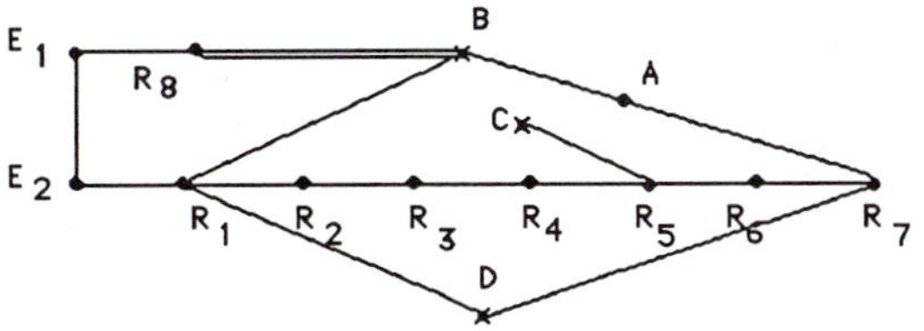

Since $F_1' \equiv 2E_1$, C and D must intersect R_1 or A. However, C enters into F_2' with multiplicity 2, hence $F_2' \cdot F_1' \geq 6$.

2.3.2.4: F_2' is of type $\tilde{D}_7$:

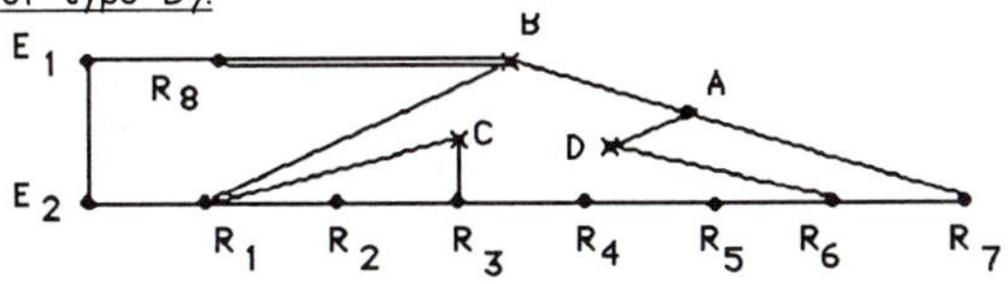

By the same reason as above , $C \cdot R_1 = D \cdot A = 1$ or $C \cdot A = D \cdot R_1 = 1$. We assume that the first case occurs. The other case is considered similarly. Since $E_1 \cdot F_2' = 2$ and neither C nor D is a component of F_1', $E_1 \cdot C = E_1 \cdot D = 1$. Let $D = R_1+R_2+R_3+C$. Then C is of canonical type and $E_1 \cdot D = E_2 \cdot D = 1$.

2.4: F_1 is of type $\tilde{E}_8$:

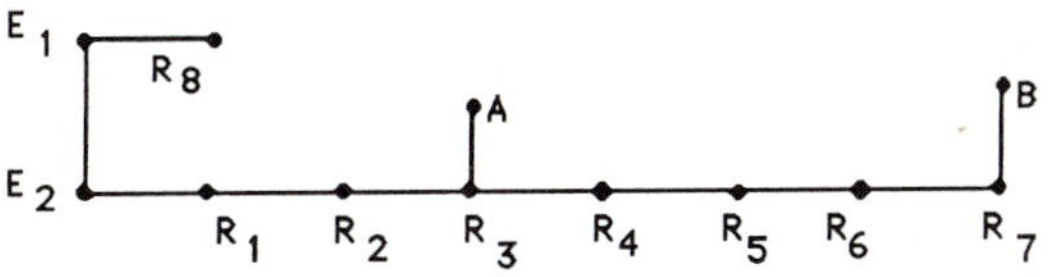

or

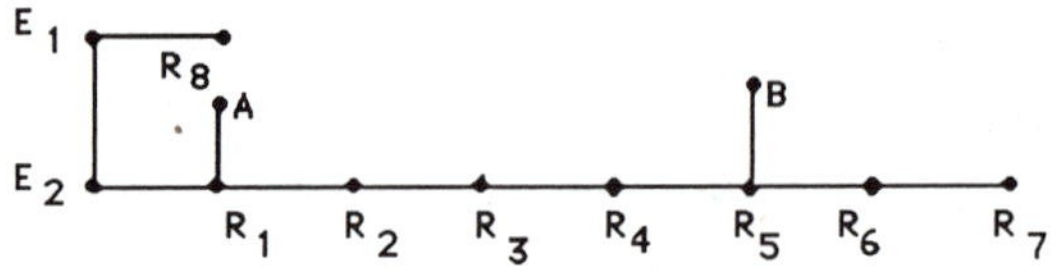

Since $E_2 \cdot F_1 = 2$, $F_1 \equiv 2E_1$, A and B are components of F_2. Since A (resp. B) is a component of F_1 of multipicity 3, R_8 must intersect B (resp. A) with multiplicity 2. This impies that F_2 is of type $\tilde{A}_1+\tilde{E}_7$:

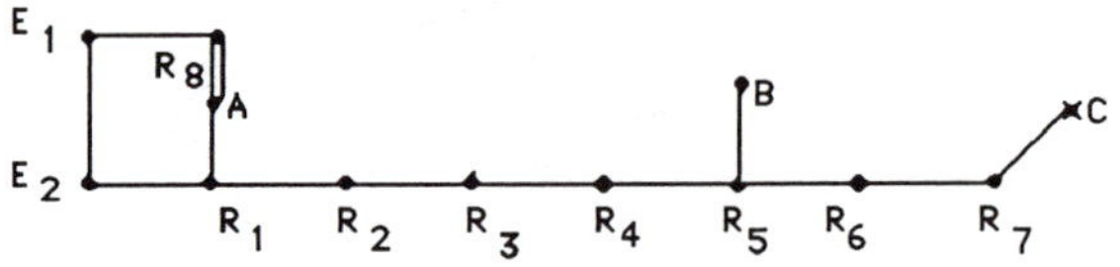

or

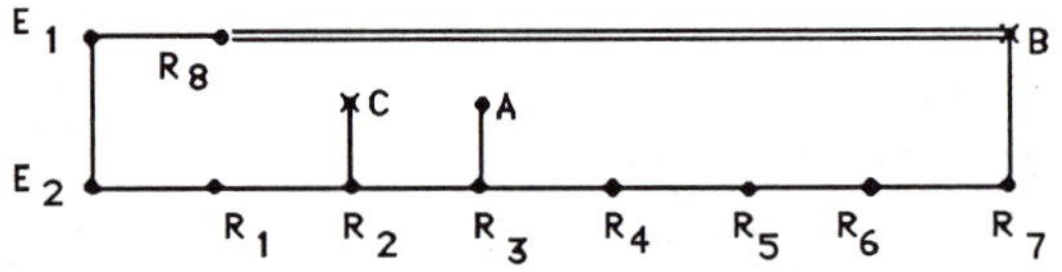

The divisor $D = R_8+A$ (resp. $D = R_8+B$) is of canonical type $\tilde{A}_1$ and has the needed property $D \cdot E_1 = D \cdot E_2 = 1$.

<u>2.5: F_1 is of type $\tilde{D}_8$:</u>

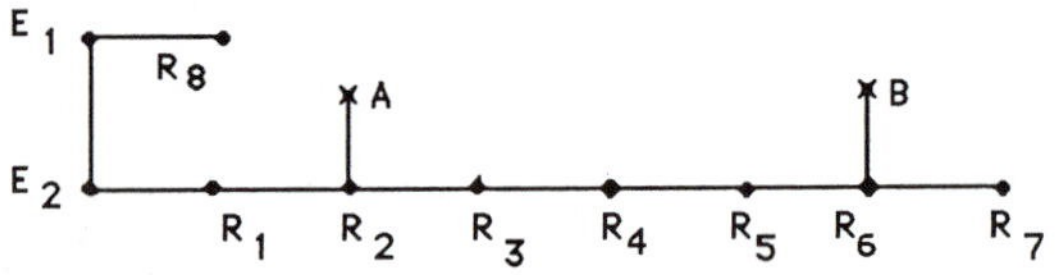

Again , as above, $R_8 \cdot (A+B) = 1$ or 2.

2.5.1: $R_8 \cdot B = 1$, $R_8 \cdot A = 0$.

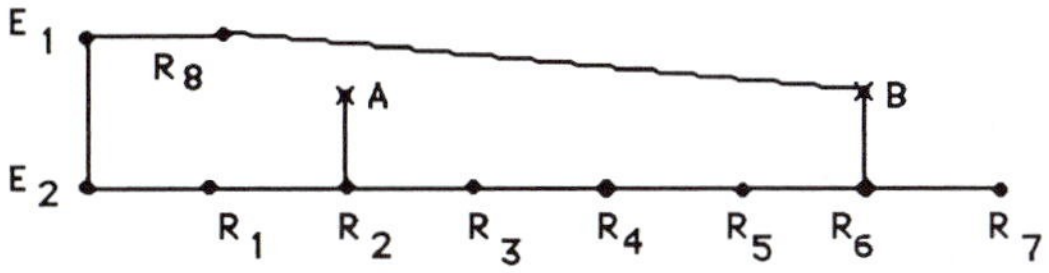

This case contradicts the assumption of the theorem (F is not $\tilde{D}_8$-special).

2.5.2: $R_8 \cdot B = 0$, $R_8 \cdot A = 1$

Similar to 2.5.1.

2.5.3: $R_8 \cdot A = R_8 \cdot B = 1$.

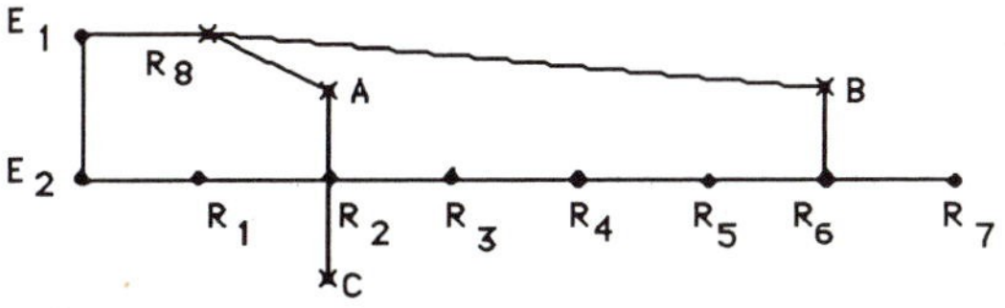

In this case $F_1 \equiv 2E_1$, A or B is a component of F_2 which is of type $\tilde{E}_8$. If A is a component of F_2, then $D = A+B+R_8+(R_2+...+R_6)$ is a divisor of canonical type such that $D \cdot E_1 = D \cdot E_2 = 1$.

2.5.4. $R_8 \cdot A = 2$, $R_8 \cdot B = 0$.

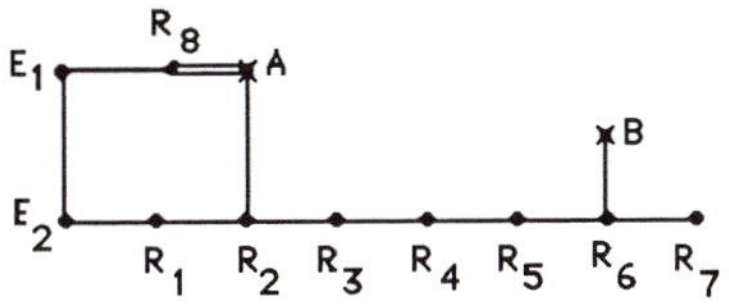

In this case, B is a component of F_2, $E_2 \bullet A = 1$ and $D = A + R_8$ is a canonical divisor with the property $D \bullet E_1 = D \bullet E_2 = 1$.

2.5.5. $R_8 \bullet A = 0$, $R_8 \bullet B = 2$.

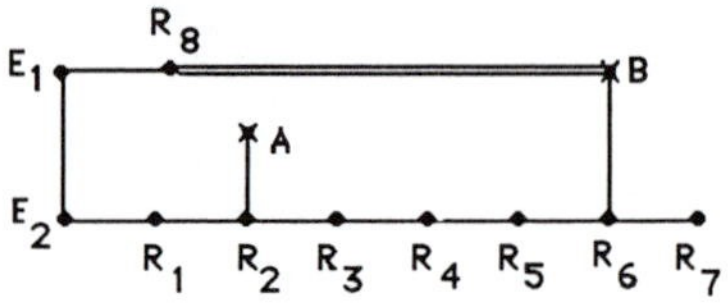

Similar to 2.5.4.

Case 3: p = 6:

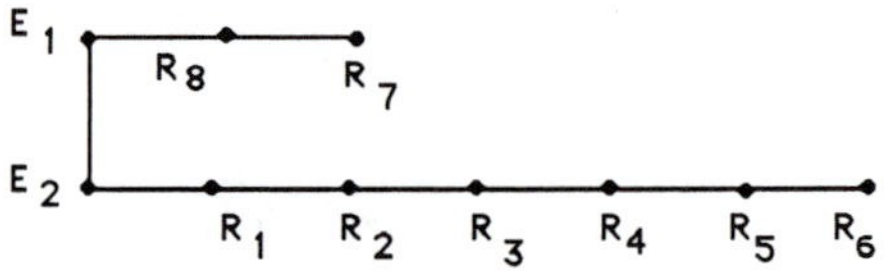

3.1. The indecomposable component F_1' of F_1 containing $R_1,\ldots,R_6$ is of type $\tilde{A}_6$:

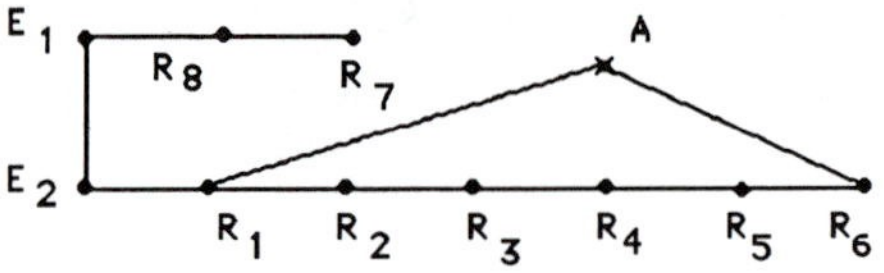

3.1.1. $R_8 \bullet A = 1$.

In this case $F_1' \equiv E_1$ and A must be a component of F_2. The curves $R_2,\dots,R_8,A$ are components of F_2. The only possibility for F_2 is to be of type $\tilde{A}_8$:

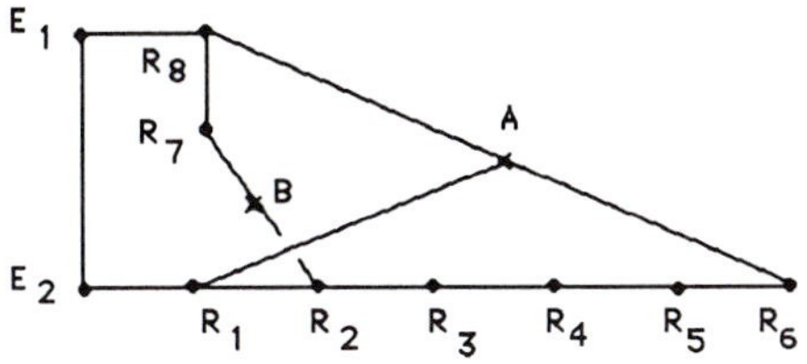

The divisor

$$D = R_7+R_8+A+B+R_1+R_2$$

is of canonical type $\tilde{A}_5$ satisfying $D \bullet E_1 = D \bullet E_2 = 1$.

3.1.2. $R_8 \bullet A = 2$.

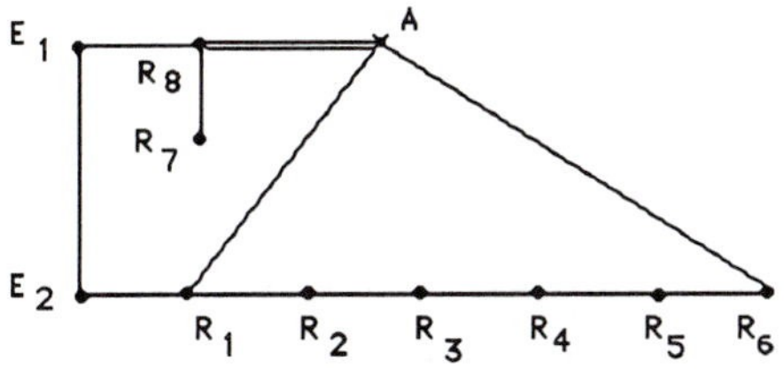

Since A is not a component of F_2, $E_2 \bullet A = 1$.The divisor $D = R_8+A$ is of canonical type with the property $D \bullet E_1 = D \bullet E_2 = 1$.

3.2. The indecomposable component F_1' of F_1 containing $R_1,\dots,R_6$ is of type $\tilde{A}_7$:

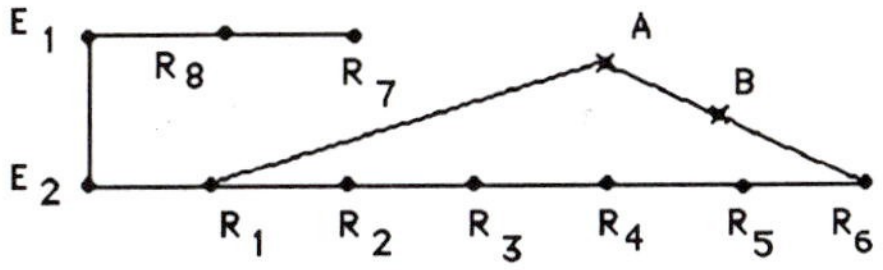

As before, $R_8 \cdot (A+B) = 1$ or 2.

3.2.1. $R_8 \cdot A = 1$, $R_8 \cdot B = 0$.

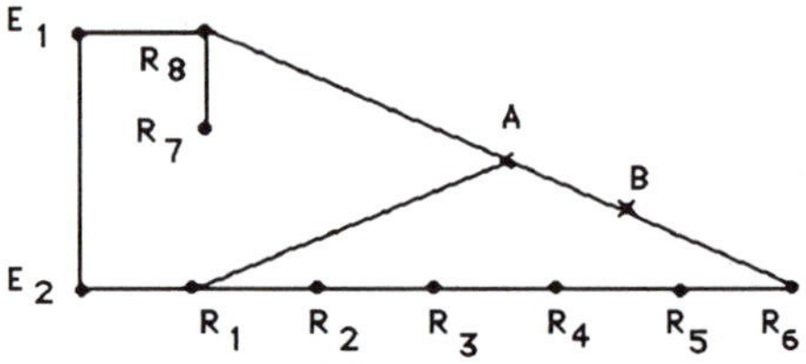

In this case $R_2, \ldots, R_8$, A, B are components of F_2. To get F_2 we must add at least one more nodal curve. However, this would contradict Lemma 3.4.1.

3.2.2. $R_8 \cdot A = 0$, $R_8 \cdot B = 1$.

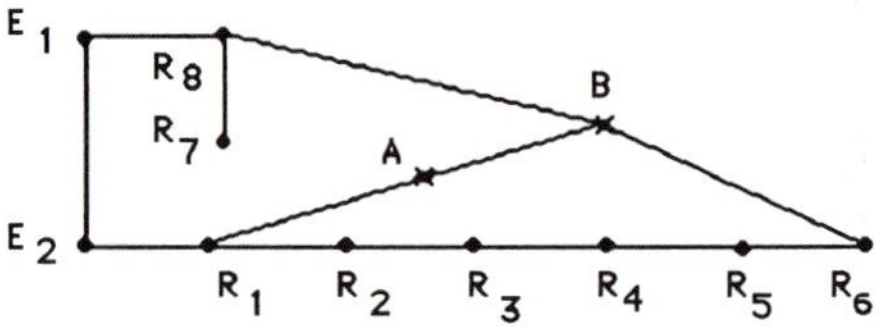

Together with A and B the curves $R_2, \ldots, R_8$ form F_2 which is of type $\tilde{E}_8$. Since $R_8 \cdot F_1' = 1$, $F_1' \equiv E_1$. However, $F_1' \cdot F_2 = 4$. This case is impossible.

3.2.3. $R_8 \cdot A = R_8 \cdot B = 1$.

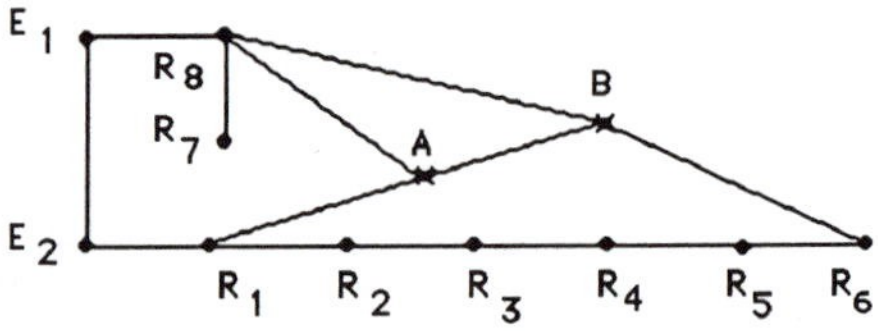

In this case $F_1' \equiv 2E_1$ and A or B must be a component of F_2. But then,

3.2.4. $R_8 \cdot A = 2$, $R_8 \cdot B = 0$.

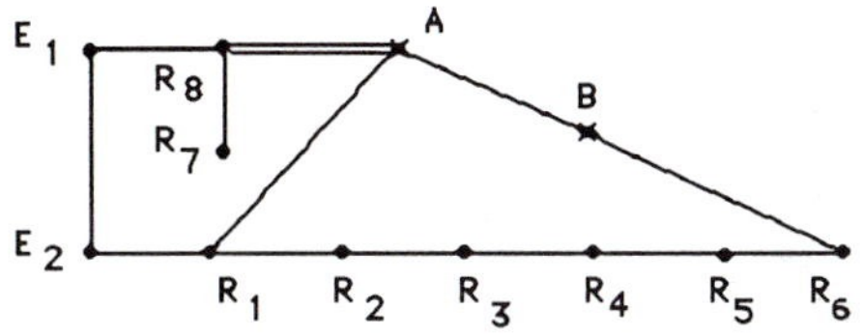

Since R_7, R_8 and A cannot all be components of F_2, A is a component of F_1 and $E_2 \cdot A = 1$. The divisor $D = R_8 + A$ is of canonical type with $D \cdot E_1 = D \cdot E_2 = 1$.

3.2.5. $R_8 \cdot B = 2$, $R_8 \cdot A = 0$.

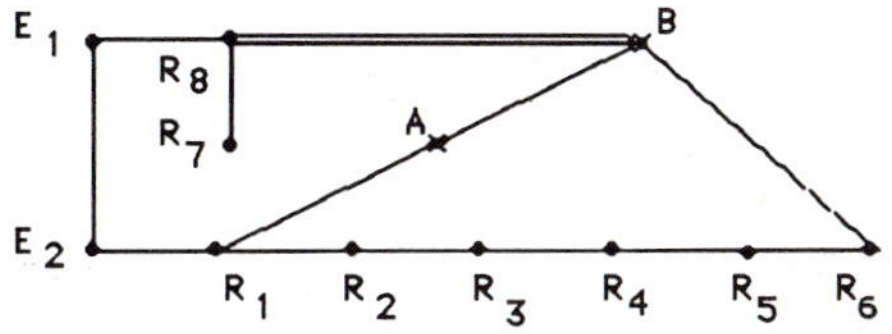

Similar to 3.2.4.

3.3. The indecomposable part F_1' of F_1 containing $R_1, \ldots, R_6$ is of type $\tilde{D}_7$:

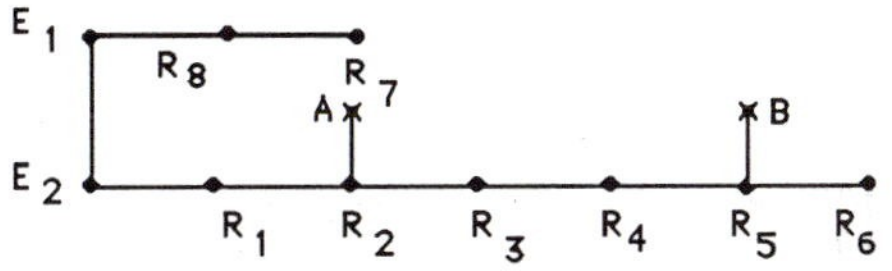

3.3.1. $R_8 \cdot A = 1$, $R_8 \cdot B = 0$:

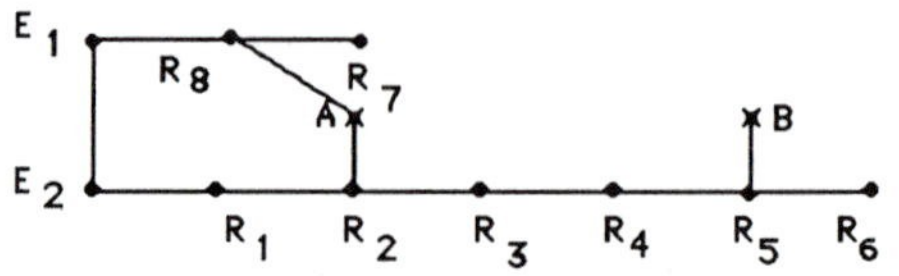

In this case the curves A and B must be components of F_2. The following picture shows the only way of completing $A, B, R_2, \ldots, R_6$ to a divisor of canonical type:

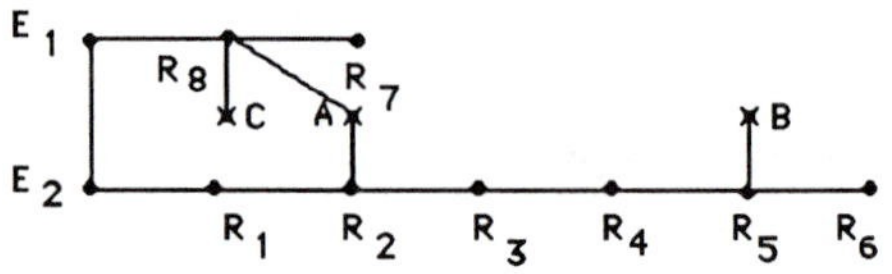

However this makes F_2 to be of type $\tilde{D}_9$ which contradicts Lemma 3.4.1.

3.3.2. $R_8 \cdot A = 0$, $R_8 \cdot B = 1$.

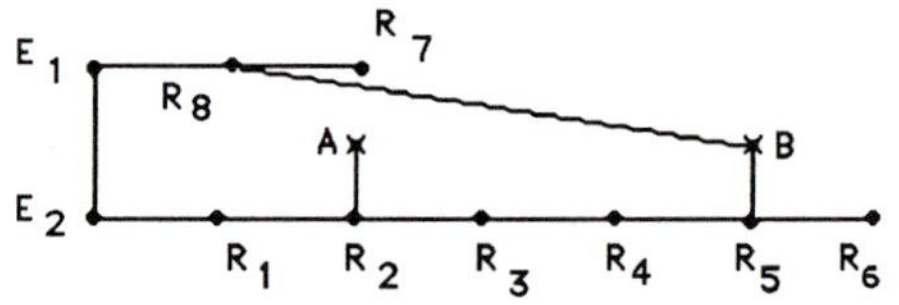

In this case the curves $A, R_2, \ldots, R_7$ cannot be completed to a divisor of canonical type F_2 with at most 9 irreducible compoments (Lemma 3.4.1).

3.3.3. $R_8 \cdot A = 2$, $R_8 \cdot B = 0$.

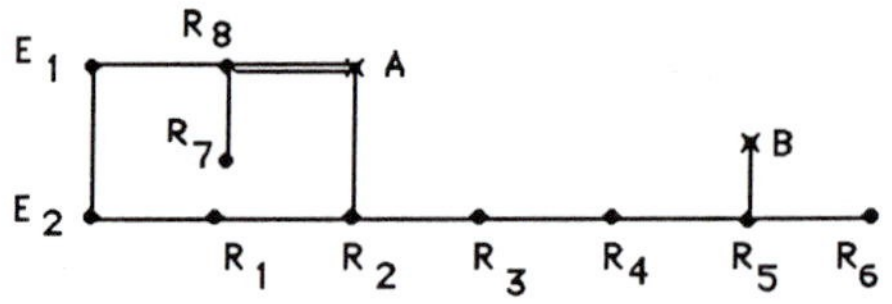

Clearly A is not a component of F_2. Hence $E_2 \bullet A = 1$ and $D = R_8+A$ is a divisor of canonical type satisfying $D \bullet E_1 = D \bullet E_2 = 1$.

3.3.4. $R_8 \bullet A = 0$, $R_8 \bullet B = 2$.

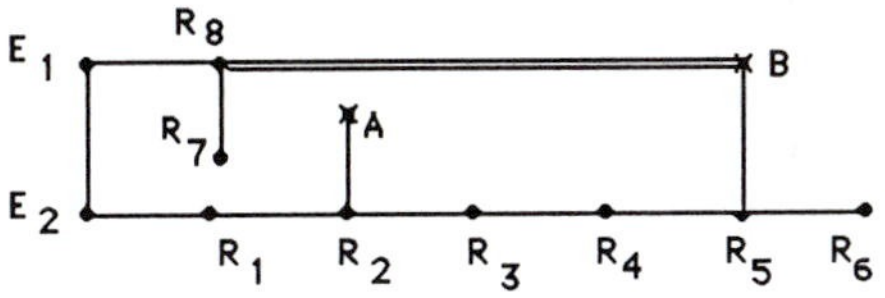

Similar to 3.3.3.

3.3.5. $R_8 \bullet A = R_8 \bullet B = 1$.

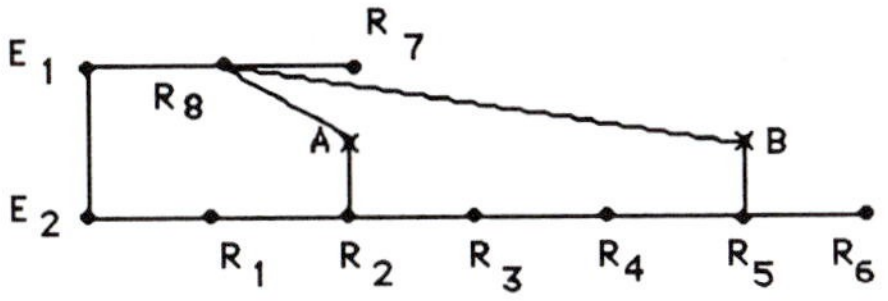

E_1 intersects only one of the curves A and B. Hence $D = R_8+A+B+R_2+\ldots+R_5$ is a divisor of canonical type satisfying $D \bullet E_1 = D \bullet E_2 = 1$.

3.4. The indecomposable part F_1' of F_1 containing $R_1,\ldots,R_6$ is of type $\tilde{E}_7$:

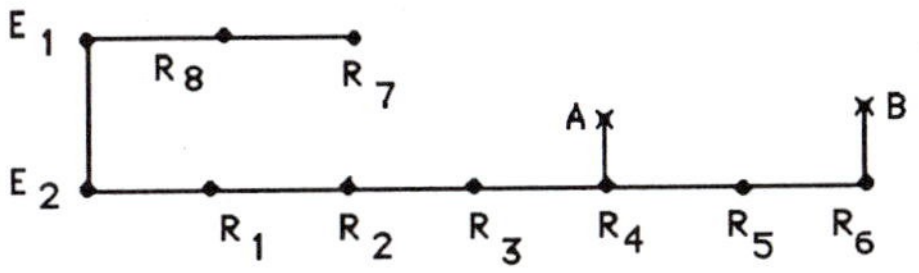

or

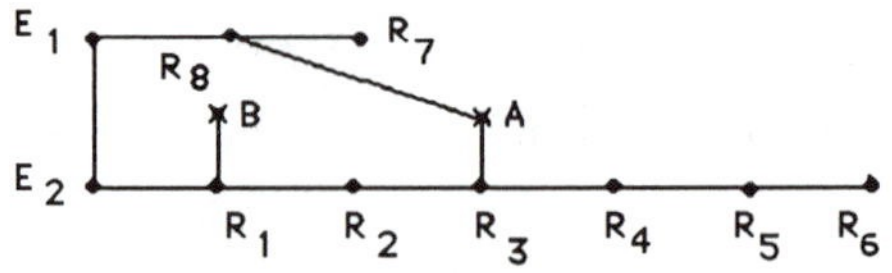

We will consider only the first possibility. In the second one E_2 does not intersect the curves $A,B,R_1,\ldots,R_8$. Hence they belong to a fibre with at least 10 components which contradicts Lemma 3.4.1.

<u>3.4.1. $R_8 \cdot (A+B) = 1.$</u>

Here A and B are components of F_2. It is easy to see that one can extend the curves, $A,B,R_2,\ldots,R_8$ to a divisor of canonical type only if $R_8 \cdot B = 1$, $R_8 \cdot A = 0$.

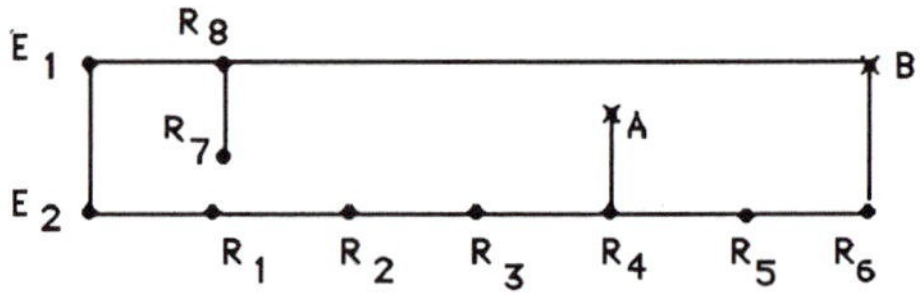

Since R_7 is a component of F_1 disjoint from F_1', it can be completed to a divisor of canonical type which is necessary of type $\tilde{A}_1$.

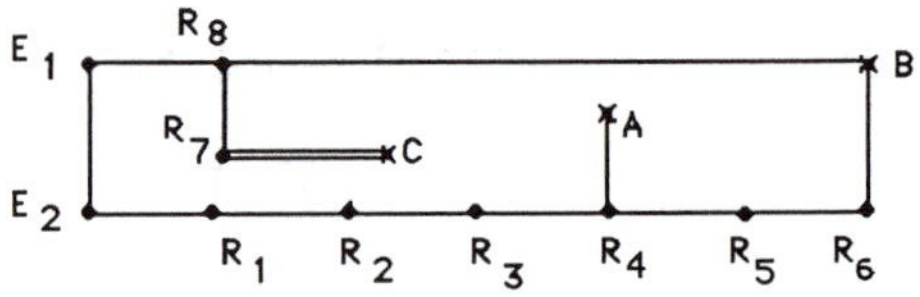

The curve R_8 being a component of F_2 of multiplicity 2 intersects R_7+C with multiplicity 1 or 2. Thus $R_8 \cdot C = 0$ or 1 and the configuration of the curves $C,A,B,R_1,\ldots,R_8$ contradicts the assumption of the theorem.

3.4.2. $R_8 \bullet A = R_8 \bullet B = 1$.

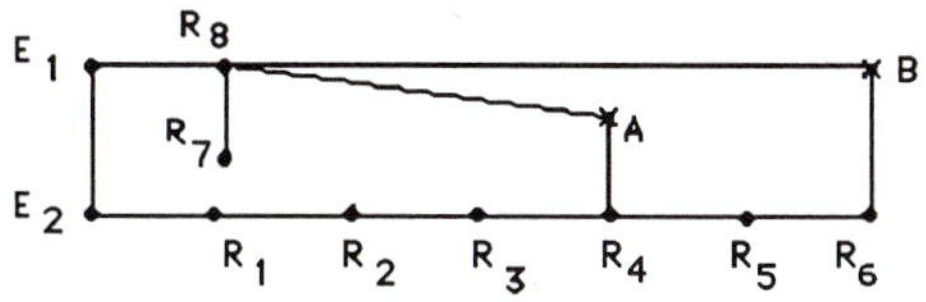

Here E_2 intersects only one of the curves A and B. Thus $D = R_8+A+B+R_4+R_5+R_6$ is a divisor of canonical type satisfying $D \bullet E_1 = D \bullet E_2 = 1$.

3.4.3. $R_8 \bullet A = 2$, $R_8 \bullet B = 0$.

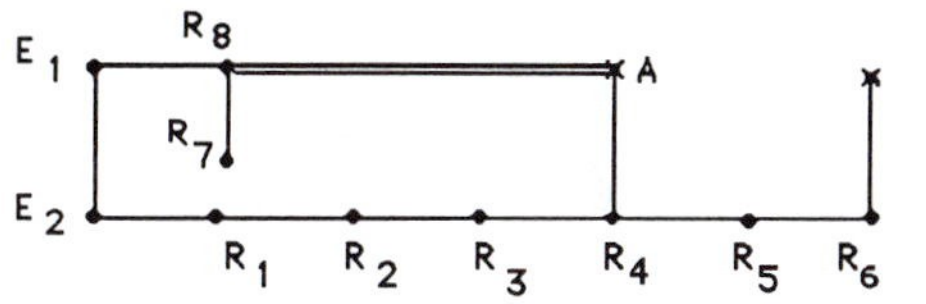

It is easy to see that A is not a component of F_2. Hence $E_2 \bullet A = 1$ and $D = R_8+A$ is a divisor of canonical type satisfying $D \bullet E_1 = D \bullet E_2 = 1$.

3.4.4. $R_8 \bullet A = 0$, $R_8 \bullet B = 2$.

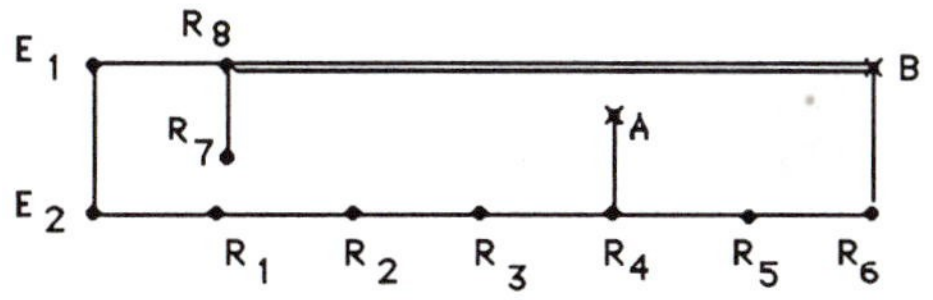

Similar to 3.4.3.

3.5. F_1 is of type $\tilde{A}_8$:

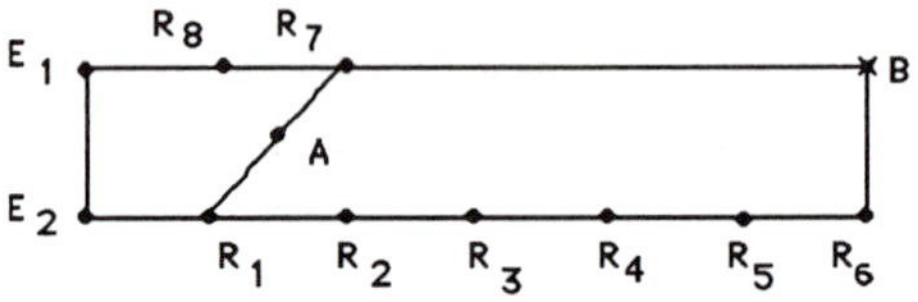

We know that $R_8 \cdot F_1 = 1$ or 2. If $R_8 \cdot F_1 = 1$, the curves A and B are components of F_2. However, F_2 must contain also the curves $R_2, \ldots, R_8$ and can be obtained only by adding at least one more component. The latter contradicts Lemma 3.4.1. This shows that R_8 must intersect A or B with multiplicity 2.

3.5.1. $R_8 \cdot A = 1$, $R_8 \cdot B = 0$.

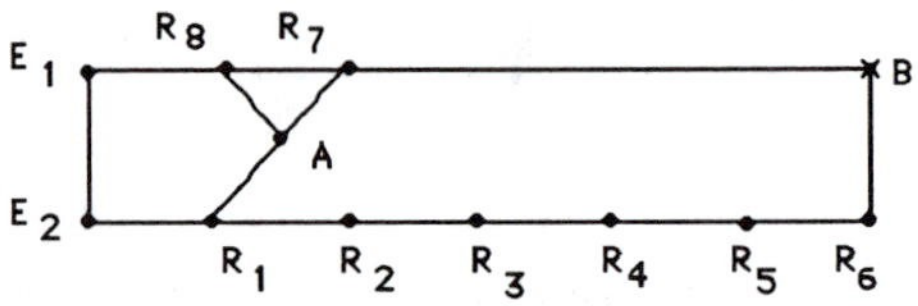

Replacing R_6 by B we obtain a configuration:

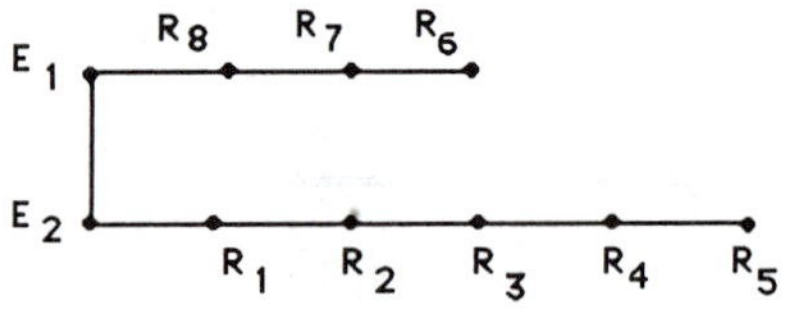

This reduces this case to the case p = 5 from below.

3.5.2. $R_8 \cdot A = 0$, $R_8 \cdot B = 1$.

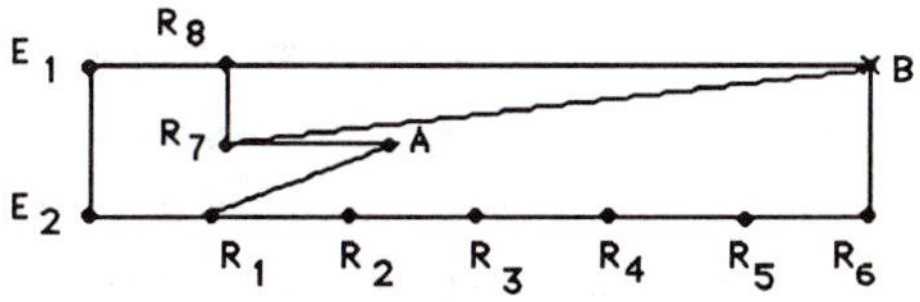

Obviously B cannot be a component of F_2. Thus A is a component of F_2 and E_2 intersects B with multiplicity 1. The divisor $D = R_8+R_7+B$ is of canonical type and satisfies $D \cdot E_1 = D \cdot E_2 = 1$.

3.6. F_1 is of type $\tilde{E}_8$.

3.6.1

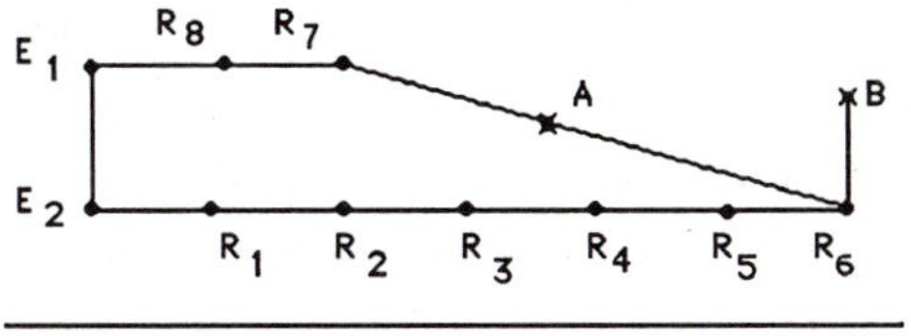

3.6.2.

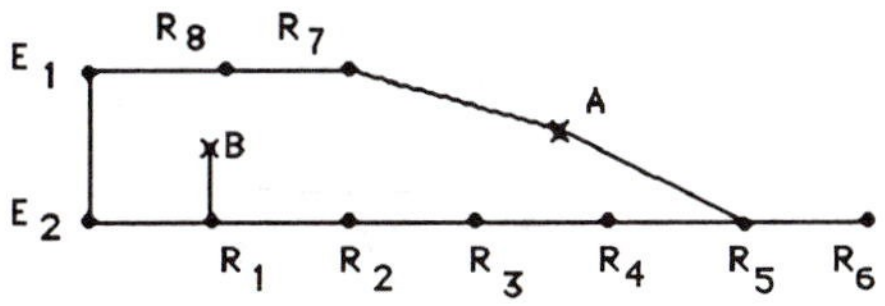

3.6.3.

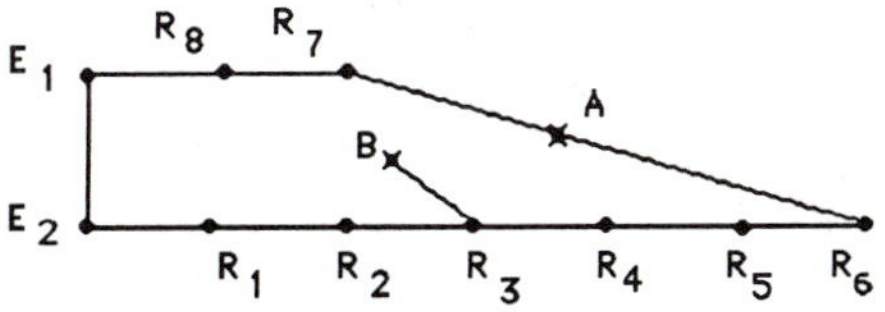

In case 3.6.1, $R \cdot F_2 = 2$ which shows that $F_2 \equiv 2F_1$. However, $E_2 \cdot F_1 = 1$ or ≥ 3. In case 3.6.2, $E_2 \cdot A = E_2 \cdot B = 0$ (otherwise $E_2 \cdot F_1 \geq 3$). This shows that A and B are components of F_2. Since F_2 has already 9 components $A, B, R_2, \ldots, R_8$ which do not form a divisor of canonical type, we have a contradiction because of Lemma 3.4.1. In case 3.6.3, as above A and B are components of F_2. Obviously, the curves $A, R_2, \ldots, R_8$ form an indecomposable part F_2' of F_2 which is type $\tilde{E}_7$. Since B is disjoint from F_2', it belongs to another part which is necessary of type $\tilde{A}_1$.

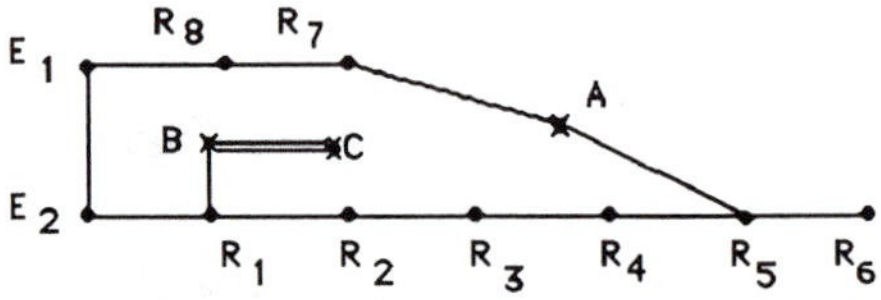

We have $R_1 \cdot C = 1$ or 0 dependent on whether $B+C \equiv 2F_2'$ or $B+C \equiv F_2'$. In any case, F is $\tilde{A}_1+\tilde{E}_7$-special, which is excluded by the assumption of the theorem.

Case 4: p = 5.

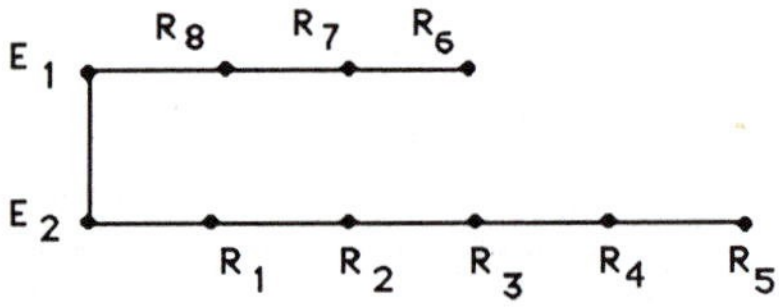

Let F_1' be the indecomposable part of F_1 containing $R_1, \ldots, R_5$. If it contains R_7 or R_6, then it consists of 8 or 9 components and cannot be of type $\tilde{A}_7$. If it does not contain R_7 and R_6, then it contains at most 7 components. Thus F_1' is of type $\tilde{A}_n (n = 5,6,8)$, $\tilde{D}_n (n = 6,7,8)$, or $\tilde{E}_n$ $(n=6,7,8)$.

<u>4.1. F_1' is of type $\tilde{A}_5$.</u>

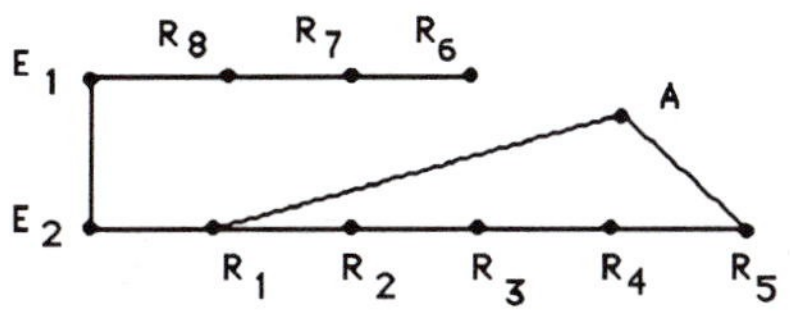

This is similar to 3.1 and is left to the reader.

4.2. F_1' is of type $\tilde{A}_6$.

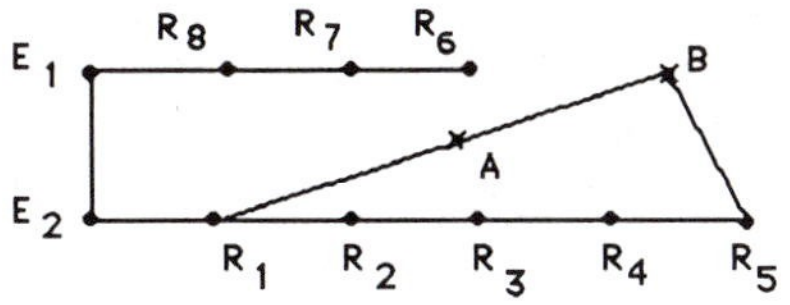

Similar to 3.2 and left to the reader.

4.3. F_1' is of type $\tilde{D}_6$.

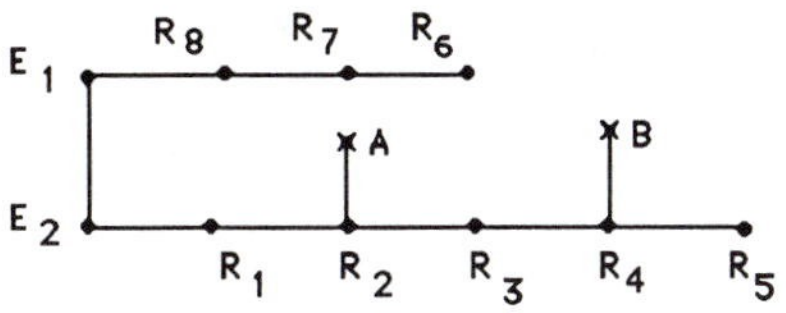

Similar to 3.3 and is left to the reader.

4.4. F_1' is of type $\tilde{E}_6$.

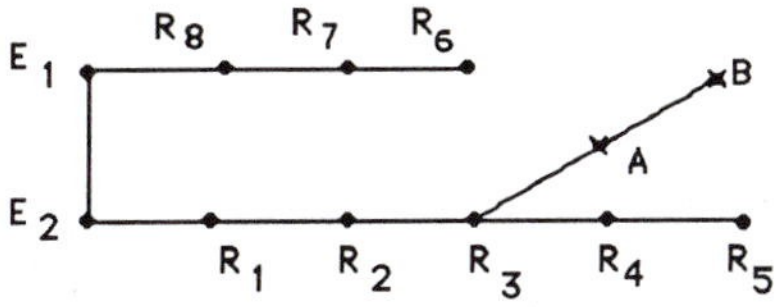

Since A is a component of multiplicity 2 in F_1', $E_2 \cdot A = 0$ Thus A is a component of F_2. The curve R_8 must intersect A or B.

<u>4.4.1. $R_8 \cdot A = 1$.</u>

In this case $F_1' \equiv 2E_1$, $E_2 \cdot B = 1$.

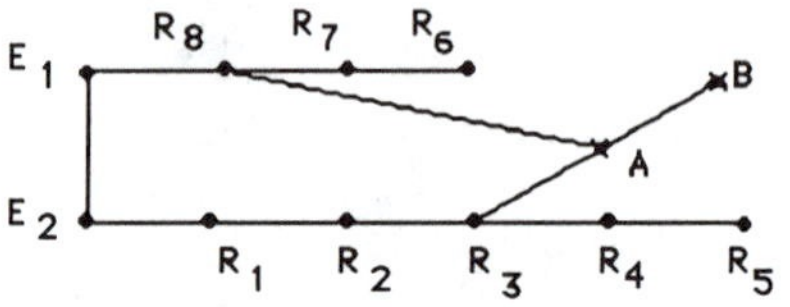

The only way to obtain F_2 is to add a component C intersecting R_6 to make F_2 of type $\tilde{E}_8$.

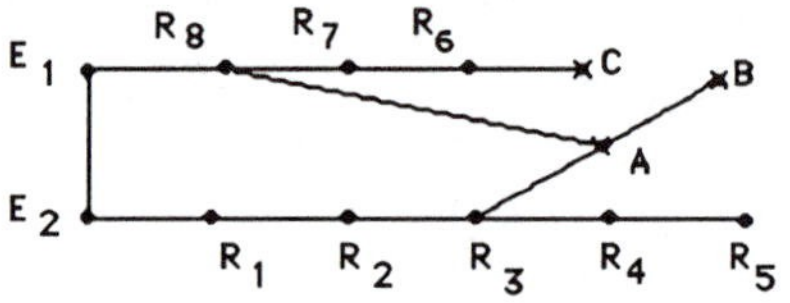

However, R_2 is of multiplicity 3 in F_2 and $R_1 \cdot F_2 = 3$. This is absurd.

<u>4.4.2. $R_8 \cdot B = 1$.</u>

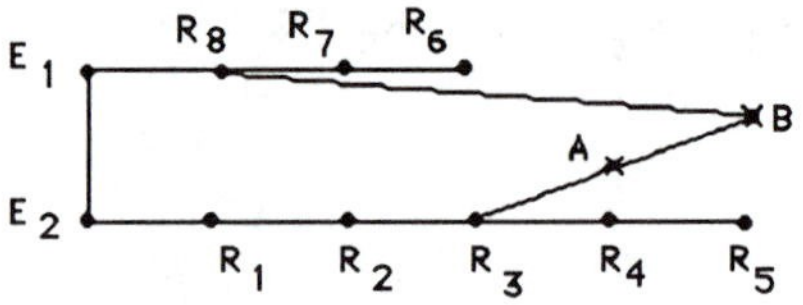

In this case $F_1' \equiv E_1$ and $E_2 \cdot A = E_2 \cdot B = 0$. The curves $R_2, \ldots, R_6$, A and B form the divisor F_2 of canonical type $\tilde{E}_8$ in which R_2 is of multiplicity 3. Thus $R_1 \cdot F_2 = 3$ which is absurd.

4.5. F_1' is of type $\tilde{D}_6$.

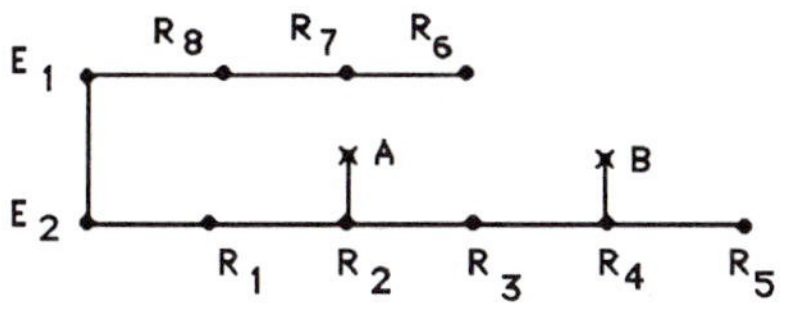

This is similar to 3.3 and is left to the reader.

4.6. F_1' is of type $\tilde{E}_7$:

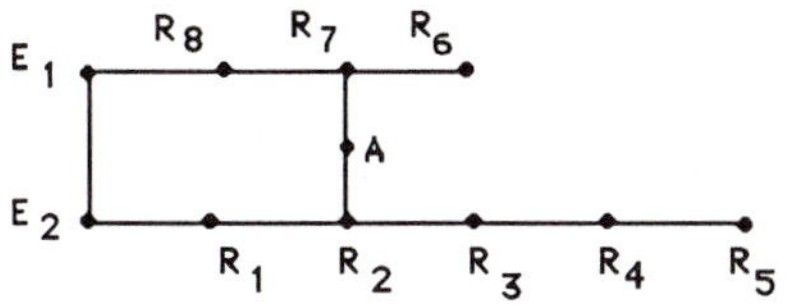

Since R_7 is a component of multiplicity 2 in F_1', $R_8 \cdot F_1' = 2$. Thus $F_1' \equiv 2E_1$ and $E_2 \cdot A = 1$, i.e. A is not a component of F_2. Since A is a component of multiplicity 3 in F_1' and intersects the components R_7 and R_2 of F_2, $F_1' \cdot F_2 \geq 3$. On the other hand, R_1 is a component fo F_1' of multiplicity 2 and cannot be a component of F_2 also. Thus $F_1' \cdot F_2 \geq 5$. Absurd.

4.6.1. F_2 is of type $\tilde{D}_8$.

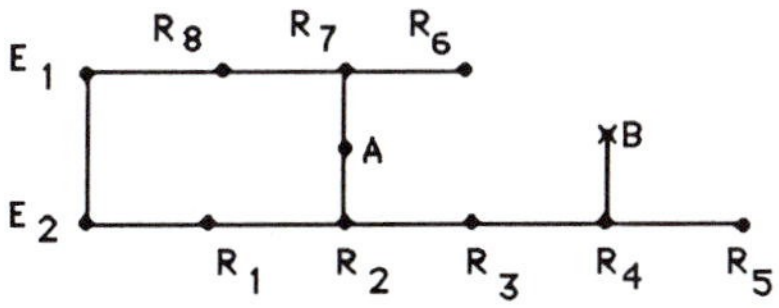

Since A is a component of F_2 of multiplicity 2, $E_1 \cdot A = 0$. Thus the divisor $D = R_8+R_6+R_1+R_3+2(A+R_7+R_2)$ is a divisor of canonical type satisfying $D \cdot E_1 = D \cdot E_2 = 1$.

4.6.2. F_2 is of type $\tilde{E}_8$.

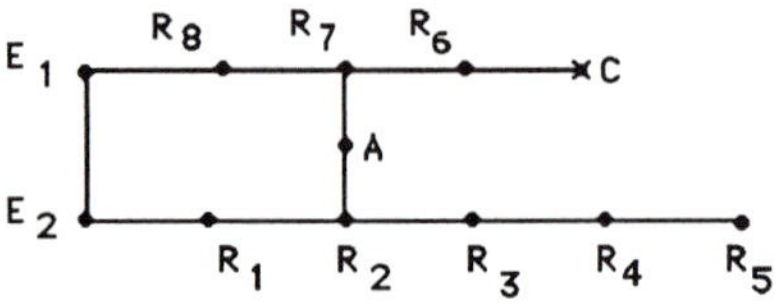

Here R_8 is a component of multiplicity 3 in F_2. Since $E_1 \cdot F_2 \leq 2$, this is absurd.

4.6.3. F_2 is of type $\tilde{E}_8$.

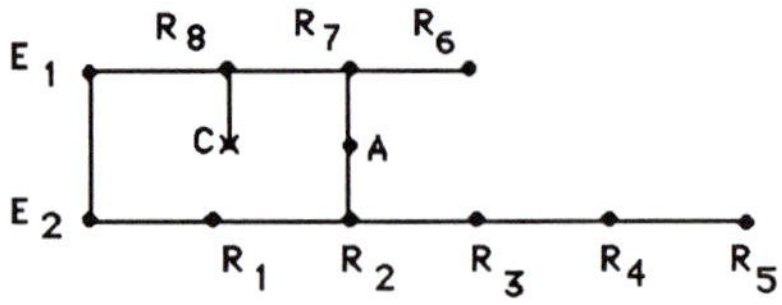

Since R_1 is a component of multiplicity 2 in F_1', $E_2 \cdot A = 0$. Hence the divisor $D = R_8+R_6+R_1+R_3+2(R_7+A+R_2)$ is of canonical type satisfying $D \cdot E_1 = D \cdot E_2 = 1$.

4.7. F_1' is of type $\tilde{E}_7$:

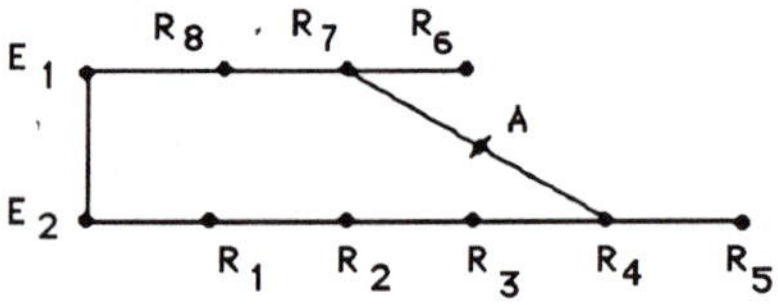

Here E_2 intersects a component R_1 of F_1' of multiplicity 1. Since A is of multiplicity 3, $E_2 \cdot A = 0$, $F_1' \equiv E_1$. However, R_8 intersects R_7 which is of multiplicity 2 in F_1'. Contradiction.

<u>4.8. F_1' is of type $\tilde{E}_7$:</u>

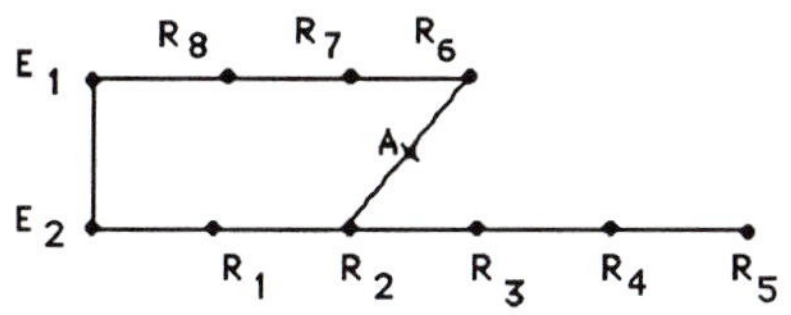

This is similar to 4.5.

<u>4.9. F_1' is of type $\tilde{E}_7$:</u>

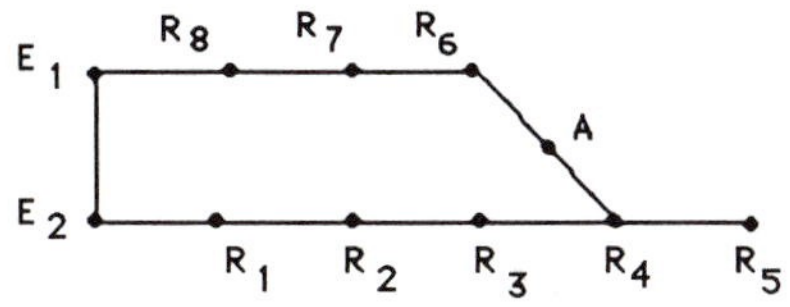

As in 4.6, $F_1' \equiv E_1$ and A is a component of F_2. The only way to extend the components $A, R_2, \ldots, R_8$ to a divisor of canonical type is to add a component C intersecting R_8.

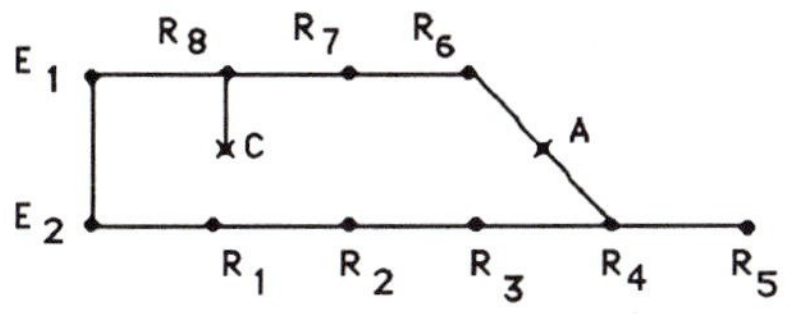

Since E_1 intersects R_8 which is a component of F_2 of multiplicity 2, $E_1 \cdot C = 0$. Thus C is a component of F_1 disjoint from F_1'. It is completed to a divisor of canonical type $\tilde{A}_1$ by adding a component D with $C \cdot D = 2$. We find that F is $\tilde{A}_1 + \tilde{E}_7$-special which contradicts the assumption of the theorem.

4.10. F_1' is of type $\tilde{A}_8$.

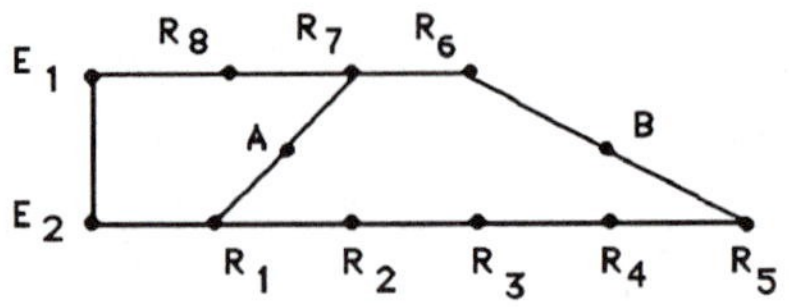

It is easy to see that both A and B cannot be components of F_2 (use Lemma 3.4.1). Thus E_2 intersects A or B, hence $F_1' \equiv 2E_1$, and R_8 intersects A or B (but not both of them).

4.10.1. $R_8 \bullet A = 1$, $R_8 \bullet B = 0$.

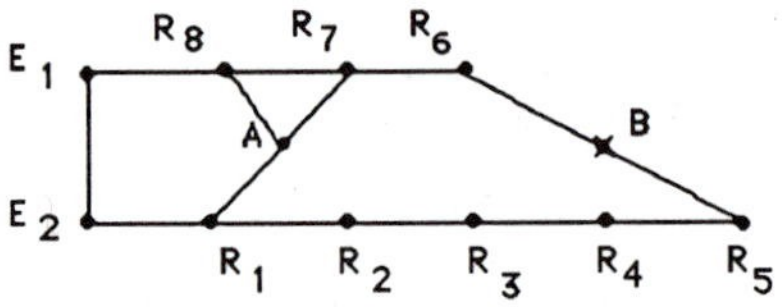

Obviously, A is not a component of F_2. Thus $E_2 \bullet A = 1$ and $D = R_8 + R_7 + A$ is a divisor of canonical type satisfying $D \bullet E_1 = D \bullet E_2 = 1$.

4.10.2. $R \bullet A = 0$, $R_8 \bullet B = 1$.

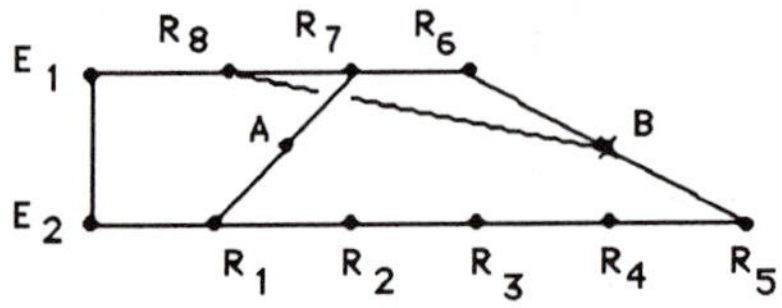

Similar to the previous case.

4.11. $F_1' = F_1$ is of type $\tilde{E}_8$.

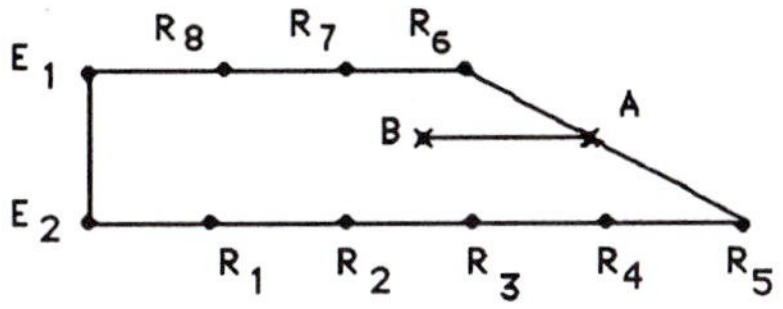

Since R_7 is of multiplicity 2 in F_1, $R_8 \cdot F_1 = 2$. Thus $F_1 \equiv 2E_1$ and $E_2 \cdot F_1 = 2$. Since R_1 is of multiplicity 1 in F_1, E_2 intersects A or B. However, A and B are components in F_1 of multiplicity > 1. Contradiction.

4.12. $F_1' = F_1$ is of type $\tilde{E}_8$.

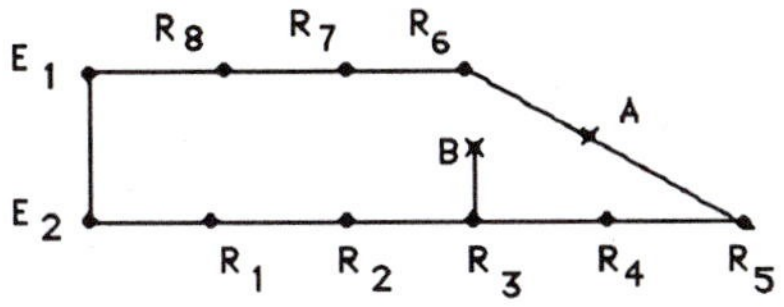

Here R_1 is of multiplicity 2. Hence E_2 does not intersect A and B. This implies that A and B are components of F_2. The latter must contain at least 10 components, which contradicts Lemma 3.4.1.

Case 5. p = 4.

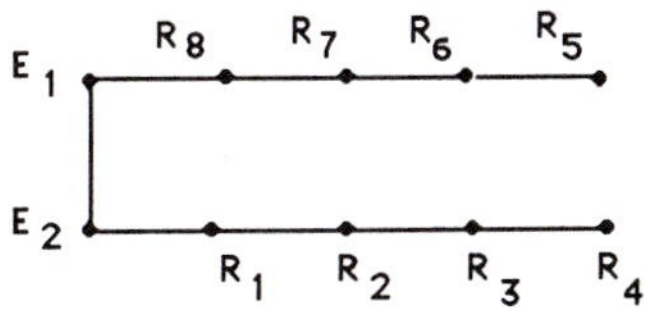

Let F_1' be the indecomposable part of F_1 containing the curves $R_1, \ldots, R_4$. If it contains one of the curves R_5, R_6, R_7, it must contain all of them and consists of 8 or 9 components. If it does not contain any of these curves, it consists of 5 or 6 components (Lemma 3.4.1).

5.1. F_1' is of type $\tilde{A}_4$.

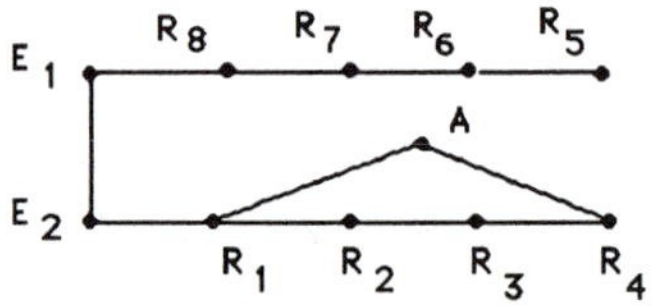

Similar to 3.1 and 4.1 and left to the reader.

5.2. F_1' is of type $\tilde{A}_5$.

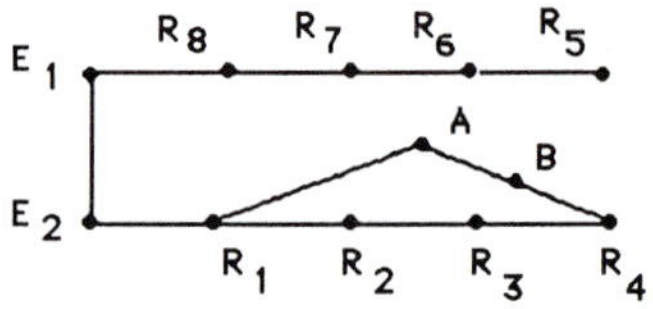

Similar to 3.2 and 4.2 and left to the reader.

5.3. F_1' is of type $\tilde{D}_5$.

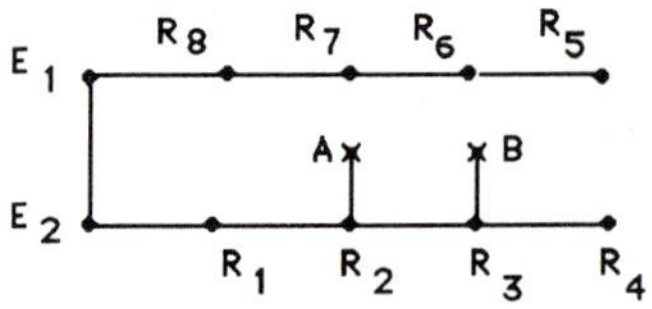

Similar to 3.3 and 4.3 and left to the reader.

5.4. $F_1' = F_1$ is of type $\tilde{A}_8$.

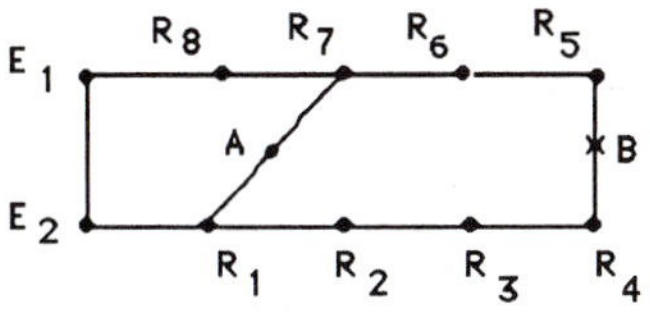

Similar to 4.9 and left to the reader.

5.5. $F_1' = F_1$ is of type $\tilde{D}_8$.

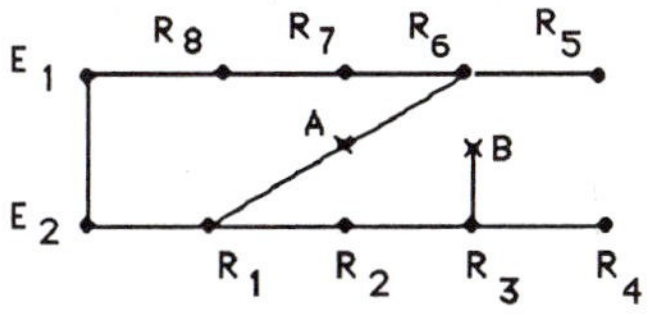

Since E_2 intersects a component of F_1 of multipicity 2, $F_1 \equiv 2E_1$ and R_8 must intersect B (since A is of multiplicity 2 in F_1).

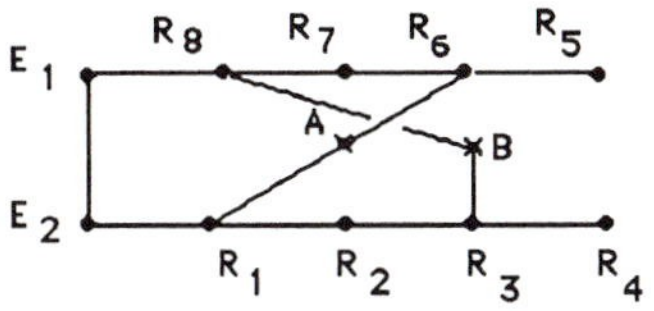

Since $E_2 \cdot A = E_2 \cdot B = 0$, the divisor $D = R_1+R_2+R_3+R_6+R_7+R\bullet+A+B$ is of canonical type and satisfies $D \cdot E_1 = D \cdot E_2 = 1$.

5.6. $F_1' = F_1$ is of type $\tilde{D}_8$

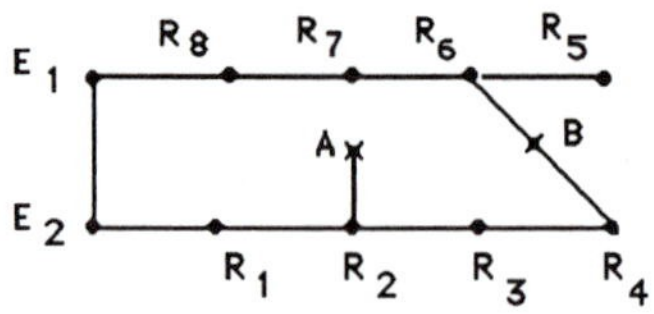

5.6.1. $F_1 \equiv E_1$.

In this case $R_8 \cdot F_1 = 1$ and F is $\tilde{D}_8$-special, which contradicts the assumption of the theorem.

5.6.2. $F_1 \equiv 2E_1$.

In this case E_2 and R_8 must intersect A, hence A is not a component of F_2. The divisor $D = A+B+R_2+R_3+R_4+R_6+R_7+R_8$ is of canonical type and satisfies $D \cdot E_1 = D \cdot E_2 = 1$.

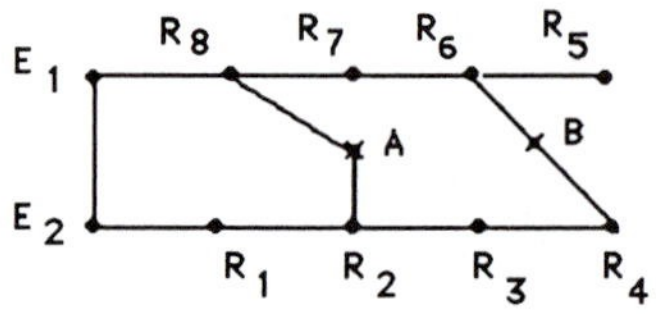

5.7. $F_1' = F_1$ is of type $\tilde{E}_8$.

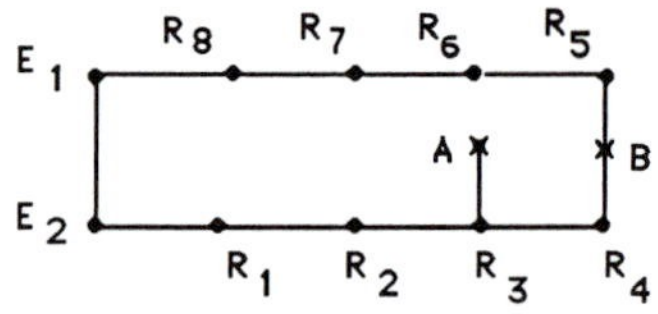

Since E_2 intersects a component of F_1 of multiplicity 2, $E_2 \cdot A = E_2 \cdot B = 0$. This shows that A and B are components of F_2. The latter must consists of at least 10 components which contradicts Lemma 3.4.1.

5.8. $F_1' = F_1$ is of type $\tilde{E}_8$

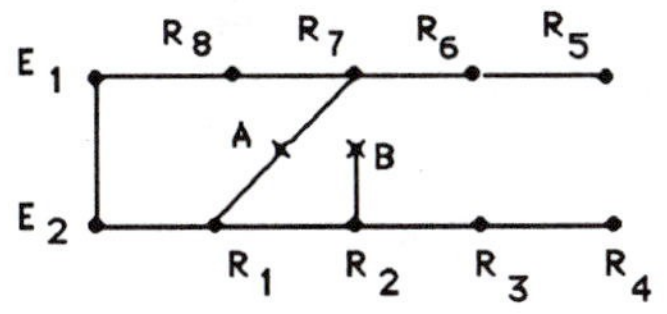

Here E_2 intersects a component of multiplicity 5. This is absurd.

This ends the proof of the theorem.

S6. Linear systems |C| with $C^2 \leq 10$.

The purpose of this section is to describe all linear systems |C| defined by an irreducible curve C on an Enriques surface F with $2 \leq C^2 \leq 10$. This will be used in the next chapter for a description of projective models of Enriques surfaces of small degree.

Lemma 3.6.1. For every C as above

$$\varphi(C) \leq 3$$

and the equality holds only if $C^2 = 10$.

PROOF. This follows immediately from Corollary 2.5.7 of Chapter 2.

Proposition 3.6.1. Let C be an irreducible curve on F with $C^2 = 2q > 0$ and $\varphi(C) = 1$. Then C is one of the following types:

(i) $|C| = |qE_1+E_2|$ for a non-degenerate U-pair $\{|2E_1|,|2E_2|\}$;

(ii) $|C| = |(q+1)E+R|$ or $|(q+1)E+R+K_F|$ for a degenerate U-pair $\{|2E|,R\}$.

PROOF. Let f_1 be an isotropic vector in Num(F) such that $[C]\cdot f_1 = 1$. By Theorem 3.2.1, $f_1 = [D]$ where $D \sim E_1 + \Sigma m_iR_i$ for some nodal curves R_i and a divisor E_1 of canonical type. Since $C\cdot D = 1$ we have $C\cdot E_1 = 1$ and we may assume that $f_1 = [E_1]$. The vector $f_2 = [C] - qf_1$ is isotropic and $f_1\cdot f_2 = 1$. Repeating this argument for f_2, we obtain $f_2 = [D']$ where $D' \sim E_2+\Sigma n_jR_j'$ with $E_1 \neq E_2$, $E_1\cdot E_2 = 1$ and $E_1\cdot R_j' = 0$ for all j or $D' \sim E_1 + R_i + \Sigma_{j\neq i}n_jR_j'$ with $E_1\cdot R_i' = 1$ and $E_1\cdot R_j' = 0$ for all $j\neq i$. For every R_j' (resp. $j\neq i$),

$$f_2\cdot R_j' = C\cdot R_j' - E_1\cdot R_j' = C\cdot R_j' \geq 0.$$

This shows that $f_2 \in C_F$ hence $f_2 = [E_2]$ (resp. $f_2 = [E_1+R]$) and $C \equiv qE_1+E_2$ (resp. $C \equiv (q+1)E_1+R$). The assertion follows from this immediately.

Corollary 3.6.1. Let C be an irreducible curve on F with $C^2 > 0$ and $\varphi(C) = 1$. There exits a unique genus 1 pencil $|2E|$ such that $E\cdot C = 1$ unless $|C| = |E_1+E_2|$ for a non-degenerate U-pair $\{|2E_1|,|2E_2|\}$. Moreover

$$H^1(F,\mathcal{O}_F(C-E)) = H^1(F,\mathcal{O}_F(C-2E)) = 0$$

If $C^2 \geq 6$,

$$H^1(F,\mathcal{O}_F(C-4E)) = 0$$

unless $|C| = |4E+R|$ for a degenerate U-pair $\{|2E|,R\}$.

PROOF. Apply the vanishing criterion from Chapter 1, §3.

The next two propositions and their corollaries can be proven in a similar way (cf.**[Cos 2]**).

Proposition 3.6.2. Let C be an irreducible curve on F with $C^2 = 4q > 0$ and $\varphi(C) = 2$. Then $|C|$ or $|C+K_F|$ is one of the following types:

(i) $|qE_1+2E_2|$, $E_1\cdot E_2 = 1$;

(ii) $|(q+2)E_1+2R_1|$, $E_1\cdot R_1 = 1$, $q \neq 1,2$,

(iii) $|qE_1+E_2|$, $E_1\cdot E_2 = 2$;

(iv) $|qE_1+E_2+R_1...+R_r|$, $E_1\cdot R_1 = E_2\cdot R_r = E_1\cdot E_2 = 1$, $E_1\cdot R_i = 0$ if $i \neq 1$, $E_2\cdot R_i = 0$ if $i \neq r$, the curves R_i are nodal and form a root base of type A_r;

(v) $|(q+1)E_1+R_1+R_2|$, $R_1\cdot R_2 = 0$, $E_1\cdot R_2 = E_1\cdot R_1 = 1$;

(vi) $|(q+1)E_1+2R_1+...+2R_r+R_{r+1}+R_{r+2}|$, $E_1\cdot R_1 = 1$, $E_1\cdot R_i = 0$ if $i \neq 1$, $R_i\cdot R_{i+1} = R_r\cdot R_{r+2} = 1$, $i = 1,...,r$, all othe intersection indices are zero;

where $|(2E_1|,|2E_2|$ are genus 1 pencils and R_i are nodal curves.

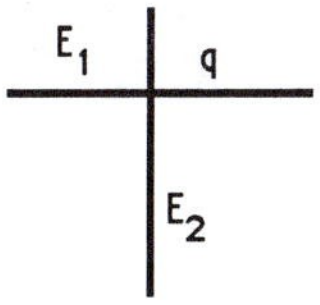

(i)

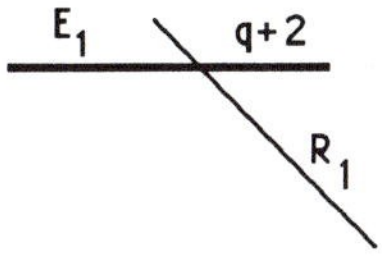

(ii)

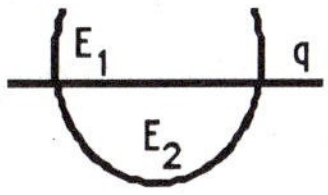

(iii)

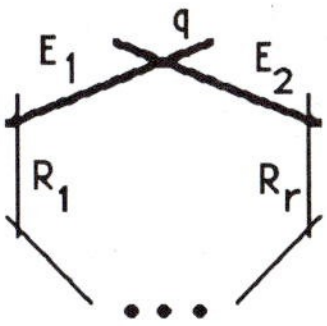

(iv)

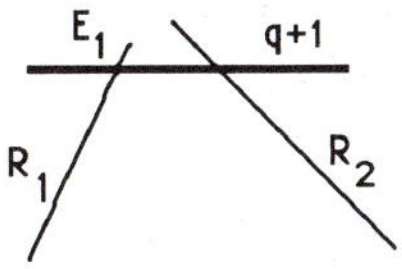

(v)

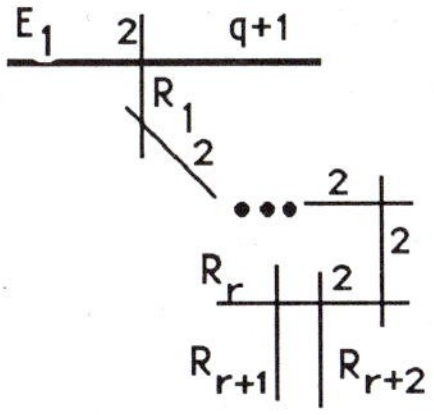

(vi)

Proposition 3.6.3. Let C be an irreducible curve on F with $C^2 = 4q+2 > 0$ and $\varphi(C) = 2$. Then $|C|$ or $|C+K_F|$ is one of the following types:

(i) $|qE_1+E_2+E_3|$, $E_i\cdot E_j = 1$, $i\neq j$;

(ii) $|qE_1+2E_2+R_1|$, $E_1\cdot E_2 = E_2\cdot R_1 = 1$, $E_1\cdot R_1 = 0$;

(iii) $|(q+1)E_1+E_2+R_1|$, $E_1\cdot E_2 = E_1\cdot R_1 = 1$, $E_2\cdot R_1 = 0$;

(iv) $|(q+2)E_1+2R_1+R_2|$, $R_1\cdot R_2 = E_1\cdot R_1 = 1$, $E_1\cdot R_2 = 0$,

where $|2E_i|$ are genus 1 pencils and R_i are nodal curves.

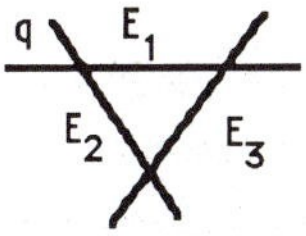

(i)

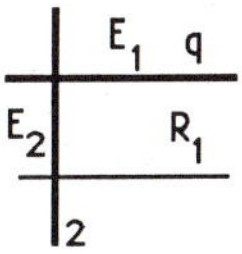

(ii)

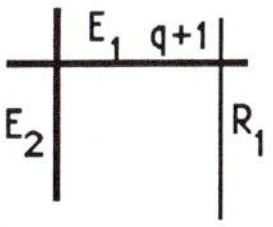

(iii)

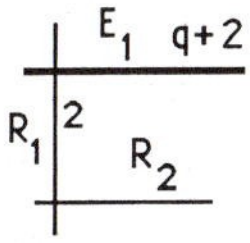

(iv)

Corollary 3.6.2. Let C be an irreducible curve on F with $C^2 > 6$ and $\varphi(C) = 2$. There is a unique genus 1 pencil $|2E|$ such that $E\cdot C = 2$ unless $|C|$ or $|C+K_F|$ is one of the following types:

(i) $|2E_1+2E_2|$, $E_1\cdot E_2 = 1$;

(ii) $|E_1+E_2+E_3|$, $E_i\cdot E_j = 1$, $i \neq j$;

(iii) $|2E_1+E_2+R_1|$, $E_1\cdot E_2 = E_1\cdot R_1 = 1$, $E_2\cdot R_1 = 0$,

where $|2E_i|$ are genus 1 pencils and R_i are nodal curves.

Corollary 3.6.3. Let C be an irreducible curve on F with $C^2 > 2$ and $\varphi(C) = 2$. Let $|2E|$ be a genus 1 pencil such that $E\cdot C = 2$. Then $H^1(F,\mathcal{O}_F(C-E)) = 0$.

Proposition 3.6.4. Let C be an irreducible curve on F with $C^2 = 10$ and $\varphi(C) = 3$. There exists a canonical isotropic sequence $(f_1,\dots,f_{10})$ such that

$$3[C] = f_1 + \dots + f_{10}.$$

PROOF. Use Corollary 2.5.5 of Chapter 2 and the proof of Lemma 3.3.1.

Bibliographical notes to Chapter III.

The application of arithmetic of hyperbolic integral quadratic forms and their reflection groups to algebraic geometry started in the fundamental work of I.Pyatetski-Shapiro and I.Shafarevich on the Torelli theorem for K3-surfaces **[PS-S]**. In the works **[B-P,Do 5, Nik 4,Nam]** the arithmetic of the Enriques lattice was applied to the study of automorphism groups of Enriques surfaces and its periods. F.Cossec was the first one who applied it to the study of linear systems on Enriques surfaces **[Cos 1,Cos 2]**.

The material of §1 is well known. The notion of a divisor of canonical type appears first in **[Mu 1]** and implicitly earlier in the works of K.Kodaira **[Kod 2]** on pluri-canonical maps of algebraic surfaces. The classification of divisors of canonical type is due to Kodaira **[Kod 1]**. The fact that this classification immediately follows from the classification of root systems of affine type was noticed later by many people (cf. **[B-P-vdV]**).

The notion of the nodal chamber from §2 appeared first in **[Lo 2]**. Theorem 3.2.1 is well known and is a generalization of the corresponding result for K3-surfaces from the work of Piytatetskii-Shapiro and Shafarevich. The observation that the existence of elliptic pencils can be obtained immediately from this theorem and the structure of the Enriques lattice is due to W. Lang **[La 2]**. The proof of the Reducibility Lemma (Corollary 3.2.2) based only on the arithmetic of the Enriques lattice is new. The question of its existence was first raised by Lang [loc.cit.]. Another proof (by lifting to characteristic 0) was given by him. The generalization of Enriques' proof to the case of characteristic 2 was given in **[B-M 2]**.

Theorem 3.5.1 is a new result. In the case char(k) $\neq 2$ it was proven earlier in **[Cos 1]**.The combinatorial arguments in its proof are similar to ones used in **[R-S 3]** for the study of genus 1 pencils on K3-surfaces in characteristic 2.

The classification of linear systems according to the values of the function Φ from §6 is due to Cossec **[Cos 2]** and S.Mukai (unpublished).

Chapter IV

PROJECTIVE MODELS

S1. Preliminaries.

As in the previous chapters F will denote an Enriques surface over an algebraically closed field of characteristic $p \geq 0$.

For every irreducible curve C on F with $C^2 \geq 2$ we will denote by

$$\varphi_C : F \to X \subset \mathbb{P}(H^0(F, \mathcal{O}_F(C))^*) \cong \mathbb{P}^n$$

the surjective rational map $\varphi_{|C|}$ defined by the linear system $|C|$. In this chapter we will describe all such maps.

As in the previous chapter we will use $\mathcal{R}(F)$ to denote the set of nodal curves on F, which will be identified with a subset of Num(F) which is a nodal root basis.

Let C be as above (we reserve this notation for such curves without further reminder), define

$$\mathcal{R}_C = \{ R \in \mathcal{R}(F) : C \cdot R = 0 \}.$$

We will also view the union of curves from $\mathcal{R}_C$ as a curve (reducible, in general) on F.

Proposition 4.1.1.

(i) $\mathcal{R}_C$ consists of irreducible curves A such that $C \cdot A = 0$;

(ii) the set $\mathcal{R}_C$ is a root basis of finite type in Num(F).

adjunction formula. Since $\mathcal{R}(F)$ is a root basis, so is $\mathcal{R}_C$. It remains to use that the sublattice spanned by $\mathcal{R}_C$ is negative definite which follows again from the Hodge index theorem.

Corollary 4.1.1. $\#\mathcal{R}_C \leq 9$ and the Dynkin diagram of $\mathcal{R}_C$ is a disjoint sum of diagrams of type A_K, D_n and E_r. The map φ_C blows down every irreducible root basis in $\mathcal{R}_C$ to a rational double point of type corresponding to its Dynkin diagram.

For every connected component E of $\mathcal{R}_C$, we can define the fundamental exceptional cycle Z_E of the corresponding double rational singularity (see Chapter 0,§2). This divisor will be called a **fundamental cycle** of $\mathcal{R}_C$.

Lemma 4.1.1. Let |D| be a non-empty complete linear system on a smooth projective surface X. Assume that |D| has no fixed component and is not composed of a pencil. Let $\varphi_{|D|}$ be the corresponding rational map. Then

$$\deg(\varphi_{|D|})\deg(\varphi_{|D|}(X)) \leq D^2.$$

Moreover, equality holds if and only if |D| has no base-points.

PROOF. Obvious.

Proposition 4.1.2. Let C be an irreducible curve on F with $C^2 \geq 4$ and $X = \varphi_C(F)$. Assume that |C| is not composed of a pencil .Then one of the following cases holds:

I) |C| has base-points and

(i) $\deg\varphi_C = 1$ or

(ii) $\deg\varphi_C = 2$, deg X = codim X + 1, |C| has two simple base-points or

(iii) $\deg\varphi_C = 3$, $C^2 = 4$, |C| has one base-point.

II) |C| has no base-points and

(i) $\deg\varphi_C = 1$ or

(ii) $\deg\varphi_C = 2$, deg X = codim X + 2 or

(i) $\deg\varphi_C = 1$ or

(ii) $\deg\varphi_C = 2$, $\deg X = \operatorname{codim} X + 2$ or

(iii) $\deg\varphi_C = 3$, $C^2 = 6$ or

(iv) $\deg\varphi_C = 4$, $C^2 = 4$.

PROOF. Apply the previous lemmas and also the facts that $\dim|C| = \frac{1}{2}C^2$ (Corollary 1.4.1) and $\deg(X) \geq \operatorname{codim}(X)+1$ for every projective variety X (Proposition 0.3.1).

Let C be an irreducible curve on F with $C^2 \geq 2$ .We say that $|C|$ (and φ_C) is **hyperelliptic** (resp. **superelliptic**) if $C^2 = 2$ or case I (ii)(resp. II (ii)) occurs.

By the adjunction formula for every curve C on F

$$\mathcal{O}_C(C) \cong \omega_C \otimes \mathcal{O}_C(K_F) = \omega_C(\varepsilon),$$

where $\varepsilon = \mathcal{O}_C(K_F) \in \operatorname{Pic}(C)_2$.

Proposition 4.1.3. Suppose that C is an irreducible curve on F and $C^2 > 0$. Then

$$\varepsilon = \mathcal{O}_C(K_F) \neq 0, \text{ if F is classical,}$$
$$= 0, \text{ otherwise}$$

In particular, $\operatorname{Pic}(C)_2 \neq 0$, if F is classical.

PROOF. Clearly, $\mathcal{O}_C(K_F) \cong \mathcal{O}_C$ if and only if $h^0(\mathcal{O}_C(K_F)) \neq 0$. By Riemann-Roch

$$h^0(\mathcal{O}_C(K_F)) = h^1(\mathcal{O}_C(C)).$$

By the vanishing theorem of Chapter 1,§ 4, $H^1(F,\mathcal{O}_F(C)) = 0$. The exact sequence

$$0 \to \mathcal{O}_F \to \mathcal{O}_F(C) \to \mathcal{O}_C(C) \to 0$$

shows that $H^0(C,\mathcal{O}_C(C)) \neq 0$ if and only if $H^2(F,\mathcal{O}_F) \neq 0$. The latter is equivalent to F being non-classical.

Corollary 4.1.2. Assume that F is classical. Let C be an irreducible member of a linear system |D| of positive dimension. Then, by restriction, the map φ_D given by |D| defines the Prym-canonical map of C given by the linear system $|\omega_C(\varepsilon)|$.

PROOF. It follows from the exact sequence from the previous proof, that the restriction map

$$H^0(F,\mathcal{O}_F(D)) \to H^0(C,\mathcal{O}_C(D)) \cong H^0(C,\omega_C(\varepsilon))$$

is surjective.

§2. Linear systems on K3 surfaces.

In this section we recall some basic properties of linear systems on a K3-surface Y. Proofs, when omitted, can be found in **[SD]**. Notations will be similar to those introduced in §1.

Let F be an Enriques surface. We assume that p = char(k) ≠ 2. Let $\pi: \bar{F} \to F$ be its K3 cover. We denote by τ the corresponding involution of $\bar{F}$.

Proposition 4.2.1. Let C be an irreducible curve on a K3 surface Y and $C^2 > 0$. Then |C| has no base-points. Moreover, the generic member of |C| is smooth and one of the following holds:

(i) deg $\varphi_C = 2$, deg $\varphi_C(Y) = p_a(C) - 1$ (in this case |C| and φ_C are said to be **hyperelliptic**);

(ii) deg $\varphi_C = 1$, deg $\varphi_C(Y) = 2p_a(C) - 2$.

Proposition 4.2.2. With the same notation as above, the following properties are equivalent:

(i) |C| is hyperelliptic;

(ii) the generic member of |C| is a smooth hyperelliptic curve;

(iii) $C^2 = 2$, or |C| = |2B| for an irreducible curve B with $B^2 = 2$, or there exists a genus 1 pencil |P| such that $P \cdot C = 2$.

Corollary 4.2.1. Let C be an irreducible curve on an Enriques surface F. Then $|\pi^{-1}(C)|$ is hyperelliptic if and only if there exists a genus 1 pencil |P| such that $C \bullet P = 2$.

PROOF. Let $\bar{C}$ be the inverse transform $\pi^{-1}(C)$ of C on $\bar{F}$. Since $\bar{C}^2 \equiv 0 \bmod 4$, $\bar{C}^2 \neq 2$. We will apply the previous theorem. First, we exclude the case $|\bar{C}| = |2B|$, where B is an irreducible curve with $B^2 = 2$. Since $\tau(\bar{C}) = \bar{C}$, $\tau(B) \sim B$. Thus, there exists a divisor D on F such that $B \sim \pi^{-1}(D)$. But, then $B^2 = 2D^2$, i.e. $D^2 = 1$, which is absurd.

Hence, there exists a genus 1 pencil $|\bar{P}|$ on F such that $C \bullet P = 2$. By the Hodge index theorem

$$4(\bar{P} \bullet \tau(\bar{P}))C^2 = (\bar{P}+\tau(\bar{P})^2\bar{C}^2 \leq ((\bar{P}+\tau(\bar{P})) \bullet \bar{C})^2 = 16.$$

Assume that $C^2 \neq 2$. Then $\bar{P} \bullet \tau(\bar{P}) < 2$, hence $\bar{P} \bullet \tau(\bar{P}) = 0$ (obviously, a pair of distinct genus 1 pencils on a K3 surface cannot have index of intersection equal to 1). Hence $\bar{P} \sim \tau(\bar{P})$. This implies that there exists a divisor E on F such that $|\bar{P}| = |\pi^{-1}(E)|$. Thus, we found a genus 1 pencil |2E| on F with $E \bullet C = 1$. If $C^2 = 2$, we know already that there exists a genus 1 pencil |P| with $P \bullet C = 2$.

The converse statement is obvious (by taking the inverse transform of $|\bar{P}|$ to $\bar{F}$).

Proposition 4.2.3. Let φ_C be a hyperelliptic map on a K3 surface Y. Then its image is singular if and only if |C| is one of the following types:

(i) |4E+2R|, where |E| is a genus 1 pencil and R is a nodal curve such that $E \bullet R = 1$;

(ii) $|2E+R_1+R_2|$, where |E| is a genus 1 pencil, R_1 and R_2 are two nodal curves such that $E \bullet R_1 = E \bullet R_2 = 1$, $R_1 \bullet R_2 = 0$;

(iii) $|2E+\Delta|$, where |E| is a genus 1 pencil and Δ is a divisor of the form $2R_1+\ldots+2R_{n-2}+R_{n-1}+R_n$, where R_i are nodal curves whose classes in Num(Y) form a root base of type D_n, $E \bullet R_1 = 1$, $E \bullet R_i = 0$, $i \neq 1$.

Corollary 4.2.2. Let C be an irreducible curve on an Enriques surface F. Assume that there exists a genus 1 pencil |P| with $C \bullet P = 2$. Let $\bar{C} = \pi^{-1}(C)$. Then the image of $\varphi_{\bar{C}}$ is singular if and only if $C^2 = 2$ and $\Phi(C) = 1$. Moreover, in this case the image of $\varphi_{\bar{C}}$ is a singular quadric in $\mathbb{P}^3$.

PROOF. The first part follows directly from the proposition. The second one will be proven later in §5.

Proposition 4.2.4. Let φ_C be a non-hyperelliptic map on a K3 surface Y. Then φ_C is a birational morphism onto a surface with isolated rational double points. Let $I(C)$ be the homogeneous ideal which is the kernel of the canonical homomorphism of graded rings:

$$\mathrm{Sym}(H^0(Y,\mathcal{O}_Y(C))) \to \bigoplus_{n\geq 0} H^0(Y,\mathcal{O}_Y(nC)).$$

Then $I(C)$ is generated by elements of degree 2 and 3. Moreover, the elements of degree 3 are needed only if $|C| = |2B+R|$, where B and R are irreducible, $B^2=2$, $R^2=-2$, $B\cdot R=1$, or there exists a genus 1 pencil $|P|$ such that $P\cdot C = 3$.

Corollary 4.2.3. Let C be an irreducible curve on an Enriques surface F. Assume that there is no genus 1 pencil $|P|$ such that $P\cdot C = 2$. Let $\bar{C} = \pi^{-1}(C)$. Then $I(\bar{C})$ is generated by elements of degree 2.

PROOF. We assume that $I(\bar{C})$ is not generated by elements of degree 2. If $|C| = |2B+R|$ as in the previous theorem, then $\bar{C}^2 = 10 \equiv 0 \bmod 4$. This shows that there exists a genus 1 pencil $|\bar{P}|$ on F such that $\bar{C}\cdot\bar{P} = 3$. Since $(\bar{P}+\tau(\bar{P}))\cdot\bar{C} = 6$, the Hodge index theorem implies that

$$(\bar{P}\cdot\tau(\bar{P}))C^2 \leq 9.$$

Assume that $\bar{P}\cdot\tau(\bar{P}) = 0$. Then $|\bar{P}| = |\tau(\bar{P})|$, hence there exists a genus 1 pencil $|2P|$ on F such that $|\pi^{-1}(P)| = |\bar{P}|$. Thus $2P\cdot C = \bar{P}\cdot\bar{C} = 3$, which is absurd.

Assume that $\bar{P}\cdot\tau(\bar{P}) \neq 0$. Since $C^2 \neq 2$ by assumption and $\bar{P}\cdot\tau(\bar{P}) \geq 2$, we must have $\bar{P}\cdot\tau(\bar{P}) = 2$, $\bar{C}^2 = 4$. Since $(\bar{C}-\bar{P}-\tau(\bar{P}))^2 = 0$, by Riemann-Roch, the moving part $|M|$ of $|\bar{C}-\bar{P}-\tau(\bar{P})|$ is non-empty. But

$$M\cdot\bar{C} \leq \bar{C}^2 - (\bar{P}+\tau(\bar{P}))\cdot\bar{C} \leq 2$$

implies $M\cdot\bar{C} = 2$. By the Hodge index theorem, $4M^2 = M^2\bar{C}^2 \leq (M\cdot\bar{C})^2 = 4$. This implies that $M^2 = 0$ and hence $|M|$ is a genus 1 pencil with $M \cdot \bar{C} = 2$. This contradicts the non-hyperellipticity of $|\bar{C}|$.

S3. Numerical connectedness.

As before, we will use F to denote an Enriques surface and use X to denote a general surface.

Recall **[Bom]** that an effective divisor D on a nonsingular projective surface X is said to be **m-connected** if for every decomposition $D = D_1+D_2$ into a sum of two effective divisors, we have $D_1 \cdot D_2 \geq m$.

A linear system $|D|$ is called m-connected if every divisor in $|D|$ is m-connected.

Proposition 4.3.1. For every 1-connected divisor D on X

$$h^0(\mathcal{O}_D) = 1.$$

Moreover, if the Frobenius map $\mathbf{F}: H^1(X,\mathcal{O}_X) \to H^1(X,\mathcal{O}_X)$ is bijective, then

$$H^1(X,\mathcal{O}_X(-D)) = H^1(X,\mathcal{O}_X(K_X+D)) = 0.$$

PROOF. See **[Ram]**.

Lemma 4.3.1. Let C be an irreducible curve on an Enriques surface F with $C^2 \geq 2$. Then $C^2 \geq 2m$ if $|C|$ is m-connected.

PROOF. By the reducibility lemma (Corollary 3.2.2 from Chapter 3), for some $D \in |C|$ there exists a decomposition $D = D_1+D_2$, where $D_i > 0$, $D_i^2 \geq 0$, $i=1,2$. Since $D_1 \cdot D_2 \geq m$, we get $C^2 \geq 2m$.

Proposition 4.3.2. Let C be an irreducible curve on F with $C^2 > 0$. Then $|C|$ is m-connected if and only if $|C+K_F|$ is m-connected.

PROOF. Suppose that $|C|$ is m-connected and $|C+K_F|$ is not m-connected. Let

$D = D_1+D_2$ be an element of $|C+K_F|$ with $D_i > 0$, $D_1 \cdot D_2 \leq m-1$. By the previous lemma $C^2 \geq 2m$,. Hence $D_i^2 > 0$ for some $i = 1,2$, say $i = 1$. By Riemann-Roch, we can find $D_1' \in |D_1+K_F|$. Then $D' = D_1'+D_2 \in |C|$, $D_1' \cdot D_2 \leq m-1$, which is a contradiction. For the converse, we note that $|C| = |C+K_F+K_F|$.

Proposition 4.3.3. Let C be an irreducible curve on F with $C^2 > 0$. Then $|C|$ is 1-connected.

PROOF. Suppose there exists a divisor $D \in |C|$ such that $D = D_1+D_2$, $D_i > 0$, $D_1 \cdot D_2 \leq 0$. We may assume that $D_1^2 > 0$. If $D_2^2 < 0$, then $D_1^2 > C^2$, and by Riemann-Roch and the vanishing theorem, $\dim |D_1| > \dim |C|$, which is absurd. Therefore, $D_2^2 \geq 0$, and again by Riemann-Roch,

$$\dim |C| = \tfrac{1}{2}C^2 \geq \dim |D_1| + \dim |D_2| \geq \tfrac{1}{2}D_1^2+\tfrac{1}{2}D_2^2 \geq \tfrac{1}{2}C^2.$$

This shows that $C^2 = D_1^2+D_2^2$, i.e. $D_1 \cdot D_2 = 0$. But this contradicts the Hodge index theorem.

Lemma 4.3.2. Let C be an irreducible curve on F with $C^2 > 0$. Then $D \in |C|$ is 2-connected unless $D = D_1+D_2$, where $D_i > 0$, $D_1^2 = 0$, $D_1 \cdot C = 1$.

PROOF. Let $D = D_1+D_2 \in |C|$, $D_i > 0$. Assume that $D_1 \cdot D_2 = 1$. Then $C^2 = D_1^2+D_2^2+2$. Suppose $D_1^2 \leq 0$, $D_2^2 \leq 0$. Then

$$C^2 = 2,\ D_1^2 = D_2^2 = 0,\ C \cdot D_1 = 1.$$

Suppose ,say $D_2^2 > 0$. By the Hodge index theorem, $D_1^2 D_2^2 < (D_1 \cdot D_2)^2 = 1$. Thus, $D_1^2 \leq 0$. If the equality holds, then $D_1 \cdot C = 1$ and we are done. If $D_1^2 < 0$, then $D_2^2 \geq C^2$. Hence, $\dim |D_2| \geq \dim |C|$, by Riemann-Roch. Since $|C|$ has no fixed components, $D_1 = 0$ which contradicts the assumption.

Proposition 4.3.4. Let C be an irreducible curve on F with $C^2 > 0$. Then $|C|$ is 2-connected if and only if $\Phi(C) \geq 2$.

PROOF. This immediately follows from the previous lemma.

Proposition 4.3.5. Let C be an irreducible curve on F with $C^2 > 0$. Assume that $|C|$ is 2-connected. Then $|C|$ is 3-connected, unless it contains a divisor D $= D_1+D_2$, $D_i > 0$, where one of the following properties holds:

(i) $D_1^2 = 0$, $D_1 \cdot C = 2$;

(ii) $D_1^2 = -2$, $D_1 \cdot C = 0$;

(iii) $D_1 \equiv D_2 \equiv M$ or $2E+R$, where M is an irreducible curve of arithmetic genus 2, $|2E|$ a genus 1 pencil and R a nodal curve.

PROOF. Let $D = D_1+D_2 \in |C|$, $D_i > 0$, $D_1 \cdot D_2 \geq 2$. Then $C^2 = D_1^2+D_2^2+4$.

Assume $D_1^2 \leq 0$, $D_2^2 \leq 0$. Then $C^2 = 4$ (it follows from the previous Lemma that $C^2 \geq 4$). Hence, $D_1^2 = D_2^2 = 0$ and $D_1 \cdot C = 2$, as wanted.

Let $D_1^2 > 0$. Then $D_1^2 D_2^2 \leq 4$ by the Hodge index theorem. If $D_1^2 > 0$, $D_1^2 = D_2^2 = D_1 \cdot D_2 = 2$, hence $D_1 \equiv D_2$. Since $2D_1 \cdot R = C \cdot R \geq 0$ for every nodal curve R, we get case (ii) by Proposition 3.1.6 of Chapter 3. If $D_1^2 = 0$, we are in case ii). If $D_1^2 < 0$, then $D_2^2 \leq C^2$ by Riemann-Roch. Hence $D_1^2 = -2$, $D_1 \cdot C = 0$ as in case iii).

Lemma 4.3.3. Let C be an irreducible curve on an Enriques surface F with $C^2 > 0$. Let Z be a fundamental cycle of $\mathcal{R}_C$. If $D = D_1+D_2 \in |C-Z|$, $D_i > 0$, then D is 1-connected unless there exists a genus 1 pencil $|2E|$ and a nodal curve R with $E \cdot R = 1$ such that

$$|C| = |2E+R+K_F|,\ Z = R,\ D_1 = E\ ,\ D_2 \sim E + K_F.$$

PROOF. Let $D \in |C-Z|$, $D = D_1+D_2, D_i > 0$, $D_1 \cdot D_2 \geq 0$. Since $|C|$ is 1-connected

$$D_1 \cdot (D_2+Z) \geq 1,\ D_2 \cdot (D_1+Z) \geq 1.$$

Therefore,

$$2D_1 \cdot D_2 + D_1 \cdot Z + D_2 \cdot Z = 2D_1 \cdot D_2 - Z^2 = 2D_1 \cdot D_2 + 2 \geq 2.$$

Thus we may assume that $D_1 \cdot D_2 = 0$, $D_1 \cdot Z = D_2 \cdot Z = 1$. Since $D_i^2 = 0$ is equivalent to $(D_i+Z)^2 = 0$, we may assume, by Lemma 4.3.2, that $D_i^2 = (D_i+Z)^2 = 0$. Since

$$C^2 = ((D_1+Z)+D_2)^2 = 2+D_2^2,$$

$D_2^2 \geq 0$. Since $D_1 \cdot D_2 = 0$, $D_2^2 \leq 0$ by Hodge Index theorem. Hence $D_2^2 = 0$. Therefore, $C^2 = 2$, $D_1^2 = D_2^2 = D_1 \cdot D_2 = 0$, $C \sim D_1+D_2+Z$. If E_1 and E_2 are two divisors of canonical type such that $D_i - E_i > 0$, then $\dim|D_1+D_2| < \dim|C| = 1$ implies $|2E_1| = |2E_2|$, $|E_1| \neq |E_2|$. Thus, $|C+K_F| = |2E| + Z$, where Z is a single nodal curve because $E \bullet Z = 1$.

Lemma 4.3.4. Let C be an irreducible curve on an Enriques surface F with $C^2 \geq 2$ and $\Phi(C) \geq 2$. Let Z be a fundamental cycle of $\mathcal{R}_C$. Then $|C-Z|$ has no fixed components unless one of the following cases occurs:

(i) $|C| = |2E+R_1+R_2|$ for a genus 1 pencil $|2E|$ and two nodal curves R_1 and R_2 with $E \bullet R_1 = E \bullet R_2 = 1$, $R_1 \bullet R_2 = 0$, $Z = R_1$;

(ii) $|C| = |2E+2R_1+\ldots+2R_n+R_{n+1}+R_{n+2}|$, where R_i are nodal curves and $Z = R_1+\ldots+2R_n+R_{n+1}+R_{n+2}$ is the fundamental exceptional cycle of a rational double point of type D_n, $E \bullet R_1 = E \bullet Z = 1$.

PROOF. Let $|C-Z| = |M| + F$ be the decomposition of $|C-Z|$ into its moving part $|M| \neq \emptyset$ and its fixed part F. We assume $F \neq 0$.

<u>Case 1:</u> $M^2 = 0$.

Then $|M| = |2E|$ is a genus 1 pencil. By Riemann-Roch, $(C-Z)^2 = C^2-2 \leq 2$. Thus $C^2 \leq 4$. If $C^2 = 2$, then $\Phi(C) = 1$ (Chapter 3,§6). Hence $C^2 = 4$ and $E \bullet C = 2$. Therefore, $C \bullet (F+Z) = C \bullet F = 0$, $(F+Z)^2 = -4$. Let $F = F_1+F_2$, $F_i \geq 0$, where F_1 and F_2 do not have common components, and every irreducible component of F_2 is contained in Z.

If $F_1 \neq 0$, we have $F_1^2 = -2$, $(F_2+Z)^2 = -2$, hence $F_2 = 0$. If $E \bullet F_1 = E \bullet Z = 1$, we are in case (i). Otherwise, $E \bullet F_1 = 2$ or $E \bullet Z = 2$ and $2 = (C-Z)^2 = (2E+F_1)^2 = 6$ or $4 \geq (C-F_1)^2 = (2E+Z)^2 = 6$ which is absurd.

If $F_1 = 0$, then $(F_2+Z)^2 = F_2^2+2F_2 \bullet Z+Z^2 = -4$ implies $F_2^2 = -2$ and $F_2 \bullet Z = 0$. As before, we must have $E \bullet F_2 = E \bullet Z = 1$. Let R be the component of F_2 such that $E \bullet R = 1$. Since E, F_2, Z and E, R, Z satisfy the same intersection properties, we must have $R = F_2$. A closer look at the possible configurations for Z, leads to case ii).

<u>Case 2:</u> $M^2 \neq 0$.

In this case

$$\dim |C| > \dim |M| = \tfrac{1}{2}M^2 = \dim |C-Z| \geq \tfrac{1}{2}(C-Z)^2 = \tfrac{1}{2}C^2 - 1$$

shows that $M^2 = C^2-2$. Since $M\cdot C = C^2-2+M\cdot(F+Z) \leq C^2$, there are 3 possible cases:

a) $M\cdot(F+Z) = 2$, $C\cdot(F+Z) = 0$, $(F+Z)^2 = -2$. Since Z is the fundamental cycle, $F = 0$, which is absurd.

b) $M\cdot(F+Z) = 1$, $C\cdot(F+Z) = 1$, $(F+Z)^2 = 0$, in which case $\Phi(C) = 1$, a contradiction.

c) $M\cdot(F+Z) = 0$, $C\cdot(F+Z) = 2$, $(F+Z)^2 = 2$, which contradicts the Hodge index theorem.

Corollary 4.3.1. Let C be an irreducible curve on F with $C^2 \geq 4$ and $\Phi(C) \geq 2$. Let Z_1 and Z_2 be two fundamental cycles of $\mathcal{R}(C)$. Then $|C-Z_1-Z_2|$ is 1-connected unless $|C| = |2E+R_1+R_2+K_F|$, where $|2E|$ is a genus 1 pencil, R_1, R_2 are nodal curves such that $E\cdot R_1 = E\cdot R_2 = 1$, $R_1\cdot R_2 = 0$, $Z_i = R_i$.

PROOF. Follows from the two previous lemmas.

Lemma 4.3.5. Let C be an irreducible curve on F with $C^2 \geq 6$ and $\Phi(C) \geq 2$. Let Z be the fundamental cycle of $\mathcal{R}(C)$. Then $\Phi(C-Z) \geq 2$ unless one of the following cases occurs

(i) $|C| = |2E+F+R|$, where $|2E|$ and $|2F|$ are genus 1 pencils, R is a nodal curve with $E\cdot F = E\cdot R = 1$, $F\cdot R = 0$, $Z = R$;

(ii) $|C| = |3E+2R_1+R_2|$, where $|2E|$ is a genus 1 pencil and R_1, R_2 are two nodal curves such that $E\cdot R_1 = R_1\cdot R_2 = 1$, $E\cdot R_2 = 0$, $Z = R_1+R_2$.

PROOF. By Lemma 4.3.4, $|C-Z|$ has no fixed components. Assume that there exists a genus 1 pencil $|2E|$ with $E\cdot(C-Z) = 1$. Two cases may occur:

There exists a genus 1 pencil $|2F|$ such that $|C-Z| = |pE+F|$ or there exists a nodal curve R such that $|C-Z| = |(p+1)E+R|$ (consider the linear systems $|C-Z-iE|$ and define p as the maximal i for which this system is non-empty).

In the first case $E\cdot F = 1$, $2p = C^2-2 \geq 4$, and

$$C^2 = (pE+F+Z)\cdot C = pE\cdot C+F\cdot C .$$

Since $\Phi(C) \geq 2$, we get $E\cdot C = F\cdot C = 2$. Therefore,

$$2 = F\cdot C = F\cdot(pE+F+Z) = p+F\cdot Z$$

implies $p = 2$, $F\cdot Z = 0$, $E\cdot Z = 1$. Let R be a component of Z such that $E\cdot R = 1$. Then $\dim|2E+F+R| = \dim|C|$ shows that $R = Z$, as in case (i).

In the second case we have $E\cdot R = 1$, $2p = C^2-2$. Then

$$C^2 = C\cdot((p+1)E+R+Z) = (p+1)E\cdot C+C\cdot R$$

and $\Phi(C)$ implies $E\cdot C = 2$, $Z\cdot E = 1$, $C\cdot R = 0$. Since

$$C\cdot Z = ((p+1)E+R+Z)\cdot Z = (p+1)+Z\cdot R - 2 = p-1+Z\cdot R,$$

and $C\cdot R = 0$, R is a component of Z hence $Z\cdot R \le 0$. The case $Z\cdot R = 0$ is impossible, because $p = \frac{1}{2}(C^2-2) > 1$. Therefore $Z\cdot R < 0$ in which case it is easy to see that there exists a nodal curve R' such that $Z = R+R'$, $R\cdot R' = 1$, $p=3$ as wanted in case (ii).

Lemma 4.3.6. Let C be an irreducible curve on F with $C^2 \ge 10$ and $\Phi(C) \ge 2$. Let Z be the fundamental cycle of $\mathcal{R}(C)$. Then $H^1(F,\mathcal{O}_F(C-2Z)) = 0$ unless there exists a genus 1 pencil $|2E|$ such that $E\cdot C = 2$, $E-Z > 0$.

PROOF. By the two previous lemmas, we know that $|\Gamma| = |C-Z|$ is irreducible and $\Phi(\Gamma) \ge 2$. Consider the decomposition of $|\Gamma-Z|$ into its moving part $|M|$ and its fixed part F. Assume that $H^1(F,\mathcal{O}_F(C-2Z)) \ne 0$. Then,

$$\dim|\Gamma-Z| = \dim|M| \ge \tfrac{1}{2}(C-2Z)^2 + 1 = \tfrac{1}{2}C^2 - 3 \ge 2.$$

Since $(C-2Z)^2 > 0$, $\dim|M| \ge 2$, $|M|$ is composed with a pencil and $\dim|M| = \frac{1}{2}M^2$. Since $|\Gamma|$ has no fixed components, $\dim|M| < \dim|\Gamma|$, hence, $M^2 = \Gamma^2-2$ or Γ^2-4.

Case 1: $M^2 = C^2-4 = \Gamma^2 - 2$.

We have

$$\Gamma^2 = M^2+M\cdot(F+Z)+\Gamma\cdot(F+Z), \quad \Gamma\cdot Z = (C-Z)\cdot Z = 2$$

which implies

$$M\cdot(F+Z) = \Gamma\cdot F = 0, \quad (F+Z)^2 = 2.$$

The latter contradicts the Hodge index theorem.

<u>Case 2:</u> $M^2 = C^2-6 = \Gamma^2-4$.

Here

$$4 = M\bullet(F+Z)+\Gamma\bullet(F+Z) = 2M\bullet(F+Z)+(F+Z)^2, \ \Gamma\bullet Z = 2.$$

By the Hodge index theorem,

$$M\bullet(F+Z) = 2, \ (F+Z)^2 = 0.$$

So, let E be a divisor of canonical type and Z_1 an effective divisor such that $F+Z = E + Z_1$. If $M\bullet E = 2$, $|C| = |M+E|$ by Riemann-Roch, $Z_1 = 0$. $E-Z > 0$, $E\bullet C = 2$ as wanted. Therefore, we can assume that $E\bullet M = 1$. Choose E_1 to be an effective divisor such that $M \sim pE+E_1$, $E\bullet E_1 = E_1{}^2 = 0$, $2p = \Gamma^2-4 \geq 4$. Then

$$\Gamma^2 = \Gamma\bullet((p+1)E+E_1+Z_1), \ \Gamma\bullet E \geq 2, \ \Gamma\bullet E_1 \geq 2.$$

This implies that $\Gamma\bullet E_1 = \Gamma\bullet E = 2$, $\Gamma\bullet Z_1 = 0$. If E_1 is of canonical type, then

$$2 = \Gamma\bullet E_1 = ((p+1)E+E_1+Z_1) = (p+1)+E\bullet Z_1$$

gives $p = 1$, a contradiction. Therefore, $E_1 \equiv E+R$ for some nodal R with $E\bullet R = 1$. Then

$$0 = \Gamma\bullet R = R\bullet((p+2)E+R+Z_1) = p+R\bullet Z_1$$

shows that R is a component of Z_1. Since

$$4p = ((p+2)E+2R)^2 \leq \Gamma^2 = 2p+4,$$

we get $p = 2$, $\Gamma \equiv 2(2E+R)$, $Z_1 = R$, $Z\bullet R = 1$. Since $C^2 = 10$, $C \equiv Z+2(2E+R)$ implies $E\bullet C = 2$. Finally, $F+Z = E+R$, $Z\bullet R = 1$ yields $E-Z > 0$, as wanted.

Let x be a closed point on F, C be an irreducible curve on F with $C^2 > 0$. We assume that $x \notin \mathcal{R}_C$. We denote by $\sigma: F' \to F$ the blowing-up of F at x and by L the exceptional curve of σ.

Lemma 4.3.7. Let $D \in |\sigma^*(C)-2L|$. Then D is 1-connected unless $D = D_1+D_2$, where $D_i > 0$, $D_1\bullet D_2 = 0$, $D_i\bullet L = 1$, in which case $\Phi(C) = 1$.

PROOF. Let $D = D_1+D_2$, $D_i > 0$. Define

$$\Delta_i = D_i + (D_i\bullet L)L, \text{ for } i =- 1,2.$$

Then $\Delta_i \geq 0$ and there exists an effective divisor C_i on F such that $\sigma^*(C_i) = \Delta_i$. We have

$$D_1 \bullet L + D_2 \bullet L = 2,$$

$$D_1 \bullet D_2 = C_1 \bullet C_2 - (D \bullet L)(D_2 \bullet L).$$

If $C_1 = 0$, $D_1 \bullet L < 0$, hence $D_1 \bullet D_2 \geq 3$. If $C_1 \neq 0$, $C_2 \neq 0$, then $D_1 \bullet D_2 \geq 1$ unless $C_1 \bullet C_2 = 1$, $D_1 \bullet L = D_2 \bullet L = 1$. In the latter case we conclude by Lemma 4.3.2.

Lemma 4.3.8. Let $D \in |\sigma^*(C) - 3L|$. Then D is 1-connected unless $D = D_1 + D_2$, where $D_i > 0$, $D_1 \bullet D_2 = 0$, $D_1 \bullet L = 1$, $D_2 \bullet L = 2$, in which case $\Phi(C) \leq 2$.

PROOF. Similar to the proof of the previous lemma.

Remark 4.3.1. With the notation of Lemma 4.3.7, if we assume $\Phi(C) \geq 2$ and that $|C|$ is not numerically equivalent to $|2M|$ for an irreducible curve M with $M^2 = 2$, then $D \in |\sigma^*(C) - 3L|$ will be 1-connected unless x does not belong to a divisor of canonical type E such that $E \bullet C = 2$. This follows from Proposition 4.3.5.

Lemma 4.3.9. Let C be an irreducible curve on F with $C^2 \geq 8$ and $\Phi(C) \geq 2$. Let Z be a fundamental cycle of $\mathfrak{R}_C$ disjoint from x. Then $D \in |\sigma^*(C-Z) - 2L|$ is 1-connected.

PROOF. This follows from Lemmas 4.3.4, 4.3.5 and 4.3.7.

Finally, we consider the blowing-up $\sigma: F' \to F$ at two distinct closed points x,y not on $\mathfrak{R}_C$. Let $L = \sigma^{-1}(x)$ and $M = \sigma^{-1}(y)$ be the corresponding exceptional curves.

Lemma 4.3.10. Let C be an irreducible curve on F with $C^2 > 0$ and $\Phi(C) \geq 2$. Then $D \in |\sigma^*(C) - 2L - 2M|$ is 1-connected unless

$$D = D_1 + D_2,\ D_i > 0,\ D_1 \bullet D_2 = 0,\ D_i \bullet L = D_i \bullet M = 1.$$

PROOF. Let $D = D_1 + D_2 \in |\sigma^*(C) - 2L - 2M|$, $D_i > 0$. Define

$$\Delta_i = D_i+(D_i\cdot L)L+(D_i\cdot M)M.$$

Then $\Delta_i \geq 0$ and $\Delta_i\cdot L = \Delta_i\cdot M = 0$, so that there exist some effective divisors C_i such that

$$D_1\cdot L+D_2\cdot L = D_1\cdot M+D_2\cdot M = 2,$$

$$D_1\cdot D_2 = C_1\cdot C_2-(D_1\cdot L)(D_2\cdot L)-(D_1\cdot M)(D_2\cdot M).$$

If $C_1 = 0$, then $D_1\cdot L$ or $D_1\cdot M<0$, hence, $D_1\cdot D_2 \geq 2$. If $C_1,C_2 \neq 0$, then $D_1\cdot D_2 = 0$, $C_1\cdot C_2 = 2$, $D_i\cdot L = D_i\cdot M = 1$ for $i = 1,2$. The case $C_1\cdot C_2 = 1$ is excluded by the assumption $\Phi(C) \geq 2$.

S4. Base-points.

This section is devoted to the proof of the following:

Theorem 4.4.1. Let C be an irreducible curve on an Enriques surface F with $C^2 > 0$. Then

(i) $|C|$ has base-points if and only if $\Phi(C) = 1$;

(ii) if $\Phi(C) =1$ and $K_F \nsim 0$, then $|C|$ has exactly two base-points of multiplicity 1;

(iii) if $\Phi(C) = 1$ and $K_F \sim 0$, then $|C|$ has exactly one base-point of multiplicity 2.

PROOF. (i) Let us consider first the case char(k) $\neq$ 2. Then we can give a geometric proof. Assume that $\Phi(C) = 1$. This means that there exists a genus 1 pencil $|2E|$ with $E\cdot C = 1$. In this case $x\in|\mathcal{O}_E(C)|$ is a base-point of $|C|$.

Conversely, assume that $\Phi(C) \geq 2$ (hence $C^2 \geq 4$). Let $\pi:\bar{F} \to F$ be the K3-cover of F and $\Gamma = \pi^{-1}(C)$. Let τ denote the involution on $H^0(\bar{F},\mathcal{O}_{\bar{F}}(\Gamma))^*$ and $\mathbb{P}(H^0(\bar{F},\mathcal{O}_{\bar{F}}(\Gamma))^*) = (\mathbb{P}^N)^*$ induced by the cover involution. Let $H_\pm$ be the two eigensubspaces of τ and $\mathbb{P}_\pm = \mathbb{P}(H_\pm)$. A hyperplane of $(\mathbb{P}^N)^*$ is τ-invariant if and only if it contains $\mathbb{P}_+$ or $\mathbb{P}_-$. These two families of hyperplanes correspond to $\pi^*(|C|)$ and $\pi^*(|C+K_F|)$. Since $\Phi(C) \neq 1$, the map $\varphi_\Gamma:\bar{F} \to Y$ is a birational morphism onto a surface with rational double points (Corollary 4.2.1). The system $|C|$ has base-points if and only if $\mathbb{P}_+\cap Y$ is not empty. Assume that it is the case, and let $x \in \mathbb{P}_+ \cap Y$. Then x must be a singular point of Y. Then the covering involution would leave a tree of nodal curves invariant. Since no involution acts freely on a tree of nodal curves, we have a contradiction. Therefore, $\mathbb{P}_+ \cap Y = \emptyset$ and $|C|$ has no base-points.

Now consider the case where char(κ) is arbitrary.

(i) As in the previous case, $|C|$ has base-points if $\Phi(C) = 1$. Assume that x is a base-point. We claim that

(4.4.1.1) x is not contained in any fundamental cycle Z of C.

Indeed, since $C^2 \geq 2$, there exists a divisor $D \in |C+K_F-Z|$. Then, by Lemma 4.3.3, we may assume that D is 1-connected (in the exceptional case $|C|$ has 2 base points, one on E and another on $E' \sim E+K_F$). By duality and the vanishing theorem of Chapter 1, §4

$$H^1(F, \mathcal{O}_F(C) \otimes \mathcal{J}_Z) = H^1(F, \mathcal{O}_F(-D)) = 0,$$

where $\mathcal{J}_Z$ is the sheaf of ideals of Z. This implies that $|C|$ has no base-points in Z. Let us now assume that $C^2 \geq 6$ or $C^2 \geq 4$ and $K_F \sim 0$. Then there exists a section s of $\mathcal{O}_F(C+K_F) \otimes \mathcal{J}_x^{\otimes 2}$, where $\mathcal{J}_x$ is the sheaf of ideals of x. Let $\sigma: F' \to F$ be the blowing-up of F at x with the exceptional curve $L = \sigma^{-1}(x)$. The zero set of $\sigma^*(s)$ is a divisor $D \in |\sigma^*(C+K_F)-2L|$. Assume that D is 1-connected. Then Serre duality, the Leray spectral sequence and the vanishing theorem imply

$$H^1(F, \mathcal{O}_F(C) \otimes \mathcal{J}_x) = H^1(F', \mathcal{O}_{F'}(-D)) = 0.$$

Hence x is not a base-point of $|C|$. By Lemma 4.3.7, there exists a genus 1 pencil $|2E|$ with $E \cdot C = 1$, $x \in E$. Therefore, we have proven

(4.4.1.2) If $C^2 \geq 6$ or $C^2 = 4$, $K_F = 0$, any base-point x of $|C|$ belongs to a divisor of canonical type E such that $E \cdot C = 1$.

Let us now consider the case $C^2 = 4$, $K_F \nsim 0$. We can assume $\Phi(C) \neq 1$. By Proposition 3.6.2 of Chapter 3, there exist two primitive effective divisors E_1 and E_2 with $E_i^2 = 0$ and $E_1 \cdot E_2 = 2$, such that $E_1+E_2 \in |C|$. By (4.4.1.1) we may assume that $x \in E_1$ and $|2E_1|$ is a genus 1 pencil. The cohomology sequence for

$$0 \to \mathcal{O}_F(E_2) \to \mathcal{O}_F(C) \to \mathcal{O}_{E_1}(C) \to 0$$

implies that x is a base point of $L = |\mathcal{O}_{E_1}(C)|$, dim $L = 1$, deg $L = 2$. This is absurd.

(ii) and (iii). Let C be an irreducible curve with $C^2 > 0$ and $\Phi(C) = 1$. Choose a genus 1 pencil $|2E|$ with $E \bullet C = 1$ and let $|D| = |2pE+C|$ for $p \geq 1$. We claim that (i) and (ii) hold for $|D|$. First, we note that $E \bullet D = 1$ and $|2E|$ is the only genus 1 pencil with this property. Since $D^2 \geq 6$, the base-point of $|D|$ is a point $x \in |\mathcal{O}_E(D)| = |\mathcal{O}_E(C)|$ if $K_F \sim 0$, and a point $q \in |\mathcal{O}_{E'}(C)|$ if $K_F \nsim 0$. This follows from (4.4.1.2). It is clear that

$$h^0(\mathcal{O}_F(D-2E)) = h^0(\mathcal{O}_F(D-2E)) - 2,$$

$$h^0(\mathcal{O}_F(D-4E))) \neq h^0(\mathcal{O}_F(D-2E)).$$

This shows that the image of $|2E|$ by φ_D is a pencil of lines. In particular, φ_D is not composed with a pencil and deg $\varphi_D \neq 1$. Our claim now follows from Proposition 4.1.2.

Since the base-points of $|D|$ are the base-points of $|C|$, (i) and (ii) hold for $|C|$ (again, applying Proposition 4.1.2).

Remark 4.4.1. Assuming char(k) $\neq 2$, one can give the following geometrical proof of Theorem 4.4.1(ii) . We use the notation of the previous paragraph. Let C be an irreducible curve with $\Phi(C) = 1$, let $|2E|$ be an elliptic pencil with $E \bullet C = 1$. There is a commutative diagram

$$\begin{array}{ccc} \bar{F} & \xrightarrow{\varphi_\Gamma} & Y \\ \pi\downarrow & & \downarrow\pi' \\ F & \xrightarrow{\varphi_C} & X \end{array}$$

where π' is obtained by projection from $\mathbb{P}_+$. The map φ_Γ is a morphism of degree two onto a surface which is ruled by the image of $|\pi^{-1}(E)|$. Let f_1 and f_2 be the fibres of this ruling which are τ-invariant. Then the base-points of $|C|$ and $|C+K_F|$ correspond to $f_i \cap \mathbb{P}_\pm$. One needs to consider separately the case when $|C|$ is a special pencil of genus 2 in the sense of the next section because in this case $|C+K_F|$ has fixed components.

S5. Hyperelliptic maps.

In this section we describe the image of Enriques surfaces by the maps associated to complete linear systems with base-points defined by an irreducible curve of arithmetic genus ≥ 3.

We recall (S1) that a map φ_C of an Enriques surface associated to an irreducible curve C (or the linear system $|C|$) with $C^2 \geq 2$ is said to be hyperelliptic if $C^2 = 2$ or it is of degree 2 onto a surface of degree n-1 in $\mathbb{P}^n$.

Proposition 4.5.1. Let C be an irreducible curve on an Enriques surface F with $C^2 \geq 2$. Then the following properties are equivalent:

(i) φ_C is hyperelliptic;

(ii) $|C|$ has base-points;

(iii) $\Phi(C) = 1$.

PROOF. By Proposition 4.1.2, $|C|$ has base-points if φ_C is hyperelliptic. Conversely, if $|C|$ has base-points, then $\Phi(C) = 1$ by Theorem 4.4.1. If $\Phi(C)= 1$, we saw in the proof of Theorem 4.4.1 that deg $\varphi_C \neq 1$. By Proposition 4.1.2 , φ_C is hyperelliptic, or has 3 base-points. The latter case is excluded by Theorem 4.4.1.

Corollary 4.5.1. Let $|C|$ be a hyperelliptic linear system with $C^2 = 2n$. Then $|C|$ is one of the following types:

(i) $|C| = |nE_1+E_2|$ for a non-degenerate U-pair $\{|2E_1|,|2E_2|\}$;

(ii) $|C| = |(n+1)E+R|$ or $|(n+1)E+R+K_F|$ for a degenerate U-pair $\{|2E|,R\}$.

We will say that $|C|$ is **non-special** (resp. **special**) if $|C|$ is of type (i) (resp. (ii)) from the previous corollary.

Theorem 4.5.1. Let $|C|$ be a hyperelliptic system on F. Assume $C^2 = 2n \neq 2$. Then $\varphi_C : F \to X$ is a map of degree 2 onto a surface X of degree $n-1$ in $\mathbb{P}^n$. Moreover, one of the following cases occurs:

(i) $X \cong \mathbb{P}^2$, $n = 2$, $|C| = |2E_1+E_2|$, $|3E_1+R|$ or $|3E_1+R+K_F|$;

(ii) $X \cong S_{0,p} \subset \mathbb{P}^{2p+1}$, $n = 2p+1$, $|C| = |(2p+1)E_1+E_2|$ or $|(2p+2)E_1+R+K_F|$ if $K_F \nsim 0$;

(iii) $X \cong S_{1,p} \subset \mathbb{P}^{2p}$, $n = 2p$, $|C| = |2pE_1+E_2|$ or $|(2p+1)E_1+R|$;

(iv) $X \cong S_{2,p-2} \subset \mathbb{P}^{2p}$, $n = 2p$, $|C| = |2pE_1+R|$,

where $|2E_1|$, $|2E_2|$ are genus 1 pencils and R is a nodal curve such that $E_1 \cdot R = E_2 \cdot R = E_1 \cdot E_2 = 1$ and $S_{a,k}$ is a rational scroll.

PROOF. We already know that φ_C is a map of degree 2 onto a surface of degree $n-1$ in $\mathbb{P}^n$, $n = \frac{1}{2}C^2$. By Proposition 0.3.2, X is isomorphic to $\mathbb{P}^2$ or to a rational scroll $S_{a,k}$, $a+k = n+1$.

If $C^2 = 4$, then $\dim |C| = 2$, hence $X \cong \mathbb{P}^2$. Assume that $C^2 \neq 4$. Clearly φ_C maps curves from $|2E_1|$ to the rulings of X. Then X is a cone (i.e. $X = S_{n-1,0}$) if and only if two distinct rulings of X span a plane, i.e.

$$h^0(\mathcal{O}_F(C-4E)) = h^0(\mathcal{O}_F(C)) - 3.$$

By Corollary 3.6.1 and Riemann-Roch, this happens only if $|C| = |4E_1+R|$, in which case $X \cong S_{2,0}$ as in (iv).

In all other cases, $X = S_{a,k}$, $k > 0$, that is $X \cong \mathbb{F}_a$, embedded by the linear system $|s+(a+k)f|$, $n = a+2k+1$ (see Chapter 0, §3). Assume, for example, that $C^2 \equiv 2 \bmod 4$ and $|C| = |(2p+1)E_1+E_2|$ is not special. The image of $|2E|$ is $|f|$, therefore $p = a+k$. Since $\dim |C| = 2p+1$, we have $n = 2p+1 = a+2k+1 = 2(a+k)+1$. Thus $a = 0$ and $k = p$, as wanted in case (ii). The remaining cases are considered similarly.

Remark 4.5.1. It follows from the proof of the theorem that $|C|$ has 2 base points $p \in E_1, p' \in E_1' \in |E_1+K_F|$ if F is classical and one base point of multiplicity 2 lying on E if F is non-classical. The rational map φ_C blows down E_1 (resp. E_1') to a point q (resp. q') lying on the exceptional section s of the scroll X. The image of the exceptional curves blown up from the points p and p' are two rulings of X if F is classical and one ruling otherwise (p' is infinitely near to p in the latter case). The exceptional section s of X is the image of the curve E_2.

The maps $F \to S_{a,\kappa}$ given by a non-special hyperelliptic system $|(2\kappa+3+a)E_1+E_2|$ with $a = 0,1$, $\kappa > 0$, are related by a commutative diagram:

$$\begin{array}{ccc} & & S_{a,\kappa+1} \\ & \nearrow & \downarrow \\ F & \rightarrow & S_{a,\kappa} \\ & \searrow & \downarrow \\ & & S_{a,1} \end{array}$$

The maps given by $|(i+2)E_1+E_2|$, $i=0,1,2$, are related by a commutative diagram:

$$\begin{array}{ccc} & & S_{1,1} \cong \mathbb{F}_1 \\ & \nearrow & \downarrow \\ F & \rightarrow & S_{0,1} \cong \mathbb{F}_0 \\ & \searrow & \downarrow \\ & & \mathbb{P}^2, \end{array}$$

where the vertical arrows are given by projection from the the contracted image of E_1. The maps defined by the special systems $|pE_1+R|$ or $|pE_1+R+K_F|$ are related by a commutative diagram

$$\begin{array}{ccc} S_{2,1} & \rightarrow & S_{a,\kappa+1} \\ \uparrow & \nearrow & \downarrow \\ F & \rightarrow & S_{a,\kappa} \\ \downarrow & \searrow & \downarrow \\ S_{2,0} & \rightarrow & S_{a,1} \end{array}$$

where the upper horizontal arrows are obtained by projection from the contracted image of E_1 and the right vertical arrows are obtained by projection from the contracted image of $E_1' \sim E_1+K_F$.

The next theorems describe the branch locus of hyperelliptic maps. Their proofs are easy and are left to the reader as an exercise.

Theorem 4.5.2. Assume $p \neq 2$. Let $\varphi'_C : F' \to X$ be the regular map obtained by the minimal resolution of the indeterminacy points of the rational map φ_C: $F \to X$ given by a non-special hyperellic system $|C| = |nE_1+E_2|$ with $n > 2$ (resp. $n = 2$). The map φ_C' decomposes into a composition of a birational morphism π: $F' \to Y$ and a double cover map: ψ: $Y \to X \cong S_{a,\kappa}$ (resp. $X \cong \mathbb{P}^2$). The map π blows down the nodal configuration $\mathfrak{R}_C$ and the curves $E_1, E_1' \in |E_1+K_F|, E_2$. The branch locus of ψ consists of two rulings l_1 and l_2 which are the images of the exceptional curves of F' (resp. two lines l_1 and l_2 meeting at a point O) and a curve $B \sim 4s+ (2n+2-4\kappa)f$ (resp. a curve of degree 6). The curve B has 2 tacnodes $P \in l_1$ and $Q \in l_2$ with tacnodal tangents l_1 at P and l_2 at Q , $P,Q \notin s$ (resp. has an additional node at O and $P,Q \neq O$). All other singularities of B are simple.

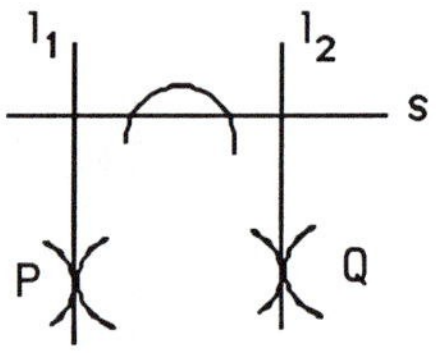

$n \neq 2$

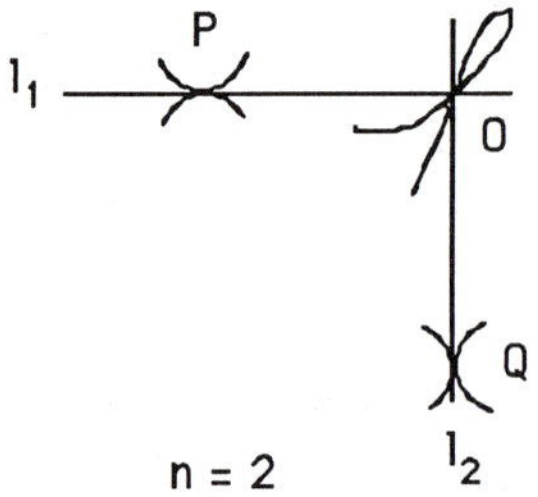

$n = 2$

Theorem 4.5.3. Assume $p \neq 2$. Let $\varphi'_C : F' \to X$ be the regular map obtained by the minimal resolution of the indeterminacy points of the rational map φ_C: $F \to X$ given by a special hyperellic system $|C| = |(n+1)E+R|$ or $|(n+1)E+R+K_F|$ with $n > 2$ and $|C| \neq |4E+R|$ (resp. $|C| = |4E+R|$, resp. $n = 2$). The map φ_C' decomposes into a composition of a birational morphism π: $F' \to Y$ and a double cover map: φ: $Y \to X \cong S_{a,k}$, $a \le 2$, $k \ne 0$ (resp. $X \cong S_{2,0}$, $X \cong \mathbb{P}^2$). The map π blows down the nodal configuration $\mathfrak{R}_C$ and the curves E , $E' \in |E+K_F|$ (and the curve R if $|C| = |4E+R|$). The branch locus of φ consists of two rulings l_1 and l_2 which are the images of the exceptional curves of F' (resp. two lines l_1 and l_2 meeting at a point 0) and a curve $B \sim 4s+ (2n+2-4k)f$ (resp. the image of such curve on $S_{2,0}$ under the blowing down s, resp. a curve of degree 6). The curve B has 2 tacnodes $P \in l_1$ and $Q \in l_2$ with tacnodal tangents l_1 at P and l_2 at Q. The tacnodes P and Q lie on the exceptional section s if $a = 0$, only one of them lies on s if $a = 1$ and none of them lie on s if $a = 2$ (resp. B has two tacnodes at nonsingular points of X not lying on the same ruling, the tacnodal tangents are the corresponding rulings, resp. B has a tacnode at a point $P \in l_1$, $P \ne 0$ with the tacnodal tangent l_1 and a singular point $Q = 0$ which becomes a tacnode after one blowing-up with the tacnodal tangent equal to the proper transform of l_2). All other singularities of B are simple.

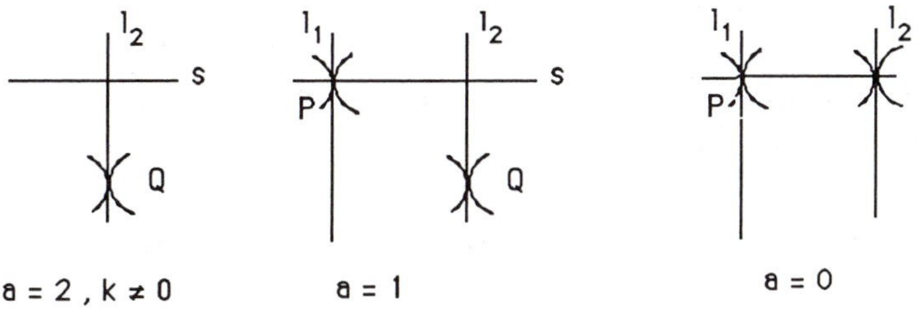

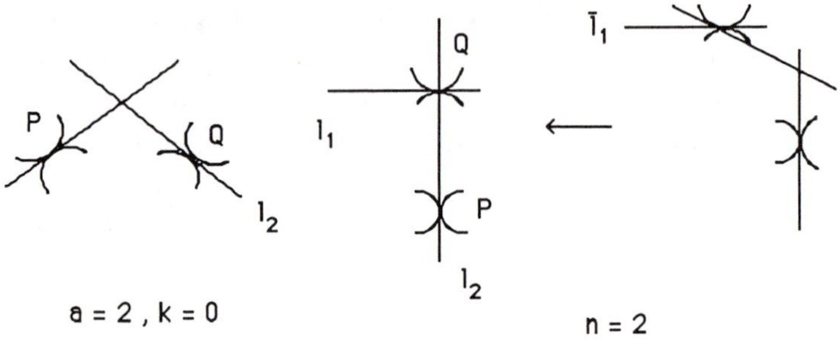

We leave to the reader to state and to prove the converse statements.

Corollary 4.5.1. Assume $p \neq 2$. The general member of a hyperelliptic system $|C|$ is a smooth hyperelliptic curve.

PROOF. The regular map φ_C': $F' \to X$ restricted the proper transform C' of a general member of $|C|$ is a double cover of a general hyperplane section H of X. Since H is a nonsingular rational curve intersecting the branch locus transversally, C' must be a smooth hyperelliptic curve.It remains to use that $|C|$ has two simple base-points or one base-point of multiplicity 2. Thus the image of C' in F is a smooth hyperelliptic member of $|C|$.

Theorem 4.5.4. Let C be an irreducible curve on a classical Enriques surface F with $C^2 > 6$. Assume that the general member of $|C|$ is a smooth hyperelliptic curve. Then $|C|$ is a hyperelliptic system.

PROOF. By Proposition 4.5.1 it suffices to prove that $|C+K_F|$ is hyperelliptic. The exact sequence of sheaves

$$0 \to \mathcal{O}_F(K_F) \to \mathcal{O}_F(C) \to \mathcal{O}_C(C+K_F) \to 0$$

together with the vanishing of $H^1(F, \mathcal{O}_F(K_F))$ shows that the restriction map

$$H^0(F,\mathcal{O}_F(C+K_F)) \to H^0(C,\mathcal{O}_C(C+K_F))$$

is surjective. Since the restriction of $|C+K_F|$ to C is a canonical map hence of degree 2, the map φ given by $|C+K_F|$ is not birational of degree >2. By Proposition 4.1.2 φ must be a superelliptic map. Thus the restriction of φ to a general member of $|C+K_F|$ is a double cover of an elliptic curve. This contradicts the property of the canonical map of a hyperelliptic map.

Remark 4.5.2. The assertion of the previous theorem is obviously true for $C^2 = 2$ and remains true if $C^2 = 4$ or 6 but its proof in the latter case is more involved. See **[Ver 1]** for the case $C^2 = 4$ and **[Cos 2]** for the case $C^2 = 6$.

S6. Birational maps.

The main result of this section is:

Theorem 4.6.1. Let C be an irreducible curve on F with $C^2 \geq 2$. Assume F is not an α_2-surface if p = 2. Then φ_C is a birational map onto a surface with rational double points as singularities if and only if $\Phi(C) \geq 3$.

Lemma 4.6.1. Let C be an irreducible curve on an Enriques surface F with $\Phi(C) \leq 2$. Then φ_C is not a birational map onto a normal surface.

PROOF. If $\Phi(C) = 1$, then $|C|$ is composed with a pencil when $C^2 = 2$ and hyperelliptic, if $C^2 > 2$ (Proposition 4.5.1). Assume $\Phi(C) = 2$. Then φ_C is a morphism of degree 4 onto $\mathbb{P}^2$ if $C^2 = 4$. If $C^2 \geq 6$ and $|2E|$ is a genus 1 pencil such that $E \cdot C = 2$, we have

$$h^0(\mathcal{O}_F(C-E)) = h^0(\mathcal{O}_F(C)) - 2$$

(Chapter I, Corollary 1.5.1). Therefore, $\varphi_C|E$ is of degree 2 onto a line.

Lemma 4.6.2. Let C be an irreducible curve on an Enriques surface F with $\Phi(C) \geq 3$. Then φ_C is a birational morphism.

PROOF. Assume that φ_C is not birational. Since $\Phi(C) \geq 3$, $C^2 \geq 10$ (Chapter 3,S6) and φ_C is a map of degree 2 onto a surface X of degree n in $\mathbb{P}^n$ (Proposition 4.1.2).

Assume that X is a projection of a scroll or an elliptic cone . Then X is ruled and the pull-back of its ruling is a complete linear system $|P|$ such that $P \cdot C = 2$. By the Hodge index theorem, $P^2 = 0$, hence $P \sim 2E$ for a curve of canonical type and $C \cdot E = 1$, i.e. $\Phi(C) = 1$. Contradiction.

By Proposition 0.3.3, X is a Del Pezzo surface in $\mathbb{P}^n$ or a reembedded quadric in $\mathbb{P}^8$, $n \leq 9$. If $n = 9$, then $X \cong \mathbb{P}^2$, embedded by the anticanonical map. Hence, X contains a two-dimensional system of cubics. Its pull-back on F is a two-dimensional linear system $|M|$ on F such that $M \cdot C = 6$. By Hodge, $M^2 \leq 2$. Since $\dim|M| = 2$, we obtain that $|M| = |2P|$, where P is a genus 1 pencil. Then $C \cdot P = 3$, which is absurd. Therefore, $n \neq 9$, and X contains a system of conics (if X is a Del Pezzo surface, this is the image of a pencil of lines passing through a fundamental point of the map $\mathbb{P}^2 \to X$). The pull-back of this system gives,as before, a genus 1 pencil $|M|$ such that $M \cdot C = 4$. A contradiction because $\Phi(C) \geq 3$.

Let $F \to F_C$ be a map obtained by contracting $\mathfrak{R}_C$ to rational double points of a normal surface F_C. The map φ_C factors through F_C. Denote by

$$\bar{\varphi}_C: F_C \to X$$

the induced map.

In view of the previous 2 lemmas Theorem 4.6.1 follows from the following one:

Theorem 4.6.2. Let C be an irreducible curve on F with $C^2 \geq 10$ and $\Phi(C) \geq 3$. Assume that F is not an $\boldsymbol{\alpha}_2$-surface if $p = 2$. Then

$$\bar{\varphi}_C: F_C \to X$$

is an isomorphism.

From now on we assume that F is not an $\boldsymbol{\alpha}_2$-surface if $p = 2$.

Proposition 4.6.1. Let C be an irreducible curve on F with $\Phi(C) \geq 3$. Assume that $K_F \sim 0$ if $C^2 = 10$. Then $\bar{\varphi}_C$ is injective.

PROOF. This will follow from the following 3 lemmas.

Lemma 4.6.3. Let C be an irreducible curve on F with $C^2 \geq 6$ and $\Phi(C) \geq 2$. Let Z_1 and Z_2 be two distinct fundamental cycles of $\mathfrak{R}_C$. Then $\varphi_C(Z_1) \neq \varphi_C(Z_2)$.

PROOF. Since $C^2 \geq 4$, there exists a divisor $D \in |C+K_F-Z_1-Z_2|$. By Proposition 4.3.2 and Corollary 4.3.1, D is 1-connected (the exceptional case does not occur, since $C^2 = 4$ in this case). Thus,

$$H^1(F,\mathcal{O}_F(-D)) = H^1(F,\mathcal{O}_F(C)\otimes\mathcal{I}_{Z_1}\otimes\mathcal{I}_{Z_2}) = 0.$$

This obviously implies that $\varphi_C(Z_1) \neq \varphi_C(Z_2)$.

Lemma 4.6.4. Let C be an irreducible curve on F with $\Phi(C) \geq 2, C^2 \geq 8$ or $C^2 \geq 6$ and $K_F \sim 0$. Let x be a closed point of F not in $\mathfrak{R}_C$. Then for any fundamental cycle Z of $\mathfrak{R}(C)$, $\varphi_C(x) \neq \varphi_C(Z)$.

PROOF. Let $\sigma:F' \to F$ be the blowing-up of F at x with the exceptional curve $L = \sigma^{-1}(x)$. Assume that $\varphi_C(x) = \varphi_C(Z)$. Then, if $C^2 \geq 8$ or $C^2 \geq 6$, $K_F \sim 0$, there exists a non-trivial section s of $\mathcal{O}_F(C+K_F)\otimes\mathcal{I}_x{}^{\otimes 2}\otimes\mathcal{I}_Z$. Clearly,

$$\mathrm{div}(\sigma^*(s)) = D \in |\sigma^*(C+K_F-Z)-2L|.$$

By Lemma 4.3.7, D is 1-connected. Thus

$$H^1(F',\mathcal{O}_{F'}(-D)) = H^1(F,\mathcal{O}_F(C)\otimes\mathcal{I}_x\otimes\mathcal{I}_Z) = 0.$$

This obviously implies that $\varphi_C(x) \neq \varphi_C(Z)$.

Lemma 4.6.5. Let C be an irreducible curve on F with $C^2 \geq 12$ or $C^2 = 10$ and $K_F \sim 0$. Assume that $\Phi(C) = 2$ .Then, for any 2 distinct closed points x and y not on any fundamental cycle of $\mathfrak{R}_C$, $\varphi_C(x) \neq \varphi_C(y)$, unless there exists a genus 1 pencil $|2E|$ such that $x,y \in E$, $E \cdot C = 2$.

PROOF. Let $\sigma:F' \to F$ be the blowing-up of F at x and y with the exceptional curves $L = \sigma^{-1}(x)$ and $M = \sigma^{-1}(y)$. Assume that $\varphi_C(x) = \varphi_C(y)$. Then, if $C^2 \geq 12$ or $C^2 = 10, K_F \sim 0$, there exists a non-trivial section s of $\mathcal{O}_F(C+K_F)\otimes\mathfrak{I}_x^{\otimes 2}\otimes\mathfrak{I}_y^{\otimes 2}$. Clearly

$$\operatorname{div}(\sigma^*(s)) = D\in|\sigma^*(C+K_F)-2L-2M|.$$

By Lemma 4.3.10, D is 1-connected. Thus

$$H^1(F',\mathcal{O}_{F'}(-D)) = H^1(F,\mathcal{O}_F(C)\otimes\mathfrak{I}_x\otimes\mathfrak{I}_Z) = 0.$$

This obviously implies that $\varphi_C(x) \neq \varphi_C(y)$.

Proposition 4.6.2. Let C be an irreducible curve on F with $\Phi(C) \geq 3$ and $K_F \sim 0$ if $C^2 = 10$. Then $\bar{\varphi}_C$ is an isomorphism.

PROOF. This follows from the next two lemmas.

Lemma 4.6.6. Let C be an irreducible curve on F with $\Phi(C) \geq 2$, $C^2 \geq 10$. Let Z be a fundamental cycle of $\mathcal{R}_C$. Then $\varphi_C(Z)$ is a rational double point unless there exists a genus 1 pencil $|2E|$ such that $E\cdot C = 2$ and $E\cdot Z > 0$.

PROOF. It is easy to see that it suffices to show that $H^1(F,\mathcal{O}_F(C)\otimes\mathfrak{I}_Z^{\otimes 2}) = 0$ (see **[Boml]**). Since $(C-2Z)^2 \geq 0$, there exists a divisor $D\in|C-2Z|$. This divisor is 1-connected by Lemma 4.3.6 unless there exists a genus 1 pencil $|2E|$ such that $E\cdot Z > 0$, $E\cdot C = 2$. We conclude as in the previous lemmas.

Lemma 4.6.7. Let C be an irreducible curve on an ordinary F with $\Phi(C) \geq 2$, $C^2 \geq 12$ or $C^2 = 10$ and $K_F \sim 0$. Let x be a closed point on F not on any curve from $\mathcal{R}_C$. Then the tangent map to φ_C at x is injective.

PROOF. As in the previous lemmas it suffices to show that $H^1(F,\mathcal{O}_F(C)\otimes\mathfrak{I}_x^{\otimes 2})=0$. The argument is similar to one used in the proof of Lemma 4.6.5 and is left to the reader.

Clearly Proposition 4.6.2 and 4.6.3 imply Theorem 4.6.2.

Theorem 4.6.3. Let C be an irreducible curve on F (could be an α_2-surface) with $C^2 \geq 4$ and $\Phi(C) = 2$. Assume that φ_C is not superelliptic. Then one of the following holds:

(i) φ_C is a morphism of degree 4 onto $\mathbb{P}^2$;

(ii) φ_C is a morphism of degree 1 onto a surface with double lines and rational double points as singularities. The double lines are the images of curves of canonical type E such that |2E| is a genus 1 pencil and $E \cdot C = 2$.

PROOF. Since $\Phi(C) = 2$, C has no base-points. By Proposition 4.1.2, φ_C is a morphism of degree 1 unless $C^2 = 6$ and φ_C is of degree 3 onto a quadric. The latter case is easily excluded. Indeed, the pull-back of a ruling of the quadric gives a pencil |D| with $D \cdot C = 3$. By the Hodge index theorem, |D| is a genus 1 pencil, which is impossible, since |D| contains a divisor of the form 2E.

§7. Superelliptic maps.

Proposition 4.7.1. Let $\varphi_C : F \to X$ be a superelliptic map of F. Then $C^2 = 6$ or 8 and $\Phi(C) = 2$.

PROOF. Since |C| is not composed with a pencil and deg $\varphi_C = 2$, $C^2 \geq 6$. Since |C| is not hyperelliptic $\Phi(C) > 1$ (Proposition 4.5.2). Using the classification of surfaces of degree n in $\mathbb{P}^n$ (Proposition 0.3.3), we obtain as in Lemma 4.6.2 that X must be an anticanonical Del Pezzo surface of degree $n \leq 8$ $(n \geq 3)$. Assume that $n \geq 5$. Again, as in the proof of Lemma 4.6.2, we consider the pull-back of a pencil of conics on X and find a genus 1 pencil |P| on F such that $P \cdot C = 4$. By Chapter 1, Corollary 1.4.1,

$$h^0(\mathcal{O}_F(C-P)) = h^0(\mathcal{O}_F(C)) - 3 \geq 3.$$

We conclude by Chapter 3, §6.

Recall from Chapter 3, §3 that a pair of genus 1 pencils $|2E_1|$ and $|2E_2|$ with $E_1 \cdot E_2 = 1$ is called a non-degenerate U-pair. A pair (|2E|,R), where |2E| is as above and R is a nodal curve with $E \cdot R = 1$ is called a degenerate U-pair.

Theorem 4.7.1. Let C be an irreducible curve on F with $C^2 = 8$. Then $|C|$ is superelliptic if and only if $|C| = |2M|$, where $|M| = |E_1+E_2|$ or $|2E+R|$ for some non-degenerate U-pair $(|2E_1|,|2E_2|)$ or a degenerate U-pair $(|2E|,R)$.

PROOF. Assume that $\varphi_C: F \to X \subset \mathbb{P}^4$ is a superelliptic map with $C^2 = 8$. As in the proof of the previous Proposition, we find a genus 1 pencil $|P|$ such that $P \cdot C = 4$ and $h^0(\mathcal{O}_F(C-P)) = h^0(\mathcal{O}_F(C)) - 3$. Again, we conclude by Proposition 3.6.2 that $|C| = |2M|$, where M is described above.

Conversely, let $|C| = |2M|$ as above and $D \in |M|$. We have an exact sequence

$$0 \to \mathcal{O}_F(M)) \to \mathcal{O}_F(C)) \to \mathcal{O}_D(C)) \to 0.$$

Since $H^1(F,\mathcal{O}_F(M)) = 0$ (Chapter 1,Theorem 1.4.1), the restriction of φ_C to D is given by the complete linear system $|\omega_D^{\otimes 2}|$. Since $D^2 = 2$, this map is of degree 2 onto a conic. This proves that deg $\varphi_C \neq 1$. However, $\Phi(C) = 2$, hence $|C|$ does not have base-points. Therefore φ_C is a superelliptic map onto a quartic Del Pezzo surface.

Theorem 4.7.2. Let C be an irreducible curve on F with $C^2 = 6$. Then $|C|$ is superelliptic if and only if one of the following cases occurs:

(i) $|C| = |E_1+E_2+E_3|$, $E_1+E_2-E_3 > 0$;

(ii) $|C| = |2E_1+E_2+R_1|$, $E_2-R_1 > 0$;

(iii) $|C| = |3E_1+2R_1+R_2|$, $E_1-R_2 > 0$,

where $|2E_i|$ are genus 1 pencils and R_i are nodal curves such that

$$E_1 \cdot E_2 = E_2 \cdot E_3 = E_1 \cdot E_3 = E_1 \cdot R_1 = R_1 \cdot R_2 = 1,\ E_2 \cdot R_1 = E_2 \cdot R_2 = 0.$$

PROOF. Let φ_C be a superelliptic map with $C^2 = 6$. By Proposition 4.7.1, $\Phi(C) = 2$ hence we can apply Proposition 3.6.3. Assume that $|C| = |E_1+E_2+E_3|$,notation as above. Let $|M| = |E_1+E_2|$. Since $h^0(\mathcal{O}_F(C-E_3)) = h^0(\mathcal{O}_F(C)) - 2$, the image of E_3 is a line on X. Therefore, the image of $|M|$ is a pencil of conics on X. Let $D \in |M|$ and consider the exact sequence

$$0 \to \mathcal{O}_F(E_1) \to \mathcal{O}_F(C) \to \mathcal{O}_D(C) \to 0.$$

Since $H^1(F,\mathcal{O}_F(E_1)) = 0$, φ_C restricted to D is given by the complete linear system $|\mathcal{O}_D(C)| = |\omega_D^{\otimes 2}|$. Therefore $\mathcal{O}_D(D-E_3)$ is trivial. Next, consider the exact sequence

$$0 \to \mathcal{O}_F(-E_3) \to \mathcal{O}_F(D-E_3)) \to \mathcal{O}_D(D-E_3) \to 0.$$

The corresponding cohomology sequence shows that $h^0(\mathcal{O}_F(D-E_3)) = h^0(\mathcal{O}_D) = 1$. Hence $D-E_3$ is effective, as wanted.

Conversely, if $E_1+E_2-E_3 > 0$, the second exact sequence shows that $\mathcal{O}_D(D-E_3) \cong \mathcal{O}_D$. Hence $\mathcal{O}_D(C)$ is isomorphic to $\omega_D^{\otimes 2}$. The first exact sequence shows that $\varphi_C|D$ is given by $|\omega_D^{\otimes 2}|$, hence it is of degree 2 onto a conic. This proves that φ_C is a superelliptic map of degree 2 onto a cubic surface.

Cases ii) and iii) are considered similarly.

Corollary 4.7.1. Let C be an irreducible curve on F with $C^2 = 6$. Then $|C|$ is superelliptic if and only if $|C|$ is one of the following types:

(i) $|Z_1+Z_2+Z_3|$, where $|2E_i| = |Z_j+Z_k|$ are genus 1 pencils, and Z_i is a fundamental cycle of $\mathcal{R}_{E_j+E_k}$ such that $Z_i \cdot Z_j = 2$ $(i \neq j \neq k)$, $Z_i^2 = -2$;

(ii) $|2Z_1+2R_1+Z_2|$, where $|2E_1| = |Z_1+Z_2|$, $|2E_2| = |2Z_1+2R_1|$ are genus 1 pencils, R_1 is a nodal curve such that Z_1 is a curve supported in $\mathcal{R}_{2E_1+R_1}$ and Z_2 is a fundamental cycle of $\mathcal{R}_{E_1+E_2}$, $Z_1 \cdot Z_2 = Z_1 \cdot R_1 = 2$, $R_1 \cdot Z_2 = 0$, $Z_1^2 = Z_2^2 = -2$;

(iii) $|3Z_1+4R_1+2R_1|$, where $|2E_1| = |2Z_1+2R_2|$ is a genus 1 pencil and Z_1 is a fundamental cycle of $\mathcal{R}_{2E_1+R_1}$ with $Z_1 \cdot R_2 = 2$, $R_1 \cdot R_2 = 1$, $Z_1 \cdot R_1 = 0$, $Z_1^2 = -2$.

PROOF. (i)In the notation of the previous theorem, let $Z_3 = E_1+E_2-E_3$. Then $Z_3 \cdot E_1 = Z_3 \cdot E_2 = 0$, $Z_3^2 = -2$. Thus Z_3 is part of a member of $|2E_1|$ and $|2E_2|$. Therefore $Z_1 = E_2+E_3-E_1 \in |2E_1-Z_3|$, $Z_2 = E_1+E_3-E_2 \in |2E_2-Z_3|$ are defined, and $|E_1+E_2+E_3| = |Z_1+Z_2+Z_3|$. Clearly, $Z_i \subset \mathcal{R}_{E_i+E_k}$ and $Z_i \cdot Z_j = 2$ for $i \neq j \neq k$. It remains to show that Z_i are fundamental cycles of $\mathcal{R}_{E_j+E_k}$. Assume that $Z_3' \geq Z_3$ is a fundamental cycle of $\mathcal{R}_{E_1+E_2}$. Let $E_3' = E_1+E_2-Z_3'$. Then $E_3'^2 = 0$, $E_3' \cdot E_1 = 1$. One easily shows that $E_3' = E_3$ and $Z_3' = Z_3$.

(ii)-(iii) are proven similarly and left to the reader.

Proposition 4.7.2. Let C be an irreducible curve on F with $C^2 = 6$. Then |C| is superelliptic if and only if |C| = |2M−Z|, where $M = E_1+E_2$ for a non-degenerate U-pair ($|2E_1|,|2E_2|$) and Z is a fundamental cycle of $\mathfrak{R}_M$ or $M = 2E_1+R$, where ($|2E_1|$,R) is a degenerate U-pair, Z is a fundamental cycle of $\mathfrak{R}_M$, $Z \neq R$.

PROOF. Assume that |C| is a superelliptic system. We use the notation of Theorem 4.7.2.

<u>Case (i)</u>: $|C| = |E_1+E_2+E_3|$.

Take $Z = E_1+E_2-E_3$, $M = E_1+E_2$.

<u>Case (ii)</u>: $|C| = |2E_1+E_2+R_1|$.

Take $Z = E_2-R_1$ ($= Z_2$ in the notation of Corollary 4.7.1), $M = E_1+E_2$.

<u>Case (iii)</u>: $|C| = |3E_1+2R_1+R_2|$.

Take $Z = E_1-R_2$ ($= Z_1$ in the notation of Corollary 4.7.1), $E=E_1, R=R_1$, $M = 2E+R$.

Conversely, assume that |C| = |2M−Z| with M and Z being as in the assertion . We consider the more difficult case M = 2E+R and leave the easier one ($M = E_1+E_2$) to the reader.

Since $Z \cdot E = 0$, there exists $Z' > 0$ such that $Z+Z' \equiv E$ or $Z+Z' \in |2E|$. In the first case, $|2(2E+R)-Z| = |3E+2R+Z'|$. Since $Z \neq R$, and Z is a fundamental cycle of 2E+R, Z' is the irreducible component of a multiple divisor from |2E| which intersects R. Thus $|C| = |3E+2R+Z'|$ satisfies Theorem 4.7.2 (iii) and hence is superelliptic. In the second case we have $R \cdot Z' = 2$, so that there exists a curve of canonical type E' and an effective divisor Z" such that $R+Z' = E'+Z''$. Assume $E \cdot E' \neq 0$. Then $2E+R \equiv E+(Z+Z'')$ implies $(2E+R)\cdot(Z''+Z) = 0$, $(Z+Z'')^2 = -2$, hence $E+R = Z+Z''$. Since R and Z are distinct fundamental cycles of $\mathfrak{R}_M$, $Z'' - R = E-Z = \bar{Z}$ for some positive divisor $\bar{Z}$. Hence $|C| = |2(2E+R)-Z| = |3E+2R+Z|$ satisfies Theorem 4.7.2 (iii) (one can easily see that $\bar{Z}$ is irreducible).

Remark 4.7.1. If $|C| = |2(2E+R)-Z|$ and Z = R , then $|C| = |3E+R|$ is hyperelliptic.

Next we describe the image of a superelliptic map. We use the notations and the definitions of Chapter 0, §4.

Theorem 4.7.3. Let $|C| = |2M|$ be a superelliptic map on F with $C^2 = 8$. Then $\varphi_C: F \to \mathcal{D}$ is a map of degree 2 onto a symmetric anticanonical quartic Del Pezzo surface $\mathcal{D}$. Moreover one of the following cases occurs:

(i) $\mathcal{D} = \mathcal{D}_1$ if $|M| = |E_1+E_2|$, F is classical;

(ii) $\mathcal{D} = \mathcal{D}_1'$ if $|M| = |2E+R|$, F is classical;

(iii) $\mathcal{D} = \mathcal{D}_2$ if $|M| = |E_1+E_2|$, F is a μ_2-surface;

(iv) $\mathcal{D} = \mathcal{D}_2'$ if $|M| = |2E+R|$, F is a μ_2-surface;

(v) $\mathcal{D} = \mathcal{D}_3$ if $|M| = |E_1+E_2|$, F is an α_2-surface;

(vi) $\mathcal{D} = \mathcal{D}_3'$ if $|M| = |2E+R|$, F is an α_2-surface;

where the notations of Theorem 4.7.1 are used.

PROOF. As we saw in the proof of Proposition 4.7.1, the image of φ_C is an anticanonical quartic Del Pezzo surface $\mathcal{D}$. The symmetric one can be distinguished by the number of lines or pencils of conics. We consider each case separately.

(i) $\mathcal{D}$ contains exactly 4 lines forming a quadrangle, these are the images of the two pairs of multiple fibres of $|2E_1|$ and $|2E_2|$. Also, $\mathcal{D}$ contains exactly 4 pencils of conics, the images of the pencils $|2E_1|$, $|2E_2|$, $|M|$ and $|M+K_F|$. By Proposition 0.4.3, $\mathcal{D} = \mathcal{D}_1$.

(ii) $\mathcal{D}$ contains exactly 2 lines, the images of the pair of multiple fibres of $|2E|$. Also, $\mathcal{D}$ contains exactly 2 pencils of conics, the images of the pencils $|2E_1|$, and $|M+K_F|$. By Proposition 0.4.3', $\mathcal{D} = \mathcal{D}_1'$.

(iii), (v) $\mathcal{D}$ contains exactly 2 lines, the images of the two unique multiple fibres of $|2E_1|$ and $|2E_2|$. Also, it contains exactly 3 pencils of conics, the images of the pencils $|2E_1|,|2E_2|$ and $|M|$. By Proposition 0.4.6, $\mathcal{D} = \mathcal{D}_2$ or $\mathcal{D}_3$. If $\mathcal{D} = \mathcal{D}_2$, then there exists a double cover $\pi: Y \to \mathcal{D}$ which is unramified over nonsingular points of $\mathcal{D}$. The pull-back of this cover to F defines a double cover ramified over at most finitely many points. By the purity theorem (Chapter 0, Proposition 0.1.10), it is unramified everywhere. Thus, F must be a μ_2-surface. Similarly, we prove that $\mathcal{D} \neq \mathcal{D}_3$ if F is a μ_2-surface.

(iv), (vi) $\mathcal{D}$ contains one line, the image of the unique multiple fibre of $|2E|$. Also, $\mathcal{D}$ contains one pencil of conics, the image of the pencil $|M+K_F|$ .By Proposition 0.4.6', $\mathcal{D} = \mathcal{D}_2'$ or $\mathcal{D}_3'$. Again, we distinguish the two cases by the argument from the proof of (iii), (v).

Remark 4.7.2. The other properties of symmetric anticanonical quartic Del Pezzo surfaces $\mathcal{D}$ can be also detected from the properties of $|C|$. For example, the singular locus of $\mathcal{D}$ is the image of the base-loci of $|M|$ and $|M+K_F|$. If $K_F \neq 0$, $\mathcal{D}$ is the intersection of two quadrics of rank 3, each of them generated by the 2-planes containing the conics. In case (i) $\mathcal{D}$ is contained in a quadric of rank 4, whose two rulings by planes contain the conics corresponding to $|2E_1|$ and $|2E_2|$.

Corollary 4.7.2. Let $p:\bar{F} \to F$ be the K3-cover of F and $|\Gamma| = |p^{-1}(M)|$. Then, we have the following commutative diagram

$$\begin{array}{ccc} \bar{F} & \xrightarrow{\varphi_\Gamma} & Y \subset \mathbb{P}^3 \\ p\downarrow & & \downarrow p' \\ F & \xrightarrow{\varphi_\Gamma} & \mathcal{D} \subset \mathbb{P}^4 \end{array}$$

where φ_Γ is a map of degree 2 of a quadric Y (singular in cases (ii) and (iv)) and $p':Y \to \mathcal{D}$ is a map of degree 2 which is a principal double cover over nonsingular points of $\mathcal{D}$. Moreover, p' and $\mathcal{D}$ is of type μ_2 (resp. $\mathbf{Z}/2$, resp. $\boldsymbol{\alpha}_2$) if and only if F is classical (resp. a μ_2-surface, resp. an $\boldsymbol{\alpha}_2$-surface).

PROOF. Take the cover $p':Y \to \mathcal{D}$ constructed in Chapter 0, §4. Its pull-back to F under the map $F \to \mathcal{D}$ defined in the theorem is a principal double cover over the pre-image of the nonsingular locus of $\mathcal{D}$ which is of type equal to the type of $\mathcal{D}$. By the purity theorem, it extends to a principal double cover of F. It remains to use Chapter 3, §2.

Theorem 4.7.4. Let $|C|$ be a superelliptic system on F with $C^2 = 6$. Then φ_C is a map of degree 2 onto a symmetric cubic $\mathcal{C}$ in $\mathbb{P}^3$. Moreover,

(i) $\mathcal{C} = \mathcal{C}_1, \mathcal{C}_1'$, or $\mathcal{C}_1''$ if F is classical;

(ii) $\mathcal{C} = \mathcal{C}_2, \mathcal{C}_2'$, or $\mathcal{C}_2''$ if F is a μ_2-surface;

(iii) $\mathcal{C} = \mathcal{C}_3, \mathcal{C}_3'$, or $\mathcal{C}_3''$ if F is an α_2-surface.

PROOF. By Proposition 4.7.2, $|C| = |2M-Z|$, where $|2M|$ is a superelliptic system with $M^2 = 2$, and Z is a fundamental cycle of $\mathcal{R}_M$. This defines a commutative diagram:

$$\begin{array}{ccc} & & \mathcal{D} \subset \mathbb{P}^4 \\ & \varphi_{|2M|} \nearrow & \\ F & & \downarrow \pi_Z \\ & \varphi_C \searrow & \\ & & \mathcal{C} \subset \mathbb{P}^3 \end{array}$$

where π_Z is the projection from the contracted image of Z. The theorem follows easily from analyzing Proposition 4.7.2 and applying Proposition 0.5.2 of Chapter 0.

Remark 4.7.3. The configuration of lines on $\mathcal{C}$ can be easily detected from the structure of $|C|$ and the following simple remark: if $|2E_i|, i=1,2,3$, are three genus 1 pencils with $E_i \cdot E_j = 1$, $i \neq j$, then $\mathcal{O}_{E_1}(E_2-E_3) \cong \mathcal{O}_{E_1}$ if and only if $E_1+E_2-E_2' > 0$, $E_2' \in |E_2+K_F|$. We use the notation of Theorem 4.7.1 and its Corollary 4.7.1.

(i) $|C| = |E_1+E_2+E_3|$, $K_F \nsim 0$. Then the images of three pairs of multiple fibres of $|2E_i|$ form a tetrahedron of lines on $\mathcal{C}_1$ (the opposite edges come from the same pencil). The images of the divisors Z_i are the other 3 lines on $\mathcal{C}_1$.

(ii) $|C| = |E_1+E_2+E_3|$, $K_F \sim 0$. Then the images of three unique multiple fibres of $|2E_i|$ are the three lines on $\mathcal{C}_1'$ intersecting at $\mathrm{Sing}(\mathcal{C}_1')$. The images of the divisors Z_i are the three coplanar lines on $\mathcal{C}_1'$ not passing through $\mathrm{Sing}(\mathcal{C}_1')$.

(iii) $|C| = |2E_1+E_2+R_1|$, $K_F \nsim 0$. Then the images of two pairs of multiple fibres of $|2E_i|$ define 4 lines on $\mathcal{B}_2$. The image of R_1 is the singular point of type A_3. The image of the divisor Z_2 is the fifth line on $\mathcal{B}_2$.

(iv) $|C| = |2E_1+E_2+R_1|$, $K_F \sim 0$. Then $\mathcal{B}_2$ contains 3 lines, the images of the unique multiple fibres of $|2E_i|$ and of Z_2.

(v) $|C| = |3E_1+2R_1+R_2|$, $K_F \not\sim 0$. The images of the two multiple fibres of $|2E_1|$ are the two lines on $\mathcal{B}_3$.

(vi) $|C| = |3E_1+2R_1+R_2|$, $K_F \sim 0$. The image of E_1 is the unique line on $\mathcal{B}_3$.

S8. The branch locus of superelliptic maps.

Let $\varphi: F \to \mathcal{D}$ be the superelliptic map given by a system $|C| = |2M|$, where F is an Enriques surface and $\mathcal{D} \subset \mathbb{P}^4$ is a symmetric anticanonical quartic Del Pezzo surface. In this section we describe the branch locus of this map.

Lemma 4.8.1. There exists a commutative diagram of morphisms

$$\begin{array}{ccccc} \tilde{F} & \xrightarrow{\tilde{r}} & \tilde{F}' & \xrightarrow{\tilde{\varphi}} & X \\ \pi\downarrow & & \tilde{\pi}'\downarrow & & p\downarrow \\ F & \xrightarrow{r} & F' & \xrightarrow{\varphi'} & \mathcal{D} \end{array} ,$$

where

$\varphi = \varphi' \circ r$,

r is the blowing-down of $\mathcal{R}_C$, φ' is a finite map of degree 2,

p is a minimal resolution of $\mathcal{D}$,

π is a birational morphism of nonsingular surfaces,

π' is a birational morphism of surfaces with double rational points,

$\tilde{\varphi}'$ is a double cover,

$\tilde{r}$ is a birational morphism.

PROOF. The lower row is the Stein decomposition of φ .Then we define $\tilde{F}$ as $(F \times_{\mathcal{D}} X)_{red}$, and use the Stein decomposition of the projection $\tilde{\varphi}:\tilde{F} \to X$ to define the upper row. The vertical arrows are defined naturally. We also apply Proposition 0.2.7 from Chapter 0 to deduce the remaining properties of the maps from the diagram.

The map $\tilde{\varphi}':\tilde{F}' \to X$ is a double cover in sense of Chapter 0,S1. Let

$$0 \to \mathcal{O}_X \to \tilde{\varphi}'_*\mathcal{O}_{\tilde{F}'} \to \mathcal{L} \to 0$$

the corresponding extension for the Algebra $\tilde{\varphi}'_*\mathcal{O}_{\tilde{F}'}$ for which

$$\tilde{F}' = \mathrm{Spec}(\tilde{\varphi}'_*\mathcal{O}_{\tilde{F}'}).$$

We denote by $\tilde{\varphi}$: $\tilde{F} \to X$ the composition $\tilde{\varphi}'\circ\tilde{r}$. We will use the notation of Theorem 4.7.3 to distinguish the 6 possible cases for the map φ and keep the notation of the previous lemma. .

Lemma 4.8.2. (i) Suppose we are in case (i) of Theorem 4.7.3. Then

a) π is the blowing up of the 4 vertices P_i of the quadrangle of the half-fibres of $|2E_1|$ and $|2E_2|$;

b) r (resp. $\tilde{r}$) is the blowing-down of $\mathcal{R}_C$ (resp. $\pi^{-1}(\mathcal{R}_C)$),

(ii) Suppose we are in case (ii) of Theorem 4.7.3. Then

a) π is the blowing up of the 4 vertices P_1,P_2,P_1',P_2' of the degenerate quadrangle of the half-fibres of $|2E_1|$ and $|2E_2|$;

b) r (resp. $\tilde{r}$) is the blowing-down of $\mathcal{R}_C \cup R$ (resp. $\pi^{-1}(\mathcal{R}_C)$),$r(R)= r(P_1) = r(P_2)$ is a double ordinary point of F' lying over the singular point of type A_3 of $\mathcal{D}'$; $r(P_1')$ and $r(P_2')$ are nonsingular points of F' lying over the singular points of type A_1 of $\mathcal{D}'$, $\tilde{\pi}^{-1}(r(R)) = \tilde{r}(R)$.

(iii), (v) Suppose we are in case (iii) or (v) of Theorem 4.7.3. Then

a) π is the blowing-up the point $P = E_1 \cap P_2$ and three points P',P'',P''' lying on the exceptional curve blown-up from P;

b) r (resp. $\tilde{r}$) is a blowing-down of $\mathcal{R}_C$ (resp. $\pi^{-1}(\mathcal{R}_C)$),$r(P)$ is a nonsingular point of F' lying over the singular point of $\mathcal{D}$.

(iv),(vi) Suppose we are in case (iv) or (vi) of Theorem 4.7.3. Then

a) π is the blowing-up the point $P = E_1 \cap R$ and three points P',P'',P''' lying on th exceptional curve blown-up from P, one of them P' corresponds to the direction at P defined by R;

b) r (resp. $\tilde{r}$) is a blowing-down of $\mathcal{R}_C \cup R$ (resp. $\pi^{-1}(\mathcal{R}_C)$), $r(R)$ is an ordinary double point of F' lying over the singular point of $\mathcal{D}$, $\tilde{\pi}^{-1}(r(R)) = \tilde{r}(R)$.

PROOF. It is easily verified case by case.

Denote by $\bar{C}$ the total transform of the points of indeterminacy of π^{-1}. We have

$$\bar{C} = \bar{C}_1+\bar{C}_2+\bar{C}_3+\bar{C}_4 \quad , \bar{C}_i^2 = -1, \text{ in cases (i) and (ii);}$$

$$\bar{C} = \bar{C}_1+2\bar{C}_2+2\bar{C}_3+2\bar{C}_4, \ \bar{C}_1^2 = -4, \ \bar{C}_i^2 = -1, \ i\neq 1, \text{ otherwise.}$$

It follows easily from the previous lemma that

$$\bar{C} = \tilde{\varphi}^*(L) \ (\text{ resp. } \bar{C} + \pi^{-1}(R) = \tilde{\varphi}^*(L)),$$

where L is the exceptional locus of p in cases (i), (iii), (v) (resp. (ii), (iv), (vi)). Moreover, in cases (iii) - (vi), $\tilde{\varphi}(L_1)$ is the central component of the exceptional curve of π.

We denote by L_0 the component $\tilde{\varphi}(\pi^{-1}(R))$ of L when it applies and understand that it is zero if it is not defined. It is the central component of the exceptional curve of the singularity A_3 in case (ii) and the furthest from the central component of the exceptional curve of the singularity D_5 in cases (iv) and (vi).

Proposition 4.8.1. Let L be the exceptional locus of p. Then

$$\mathcal{I} \cong \mathcal{O}_X(-L+L_0)\otimes p^*(\mathcal{O}_{\mathcal{D}}(-1)) \text{ if } K_F \sim 0,$$

$$\mathcal{I}^2 \cong \mathcal{O}_X(-L+L_0)\otimes p^*(\mathcal{O}_{\mathcal{D}}(-2)) \text{ if } K_F \not\sim 0.$$

PROOF. By Proposition 0.1.3,

$$\omega_{F'} = \tilde{\varphi}'^*(\omega_{Y'}\otimes \mathcal{I}^{-1}).$$

Since F' and $\mathcal{D}$ have only rational double singularities,

$$\omega_F \cong \tilde{r}^*(\omega_{F'}), \ p^*(\omega_{\mathcal{D}}) \cong \omega_X.$$

Hence

$$\tilde{\varphi}^*(\mathcal{I}) = \tilde{r}^*(\tilde{\varphi}'^*(\mathcal{I})) = \tilde{r}^*(\omega_{F'}^{-1}\otimes\tilde{\varphi}'^*(\omega_X)) = \omega_F^{-1}\otimes\tilde{\varphi}^*(\omega_X) =$$

$$= \omega_{\bar{F}}^{-1} \otimes \varphi'^*(p^*(\omega_{\mathcal{D}})) = \omega_{\bar{F}}^{-1} \otimes \varphi'^*(p^*(\mathcal{O}_{\mathcal{D}}(-1))).$$

Obviously

$$\omega_{\bar{F}} = \pi^*(\omega_F) \otimes \mathcal{O}_{\bar{F}}(\bar{L}).$$

Thus

$$\tilde{\varphi}^*(\mathcal{I}) \cong \pi^*(\omega_F^{-1}) \otimes \mathcal{O}_{\bar{F}}(-\bar{L}) \otimes \tilde{\varphi}^*(p^*(\mathcal{O}_{\mathcal{D}}(-1))) \cong$$

$$\cong \pi^*(\omega_F^{-1}) \otimes \tilde{\varphi}^*(\mathcal{O}_X(-L+L_0)) \otimes \tilde{\varphi}^*(p^*(\mathcal{O}_{\mathcal{D}}(-1)).$$

Since $\tilde{\varphi}^*$: Pic(X) $\to$ Pic($\bar{F}$) is injective, we find

$$\mathcal{I} \cong \mathcal{O}_X(-L+L_0) \otimes p^*(\mathcal{O}_{\mathcal{D}}(-1)) \text{ if } K_F \sim 0,$$

$$\mathcal{I}^2 \cong \mathcal{O}_X(-L+L_0) \otimes p^*(\mathcal{O}_{\mathcal{D}}(-2)) \text{ if } K_F \nsim 0.$$

Lemma 4.8.3. Let $q:X \to \mathbb{P}^2$ be the representation of X as the blowing up of 5 points $p_1,\dots,p_5$ on $\mathbb{P}^2$ in special position (see Chapter 0, §4). Let e_0 be the class in Pic(X) of the inverse transform of a line, and e_i be the classes of $q^{-1}(p_i)$, $i = 1,\dots,5$. Then

$$\mathcal{I}^{-1} = 4e_0-2e_1-2e_3-2e_5-e_2-e_4$$

PROOF. We know that the curves L_i, $i \neq 0$, on X can be given by (in the corresponding order:

$\mathcal{D}_1$: $e_0-e_1-e_2-e_3$, $e_0-e_1-e_4-e_5$, e_2-e_3, e_4-e_5,

$\mathcal{D}_1'$: $e_0-e_1-e_4-e_5$, $e_0-e_1-e_2-e_3$, e_2-e_3, e_4-e_5,

$\mathcal{D}_2,\mathcal{D}_3$: $e_0-e_1-e_2-e_3$, e_2-e_3, e_3-e_4, e_4-e_5

$\mathcal{D}_2',\mathcal{D}_3'$: $e_0-e_1-e_2-e_3$, e_2-e_3, e_3-e_4, e_4-e_5.

Adding these classes up, we get

$$L-L_0 = 2(e_0-e_1-e_3-e_5) \text{ if } \mathcal{D} = \mathcal{D}_1, \mathcal{D}_1',$$

$$= e_0-e_1-e_3-e_5 \text{ otherwise.}$$

Since

$$p^*(\mathcal{O}_{\mathcal{D}}(1)) = -K_X = 3e_0-e_1-e_2-e_3-e_4-e_5,$$

we obtain from Proposition 4.8.1

$$\mathcal{L}^{-1} = 4e_0-2e_1-2e_3-2e_5-e_2-e_4.$$

Corollary 4.8.1.

$$H^1(X,\mathcal{L}^{-1}) = 0, \text{ if } \mathcal{D} = \mathcal{D}_1,\mathcal{D}_1';$$
$$\cong k \text{ otherwise.}$$

PROOF. Assume $\mathcal{D} = \mathcal{D}_1$(resp. $\mathcal{D}_1'$). Then

$$4e_0-2e_1-2e_3-2e_5-e_2-e_4 =$$
$$= (2e_0-e_2-e_4)+(e_0-e_1-e_2-e_3)+(e_0-e_1-e_4-e_5)+(e_2-e_3)+(e_4-e_5).$$

(resp.

$$4e_0-2e_1-2e_3-2e_5-e_2-e_4 = (2e_0-e_1-e_2)+(e_1-e_4)+(e_0-e_1-e_2-e_3)$$
$$+(e_0-e_1-e_4-e_5)+(e_2-e_3)+(e_4-e_5).$$

This shows that the fixed part of $|\mathcal{L}^{-1}|$ consists of L and its moving part is of dimension 3. The latter is represented by the linear system of conics passing through p_2 and p_4 (resp. p_1 and p_2). It follows from Riemann-Roch that $H^1(X,\mathcal{L}^{-1}) = 0$ in this case.

Assume $\mathcal{D} \neq \mathcal{D}_1,\mathcal{D}_1'$, then

$$4e_0-2e_1-2e_3-2e_5-e_2-e_4 = (3e_0-e_1-e_2-e_3-e_4-e_5)+$$
$$+(e_0-e_1-e_2-e_3)+(e_2-e_3)+(e_3-e_4)+(e_4-e_5).$$

This shows that the fixed component of $|\mathcal{L}^{-1}|$ is represented by $L-L_0$ and its moving part is of dimension 4 . The latter is represented by the linear system of cubic curves passing through p_1,p_2,p_3,p_4 and p_5. By Riemann-Roch, again $H^1(X,\mathcal{L}^{-1}) = k$.

Theorem 4.8.1. Let $\tilde{\varphi}':\tilde{F}' \to X$ be a double cover defined by a superelliptic map φ_{2M} of an Enriques surface F. Then (in the notation of Lemma 4.8.3)
(i) Assume char(k) $\neq$ 2. Then $\tilde{\varphi}'$ is a separable cover branched along the curve

$L-L_0$ and a curve $W \in |6e_0-2e_1-2e_2-2e_3-2e_4-2e_5| = |p^*(\mathcal{O}_{\mathcal{D}}(2)|$ with at most simple singularities which does not intersect L.

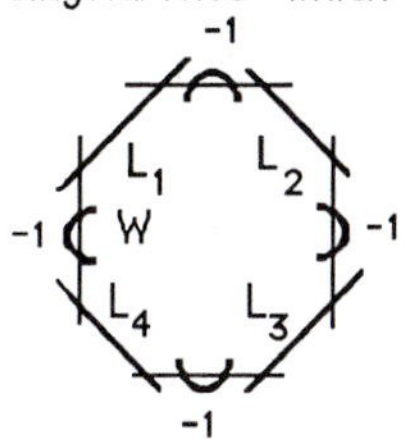

$\mathcal{D}_1$

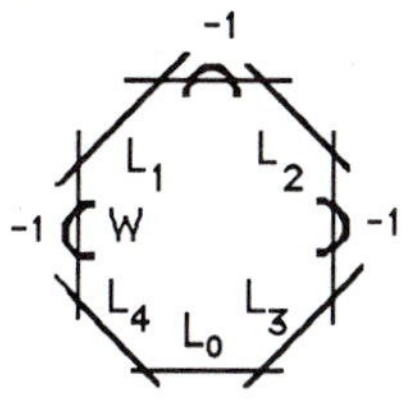

$\mathcal{D}_1'$

(ii) Assume char(k) = 2 and F is classical. The following two cases may occur:

(a) $\tilde{\varphi}'$ is a separable double cover branched along the curve L and a curve $W \in |2e_0-e_2-e_4|$ if $\mathcal{D} = \mathcal{D}_1$ (resp.along L and a curve $W \in |2e_0-e_1-e_2|$ if $\mathcal{D} = \mathcal{D}_1'$) which intersects L as shown in the picture:

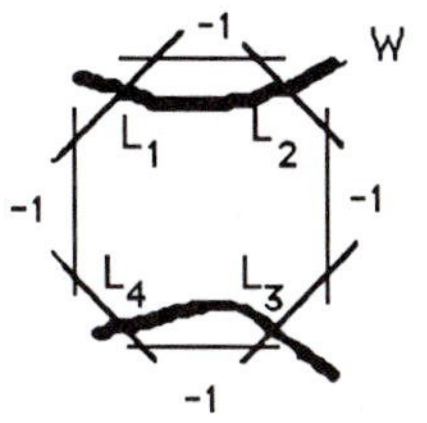

$\mathcal{D}_1$

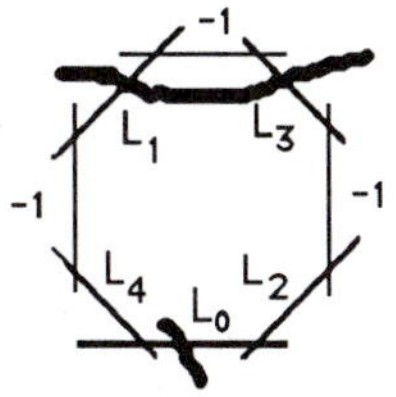

$\mathcal{D}_1'$

(b) $\tilde{\varphi}'$ is an inseparable double cover with splittable admissible triple branched along the curve $L-L_0$ and a curve curve $W \in |6e_0-2e_1-2e_2-2e_3-2e_4-2e_5| = |p^*(\mathcal{O}_{\mathcal{D}}(2)|$ which does not intersect L.

(iii) Assume F is a μ_2-surface. Then $\tilde{\varphi}'$is a separable double cover which ramifies along the curve $L-L_0$ and a curve $W \in |3e_0-e_1-e_2-e_3-e_4-e_5| = |p^*(\mathcal{O}_{\mathcal{D}}(1)|$ which is disjoint from $L-L_0$.

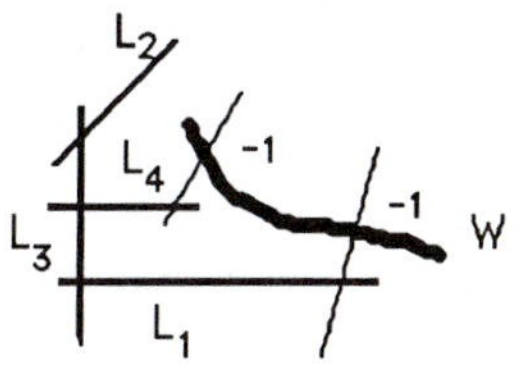

$\mathcal{D}_2$

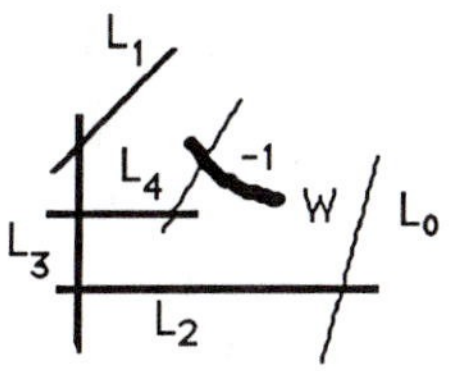

$\mathcal{D}_2'$

(iv) Assume F is an α_2-surface. The following two cases may occur:

(a) $\tilde{\varphi}'$ is a separable double cover with the branch divisor $D = 2L+L_3+W$, where $W \in |2e_0-e_2-e_3|$ if $\mathcal{D} = \mathcal{D}_3$ (resp. $D = 2L-L_0+2L_2+2L_3+W$, where $W \in |2e_0-e_1-e_2|$ if $\mathcal{D} = \mathcal{D}_3'$) and W intersects L as shown in the picture:

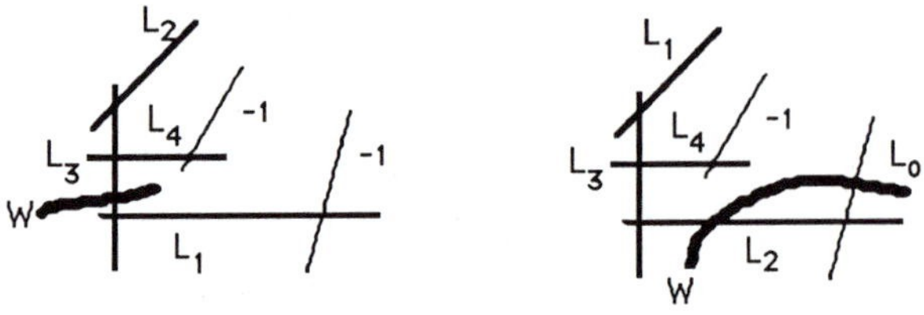

$\mathcal{D}_3$ $\qquad\qquad$ $\mathcal{D}_3'$

(b) $\tilde{\varphi}'$ is an inseparable double cover with non-splittable admissible triple.

PROOF. We will consider only the case $\mathcal{D} = \mathcal{D}_i$ and leave the case $\mathcal{D} = \mathcal{D}_i'$ to the reader.

(i) We know that $\tilde{\varphi}$ is a separable double cover with the branch divisor $D \in |\mathcal{L}^{-2}|$. Since $\tilde{\varphi}$ ramifies over L, $D = L+W$ for some curve W. We saw in the proof of Lemma 4.8.3 that $L \in |2(e_0-e_1-e_3-e_5)|$ and $D \in |8e_0-4e_1-2e_2-4e_3-2e_4-4e_5|$. Hence

$$W \in |6e_0-2e_1-2e_2-2e_3-2e_4-2e_5|.$$

and $W \cdot L = 0$.

(ii) Assume $\tilde{\varphi}$ is separable. By Lemma 4.8.2, the branch divisor D of $\tilde{\varphi}$ belongs to the linear system $|4e_0-2e_1-e_2-2e_3-e_4-2e_5|$. We know that $\tilde{\varphi}(L_i) = 2\bar{L}_i$ for all $L_i \neq L_0$. By Corollary 0.1.1, L must enter into the branch divisor. Thus, $D = L+W$, and since $L \sim 2(e_0-e_1-e_3-e_5)$, we obtain

$$W \in |2e_0-e_2-e_4|$$

and $W \cdot L_i = 1$ for $i \neq 0$.

Assume $\tilde{\varphi}$ is inseparable. By Corollary 4.8.1, $H^1(X,\mathcal{L}^{-1}) = 0$. This shows that $\tilde{\varphi}$ corresponds to a splittable admissible triple and its branch divisor D $\in |\mathcal{L}^{-2}|$. as in the previous case we see that $D = L+W$ for some curve W. Knowing the classes of D and of $L-L_0$ in Pic(X), we easily find $W \in |6e_0-2e_1-2e_2-2e_3-2e_4-2e_5|$. Since $\tilde{F}'$ is normal, W does not contain common components with L. Counting the intersection number $W \cdot L$, we see that it is zero.

(iii) Assume $\tilde{\varphi}$ is inseparable. Then φ is inseparable and F contains a genus 1 pencil which is the inverse transform of a pencil of conics on $\mathcal{D}$ (see the proof of Theorem 4.7.3). This implies that F contains a quasi-elliptic pencil (since a smooth elliptic curve is not homeomorphic to a smooth rational curve). We will see in Chapter 5, §7 that a μ_2-surface does not contain quasi-elliptic pencils. This shows that $\tilde{\varphi}$ is separable. Let D be the branch divisor. Let L_1 be the central component of L. Since $\tilde{\varphi}^{-1}(L_i) = 2\bar{L}_i$ for $i \neq 1$, the components L_2, L_3 and L_4 enter in D. Since L_1 intersects the branch divisor at at least three points and $\tilde{\varphi}^{-1}(L_1)$ is a smooth rational curve, L_1 must enter into D also. Thus D = L+W, where

$$W \in |D-L| = |(4e_0-2e_1-e_2-2e_3-e_4-2e_5)-(e_0-e_1-e_3-e_5)| =$$

$$= |3e_0-e_1\ldots-e_5|.$$

Since the local fundamental group of the singularity of $\mathcal{D}_2$, $\mathcal{D}_2'$ is $\mathbb{Z}/2$, W does not intersect L. In particular, it has no common components with L.

(iv) Assume $\tilde{\varphi}$ is separable. Then, as in the previous case, the branch divisor $D = L+W'$, $W' \in |3e_0-e_1\ldots-e_5|$. However, in this case W' must intersect L, since the singularity of $\mathcal{D}$ has trivial fundamental group. Since $W' \cdot L = 0$, this can happen only if W' and L have common components. One easily checks that this is possible only if $W' = W+L+L_2$, where $W \in |2e_0-e_2-e_3|$.

Assume $\tilde{\varphi}$ is inseparable. By Corollary 4.8.1, $H^1(X, \mathcal{L}^{-1}) \neq 0$. Thus $\tilde{\varphi}$ corresponds to nonsplittable admissible triple.

Theorem 4.8.2. Assume $char(k) \neq 2$. Let $\mathcal{D} = \mathcal{D}_1$ or $\mathcal{D}_1'$ and $W \in |\mathcal{O}_{\mathcal{D}}(2)|$ with simple singularities which does not pass through $Sing(\mathcal{D})$. Then there exists a superelliptic map $\varphi: F \to \mathcal{D}$ which is a non-ramified double cover outside W and $Sing(\mathcal{D})$.

PROOF. Left to the reader.

The branch locus of a superelliptic map φ_C with $C^2 = 6$ is described similarly. We use the fact that it is equal to the projection of the branch locus described above. We omit the lengthy statements of the corresponding results.

§9. Projective models of degree ≤ 10.

The purpose of this paragraph is to discuss some projective models of Enriques surfaces obtained by complete linear systems |C| with $4 \le C^2 \le 10$.

The following table summarizes our results:

C^2	$\Phi(C)$	deg φ_C	$\varphi_C(F)$	type
4	1	2	$\mathbb{P}^2$	double plane
	2	4	$\mathbb{P}^2$	quadruple plane
6	1	2	$S_{0,0}$ or $S_{2,0}$	double quadric
	2	2	$\mathcal{C}_i$ or $\mathcal{C}_i'$	double cubics
	2	1		Enriques sextic
8	1	2	$S_{1,0}$	double $S_{1,0}$
	2	2	$\mathcal{D}_i$ or $\mathcal{D}_i'$	double quartic Del Pezzo surface
	2	1	non-normal octic	
10	1	2	$S_{0,2}$	double $S_{0,2}$
	2		non-normal surface of degree 10	
	3	1	normal surface of degree 10	Fano model (F is not an α_2-surface)

Recall that the existence of curves C with the given values of $\varphi(C)$ follows from the classification of O(Num(F))-orbits of vectors x in Num(F) with $x^2 \le 10$ (Chapter 2, §5). The intersection of these orbits with the nodal chamber C_F defines the class [C]. The properties of the corresponding map φ_C follow from the previous paragraphs in this chapter.

We proceed to make some comments about this table.

A) Double planes.

Enriques' double plane construction corresponds to hyperelliptic systems $|C|$ with $C^2 = 4$. There exist two types of double planes: the **non-special type** when $|C| = |2E_1+E_2|$ is non-special and the **special type** when $|C| = |3E+R|$ is special.

<u>$p \neq 2$</u>.

Let $|C| = |2E_1+E_2|$ be a non-special hyperelliptic system with $C^2 = 4$. Assume $p \neq 2$. Recall from S5 that the rational map $\varphi_C = \varphi_1 \circ \varphi_2 \circ \varphi_3$, where φ_1 blows up the two points $E_1 \cap E_2$ and $E_1' \cap E_2$ ($E_1' \in |E_1+K_F|$), φ_2 blows down the proper transforms of curves E_1, E_1', and E_2, and φ_3 is a cover of $\mathbb{P}^2$ with the branch curve of degree 8 which is the union of two lines l_1 and l_2 and a curve of degree 6 with 2 tacnodes P and Q, a node O and some other simple singularties.

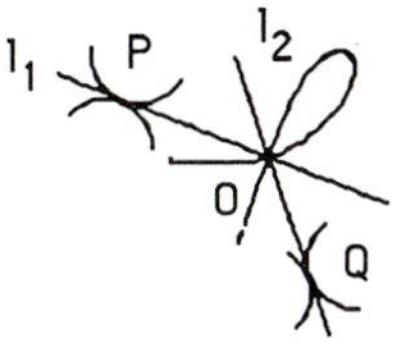

The curves E_1 and E_1' are sent to the tacnodes P and Q. The curve E_2 is mapped to O. The exceptional curves blown up from $E_1 \cap E_2$ and $E_1' \cap E_2$ are mapped to the branch lines l_1 and l_2.

Let $|C| = |3E+R|$ be a special hyperelliptic system with $C^2 = 4$. Recall from S5 that the rational map $\varphi_C = \varphi_1 \circ \varphi_2 \circ \varphi_3$, where φ_1 blows up the two points $p \in E_1, p \notin R$ and $q = E_1' \cap R$ ($E_1' \in |E_1+K_F|$), φ_2 blows down the proper transforms of curves E_1, E_1', and R, and φ_3 is a cover of $\mathbb{P}^2$ with the branch curve of degree 8 which is the union of two lines l_1 and l_2 and a curve of degree 6 with 3 tacnodes, where two of them are infinitely near and some other simple singularities.

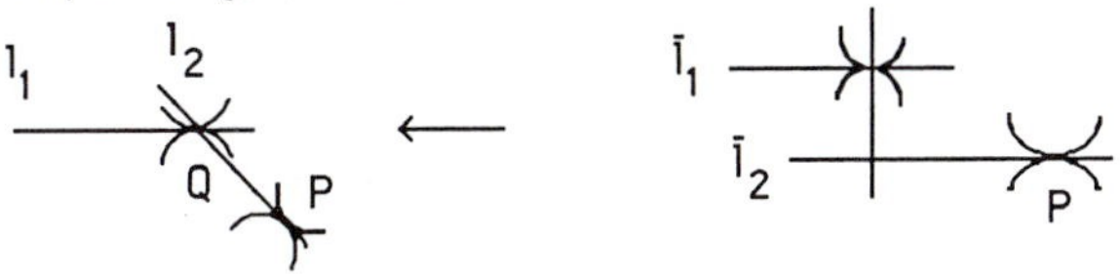

The curve E_1 is carried to the tacnode P. The curve E_1' is mapped to Q. The exceptional curves blown up from p is mapped to l_1 and the one which is blown up from q is mapped to l_1.

p = 2.

We leave to the reader to modify the previous discussion in this case by using the description of the branch locus of superelliptic maps given in Theorem 4.8.1. The following pictures describe the corresponding branch loci in the case where φ_C is separable:

F is classical

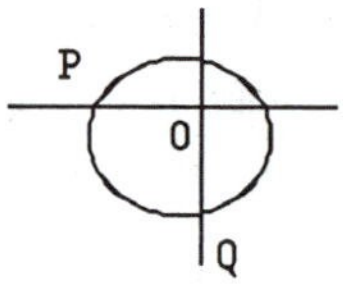

non-special

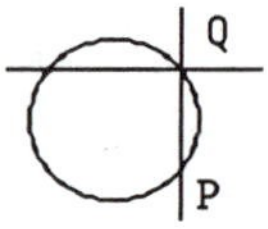

special

F is a μ_2-surface:

non-special

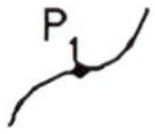

special

F is an α_2-surface:

non-special

special

Applying Theorem 3.4.1, we obtain

Theorem 4.9.1. Assume char(k) ≠ 2 or F is not E_8-special. Then F is isomorphic to a double plane of non-special type.

B) Quadruple planes.

By Propositions 4.5.1 and 4.7.1 they are defined by the linear systems |C| with

$C^2 = 4$ and $\varphi(C) = 2$. The linear systems with this property are described by Proposition 3.6.2. The corresponding map φ_C is a 4-fold cover of $\mathbb{P}^2$. If $p \neq 2$, the branch locus of this map was studied by A. Verra **[Ve 1]**. In the generic case, this is a curve of degree 12, geometric genus 19 with exactly 36 ordinary cusps which lie on a sextic curve. This description is intimately related to the following theorem:

Theorem 4.9.2. Assume $p \neq 2$. Let $\pi: \bar{F} \to F$ be the K3-cover of F and $|\Gamma| = |\pi^*(C)|$, where $|C|$ as above. Then $\varphi_\Gamma(\bar{F})$ is the intersection of 3 quadrics in $\mathbb{P}^5$.

PROOF. Since φ_C is not hyperelliptic, φ_Γ is not hyperelliptic (Corollary 4.2.1). Therefore, it is a morphism of degree 1 onto a surface of degree 8 in $\mathbb{P}^5$ (see §2). Moreover, we can apply Corollary 4.2.3 to deduce that this surface is given by quadric equations.

Remark 4.9.1. Let τ be the involution of $\bar{F}$ defined defined by the K3-cover π. We can choose projective coordinates $(x_0,x_1,x_2,y_0,y_1,y_2)$ in $\mathbb{P}^5$ such that any quadric $Q(\lambda)$ containing $\varphi_\Gamma(F)$ is given by an equation of the form:

$$Q(\lambda):\ q(\lambda)(x)+q'(\lambda)(y) = 0$$

and the involution τ is induced by the projective involution

$$\tau(x,y) = (x,-y).$$

Conversely, given a net of quadrics in $\mathbb{P}^5$ with smooth base locus F and an involution τ in $\mathbb{P}^5$ which acts freely on F, then we can reduce the equations of the quadrics and the formula for the involution to the the form above. The quotient surface $\bar{F} = F/(\tau)$ is an Enriques surface.

C) Double quadrics.

They correspond to hyperelliptic systems $|3E_1+E_2|$ and $|4E+R|$. We refer to §5 for the description of these maps in the case $p \neq 2$ and leave to the reader the corresponding statements in the case $p = 2$ (compare A)).

D) Double cubic surfaces ($p \neq 2$).

They correspond to superelliptic systems of degree 6. We know that there are 3 cases described by Theorem 4.7.2 or Corollary 4.7.1.

Case $\mathcal{C}_1$.

In this case $|C| = |E_1+E_2+E_3|$, where $|2E_i|$ are genus 1 pencils with $E_i \cdot E_j = 1$, $i \neq j$, $|E_1+E_2-E_3| \neq \emptyset$. The system $|E_1+E_2-E_3|$ consists of a connected nodal configuration Z and $|C| = |2M-Z|$, where $M = E_1+E_2$. The map $\varphi_{2M}: F \to \mathcal{D}_1$ blows down Z to a simple singularity x_0 of the branch curve B not lying on the the standard quadrangle. The rational map φ_C is the composition of φ_{2M} and the projection of $\mathcal{D}_1$ to $\mathcal{C}_1$ from x_0.

Comparing this construction with the construction of the double plane, we see that the sextic branch curve B of the double plane construction corresponding to the linear system $|2E_1+E_2|$ has a simple singularity S besides the singular points O, P and Q which does not lie on the branching lines. The standard Cremona transformation defined by the net of conics passing through P, Q and S transforms B into a Wirtinger sextic W with 6 nodes lying on the vertices of the complete quadrilateral. The anti-canonical model of the blowing-up of these vertices(given by the linear system of cubic curves passing through the nodes of W) defines the cubic surface $\mathcal{C}_1$. The image of W (and B) is a smooth curve X of genus 4 canonically embedded into $\mathbb{P}^3$. The double cover of $\mathcal{C}_1$ unramified outside the nodes of $\mathcal{C}_1$ defines an etale double cover of X. The corresponding Prym canonical model is the Wirtinger sextic W (cf.**[Co 3,Cat]**).

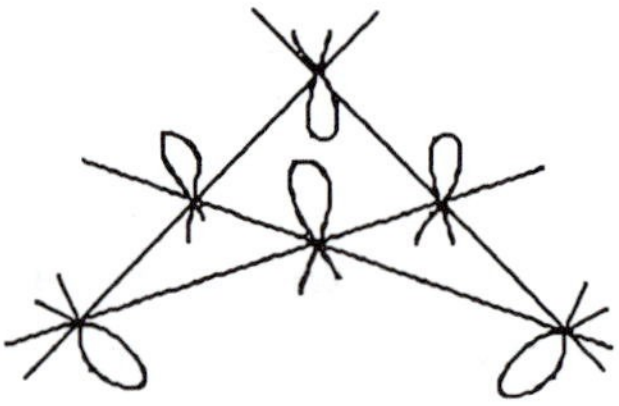

Case $\mathcal{B}_1'$:

In this case $|C| = |2E_1+E_2+R_1|$, $R_1 \bullet E_1 = E_1 \bullet E_2 = 1$, $R_1 \bullet E_2 = 0$, R_1 is a nodal curve, $|2E_1|$, $|2E_2|$ are genus 1 pencils, $|E_2-R_1| \neq \emptyset$ and consists of a connected nodal configuration Z. The linear system $|2M|$, where $M = E_1+E_2$, defines a map $\varphi_{2M}: F \to \mathcal{D}_1$ which blows down Z to a simple singularity x_0 of the branch curve which lies on the line $\varphi_{2M}(E_2)$.The rational map φ_C is the composition of φ_{2M} and the projection of $\mathcal{D}_1$ to $\mathcal{B}_1$ from x_0.

We leave to the reader the study of the corresponding branch locus.

Case $\mathcal{B}_1''$:

In this case $|C| = |3E_1+2R_1+R_2|$, $R_1 \bullet E_1 = R_1 \bullet R_2 = 1$, $E_1 \bullet R_2 = 0$, R_1 and R_2 are nodal curves, $|2E_1|$ is a genus 1 pencil, $|E_1-R_2| \neq \emptyset$ and consists of a connected nodal configuration Z. The linear system $|2M|$, where $M = 2E_1+R_1$, defines a map $\varphi_{2M}: F \to \mathcal{D}_2$ which blows down Z to a simple singularity x_0 of the branch curve which lies on the conic $\varphi_{2M}(R_2)$. The rational map φ_C is the composition of φ_{2M} and the projection of $\mathcal{D}_1$ to $\mathcal{B}_1$ from x_0.

We leave to the reader the study of the corresponding branch locus .

E) Enriques sextics (F is classical).

Theorem 4.9.3. Let $|C| = |E_1+E_2+E_3|$, where $|2E_i|$ are a genus 1 pencils with $E_i \bullet E_j = 1$, $i \neq j$. Assume that $|C|$ is not superelliptic. Then φ_C is a morphism of degree 1 onto a sextic surface in $\mathbb{P}^3$ isomorphic to a surface defined by an equation:

$$x_0 l_1 l_2 l_3 \varphi(x_0,x_1,x_2,x_3)+\psi(x_0 l_1 l_2, x_0 l_2 l_3, x_0 l_3 l_1, l_1 l_2 l_3) = 0 ,$$

where φ and ψ are homogeneneous polynomials of degree 2, l_i are linear forms in projective coordinates x_0,x_1,x_2,x_3 in $\mathbb{P}^3$.

PROOF. Since $|C|$ is not superelliptic, the map φ_C is of degree 1 onto a sextic surface with double lines $\varphi_C(E_i),\varphi_C(E_i')$, where $|2E_i'| = |2E_i|$. To find the equation of such a sextic we consider the rational functions $l_i \in H^0(F,\mathcal{O}_F(C))$ with the divisors

$$(l_1) = E_2'+E_3'-E_2-E_3, \quad (l_2) = E_1'+E_3'-E_1-E_3, \quad (l_3) = E_2'+E_1'-E_2-E_1$$

Let $1, x_1, x_2, x_3$ be a basis of $H^0(F, \mathcal{O}_F(C))$. By Riemann-Roch, dim $H^0(F, \mathcal{O}_F(2C)) = 13$. Therefore the following 14 functions from $H^0(F, \mathcal{O}_F(2C))$ are linearly dependent:

$$1, x_1, x_2, x_3, x_2 x_3, x_1 x_2, x_1 x_3, x_1^2, x_2^2, x_3^2, l_1 l_2 l_3^{-1}, l_2 l_3 l_1^{-1}, l_1 l_3 l_2^{-1}, l_1 l_2 l_3 \quad .$$

Multiplying a linear dependence relation by the product $l_1 l_2 l_3$ and homogenizing leads to the required equation.

Remark 4.9.2. The linear function $1, l_1, l_2, l_3$ are linearly independent if and only if the corresponding planes H_i in $\mathbb{P}^3$ associated to the divisors

$$E_1+E_2+E_3, \quad E_2'+E_3'+E_1, \quad E_3'+E_1'+E_2, \quad E_2'+E_1'+E_3$$

do not have a common point. Since H_1 intersects H_2 along a line, the image of E_1, and H_3 intersects H_4 along the image of E_1', we see that the condition of linear independence is equivalent to $|C-E_1-E_1'| = \emptyset$. The latter is equivalent to non-superellipticity of $|C+K_F|$. In this case, $\varphi_C(F)$ is isomorphic to a surface with an equation of the form:

$$x_0 x_1 x_2 x_3 \varphi(x) + \psi(x_1 x_2 x_3, x_0 x_2 x_3, x_0 x_1 x_3, x_0 x_1 x_2) = 0.$$

Clearly one can rewrite this equation as follows:

$$x_0 x_1 x_2 x_3 \varphi(x) + \lambda_1 (x_1 x_2 x_3)^2 + \lambda_2 (x_0 x_2 x_3)^2 + \lambda_3 (x_0 x_1 x_3)^2 + \lambda_4 (x_0 x_1 x_2)^2 = 0.$$

for some $\lambda_i \in K$. This is the classic form of an Enriques surface mentioned in the Introduction: a sextic surface passing doubly through the edges of the coordinate tetrahedron in $\mathbb{P}^3$.

A sextic surface given by the equation from Theorem 4.9.2 which has only rational double points besides the double lines will be called a **classical Enriques sextic** It is said to be **non-degenerate** if the linear functions $1, l_1, l_2, l_3$ are linearly independent hence the equation can be written in the form given in Remark 4.9.3. It is called **degenerate** otherwise.

Corollary 4.9.1. Let F be a classical Enriques surface which is not extra-special. Then F is birationally isomorphic to a classical non-degenerate Enriques sextic. .

PROOF. We know from Theorem 3.5.1 that F contains three genus 1 pencils $|2E_i|$ with $E_i \cdot E_j = 1$, $i \neq j$. It follows from Theorem 4.7.2 that either $|E_1+E_2+E_3|$ or $|E_1+E_2+E_3+K_F|$ is not superelliptic.

Proposition 1.9.1. A minimal nonsingular model of a classical Enriques sextic surface is a classical Enriques surface.

PROOF. By adjunction, the canonical system of an Enriques sextic Y is cut out by quadrics passing through the double lines. Obviously, it is empty. The bicanonical system is cut by the linear system of quartic surfaces passing doubly through the double lines. It is isolated and consists of one quartic, the union of the four coordinate planes. Therefore, $p_g(F) = 0$, $2K_F = 0$ for a minimal nonsingular model F of Y.

Note also that the proper transforms of three coplanar edges of the tetrahedron define three curves on F which are not numerically equivalent (consider their intersection with the inverse transforms of the oppositie edges). This shows that $b_2(F) \geq 3$ and excludes the possibility of F being a hyperelliptic surface (see Chapter 1, §1). Thus, F is an Enriques surface.

Remark 4.9.3. The standard cubic Cremona transformation in $\mathbb{P}^3$:

$$(x_0,x_1,x_2,x_3) \rightarrow (x_1x_2x_3, x_0x_2x_3, x_0x_1x_3, x_0x_1x_2)$$

composed with the map φ_C defines the map $\varphi_{C'}$, where $|C'| = |C+K_F|$. Its image is isomorphic to the surface

$$x_0x_1x_2x_3\psi(x) + \varphi(x_1x_2x_3, x_0x_2x_3, x_0x_1x_3, x_0x_1x_2) = 0.$$

When $\psi = \varphi$, the two surfaces are projectively equivalent.

Remark 4.9.4. The pencil of planes passing through an edge of the tetrahedron defines a pencil of curves of genus 2 on F. Taking a face of the tetrahedron, we see that the corresponding curve decomposes into a sum C_1+C_2, where C_i are the proper transforms of two edges. This obviously imply that $C_1^2 = C_2^2 = 0$, $C_1 \cdot C_2 = 1$, i.e. $\{|2C_1|, |2C_2|\}$ is a non-degenerate U-pair. In this way we reconstruct the three elliptic pencils $|2E_i|$.

Theorem 4.9.4. Let $|C| = |2E_1+E_2+R|$, where $|2E_i|$ are genus 1 pencils with $E_1 \cdot E_2 = 1$, and R is a nodal curve with $E_1 \cdot R = 1$, $E_2 \cdot R = 0$. Assume that $|C|$ is not superelliptic. Then φ_C is a morphism of degree 1 onto a sextic surface in $\mathbb{P}^3$ isomorphic to a surface defined by an equation:

$$x_0^2 l_1 l_2 \varphi(y)+a_1 x_0^4 l_1^2+a_2 x_0^4 l_2^2 + a_3 x_0 l_1^2 l_2^2 l_3 + a_4(l_1 l_2 l_3)^2 = 0,$$

where φ is a homogeneneous polynomial of degree 2, l_i are linear forms in projective coordinates x_0,x_1,x_2,x_3 in $\mathbb{P}^3$ and $a_i \in K$.

PROOF. Since $|C|$ is not superelliptic, the map φ_C is of degree 1 onto a sextic surface with 4 double lines $\varphi_C(E_i), \varphi_C(E_i')$, where $|2E_i'| = |2E_i|$. Let $C = E_1+E_1'+E_2'+R$ and consider the rational functions $l_i \in H^0(F,\mathcal{O}_F(C))$ with the divisors

$$(l_1) = E_2+E_1'-E_2'-E_1,\ (l_2) = E_1+E_2-E_1'-E_2',\ (l_3) = M-E_1-E_2-R,$$

where $M \in |E_1+E_1'+R|$ is disjoint from $E_1+E_1'+R$. Let $1,x_1,x_2,x_3$ be a basis of $H^0(F,\mathcal{O}_F(C))$. By Riemann-Roch, $\dim H^0(F,\mathcal{O}_F(2C)) = 13$. Therefore the following 14 functions from $H^0(F,\mathcal{O}_F(2C))$ are linearly dependent:

$$1,x_1,x_2,x_3,x_2x_3,x_1x_2,x_1x_3,x_1^2,x_2^2,x_3^2,l_1l_2^{-1},l_2l_3,l_1l_3,l_1l_2l_3.$$

Multiplying a linear dependence relation by $l_1l_2l_3$ and homogenizing leads to the required equation.

Corollary 4.9.2. Let F be a classical Enriques surface which is not E_8-special. Then F is birationally isomorphic to a sextic surface given by the equation from Theorem 4.9.3.

PROOF. Similarly to the proof of Corollary 4.9.1 where we refer to Theorem 3.5.1 instead of Corollary 3.4.1.

E') Enriques sextics (F is not classical).

Not much is known in this case. If F is unnodal, the linear system $|E_1+E_2+E_3|$ is not-superelliptic and the corresponding map defines a birational isomorphism between F and a sextic surface in $\mathbb{P}^3$. The images of the curves E_i are double lines of the sextic surface. What is its equation ? If F is nodal, we do not know whether there exist non-superelliptic linear systems $|E_1+E_2+E_3|$, even we do know that one can find a non-degenerate canonical isotropic 3-sequence provided F is not an extra-special α_2-surface.

F) Double quartic Del Pezzo surfaces.

They are defined by superelliptic linear systems $|C| = |2E_1+2E_2|$ or $|4E+2R|$. We refer to the previous section for the description of the corresponding maps. It is closely related to the double plane construction.

Let $C = |2E_1+E_2|$ be a non-special hyperelliptic system defining a non-special double plane construction. Let $M = |2E_1+2E_2|$. We have the following diagram of rational maps:

$$\begin{array}{ccc} & & \mathcal{D}_1 \\ & \varphi_{2M} \nearrow & \\ F & & \downarrow \pi_{E_2} \\ & \varphi_C \searrow & \\ & & \mathbb{P}^2 \end{array}$$

where $\pi_{E_2}:\mathcal{D}_1 \to \mathbb{P}^2$ is obtained by projection from the line $\varphi_{2M}(E_2)$ on $\mathcal{D}_1$.

Let $|C| = |3E+R|$ be a special hyperelliptic system defining a special double plane construcion. Let $|M| = |2E+R+K_F|$. We have the following diagram:

$$\begin{array}{ccc} & & \mathcal{D}_1' \\ & \varphi_{2M} \nearrow & \\ F & & \downarrow \pi_{E} \\ & \varphi_C \searrow & \\ & & \mathbb{P}^2 \end{array}$$

where $\pi_E:\mathcal{D}_1' \to \mathbb{P}^2$ is obtained by projection from the line $\varphi_{2M}(E)$ on $\mathcal{D}_1'$.

Assume that F is classical and $|M| = |E_1+E_2|$. Let $\pi:\mathcal{D}_1 \to \mathbb{F}_0$ be the rational map which is the composition of the blowing up of the vertices of the non-degenerate quadrangle with the blowing down the proper transform of the sides of the quadrangle. The composed rational map $\varphi = \pi\circ\varphi_{2M}: F \to \mathbb{F}_0$ is a double cover branched along a quadrangle of the rulings and a curve B of bidegree (4,4) with 4 nodes at the vertices of the quadrangle.

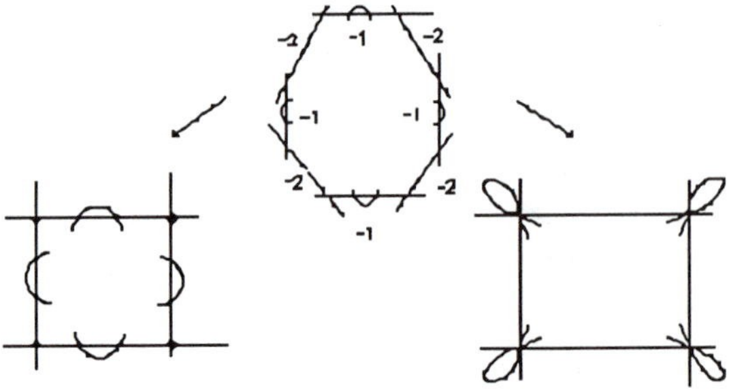

The image of the four multiple fibres of $|2E_1$ and $|2E_2|$ are the vertices of the quadrangle. The curve B is the image of the branch curve of φ_{2M} under π. It is a Prym-canonical model of a curve of arithmetic genus 5.

We leave to the reader the statement of the corresponding result in the case $|C| = |4E+2R|$. The corresponding picture is as follows:

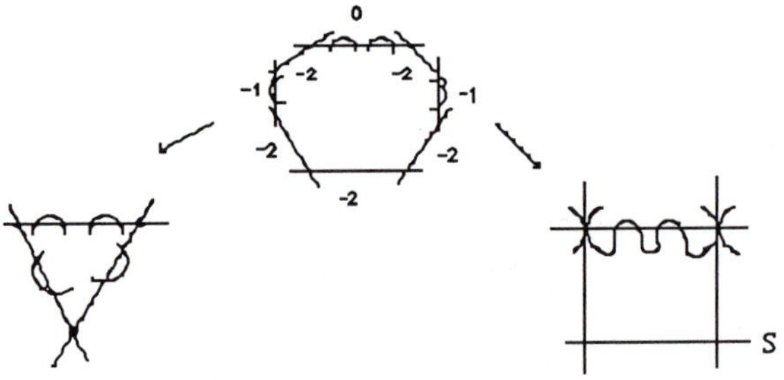

F) Non-normal octics.

These models correspond to non-superelliptic and linear systems $|C|$ with $C^2 = 8$ and $\Phi(C) = 2$. By Proposition 4.7.1, $|C|$ is of types (iii)-(v) from Propositions 3.6.2. This model has not been studied yet. In the case $|C| = |2E_1+E_2|$ and $p\neq 2$, the octic surface $V = \varphi_C(F)$ in $\mathbb{P}^4$ has 2 double lines, the images of the curves E_1 and $E_2\in|E_1+K_F|$ The hyperplane sections of V are curves of degree 8 with 2 nodes. They are Prym-canonical models of curves of genus 5 from $|C|$.

G) Double F_1.

This was discussed in S5.

H) Non-normal surfaces of degree 10 in $\mathbf{P}^5$.

They correspond to linear systems $|C|$ described in Proposition 3.6.3 ($q = 2$). Very little is known about these models. For example, if $|C| = |2E_1+E_2+E_3|$ and $p \neq 2$, we can say only that the image V of φ_C has two double lines which are the images of the curves E_1 and $E_1' \in |E_1+K_F|$. The curves E_2, E_2', E_3, E_3' are mapped to plane cubics. The projections from the corresponding planes is the non-special double plane construction.

I) Double $S_{0,2}$.

They corrspond to hyperelliptic maps defined by the linear systems $|5E_1+E_2|$ and $|6E_1+R+K_F|$. We refer to S5 for the description of these maps in the case $p \neq 2$.

J) Fano models.

These are normal surfaces of degree 10 in $\mathbb{P}^5$. We will study these models later in Part 2.

S10. Applications to linear systems.

Theorem 4.10.1. Let D an ample divisor on an Enriques surface F. Then there exists an irreducible curve C such that $C \in |D|$.

PROOF. Since $D^2 > 0$, $\dim|D| > 0$. Let $|D| = |M|+Z$ be the decomposition of $|D|$ into its moving part and the fixed part. Assume $M^2 = 0$. Then $|M|$ is composed of a genus 1 pencil and $D \cdot M \geq 2$. This implies that $D^2 = D \cdot M + D \cdot Z > 2$. This contradicts Corollary 3.1.2 of Chapter 3.

Therefore, we may assume that $M^2 > 0$ and $|M|$ is not composed with a pencil. If $Z \neq 0$,

$$\dim|M| = \tfrac{1}{2}M^2 = \dim|M+F| \geq \tfrac{1}{2}(M+F)^2 = \tfrac{1}{2}M^2 + \tfrac{1}{2}(2M\cdot F+F^2).$$

Hence $D\cdot F = D\cdot M + D\cdot Z = (2M\cdot F+F^2) - M\cdot F \leq 0$. This contradicts the ampleness of D.

Corollary 4.10.1. Let $\mathcal{L}$ be an ample invertible sheaf on F. Then

$$H^1(F,\mathcal{L}) = H^1(F,\mathcal{L}^{-1}) = 0.$$

PROOF. This follows from the vanishing theorem 1.3.1 of Chapter 1.

Corollary 4.10.2. Let $\mathcal{L}$ be an ample invertible sheaf on F. Then $\mathcal{L}^2$ is generated by its sections and $\mathcal{L}^{\otimes 3}$ is very ample.

PROOF. This follows from the previous results by standard arguments.

Theorem 4.10.2. Let C be an irreducible curve on F with $C^2 \geq 0$, $\dim|C|>0$. Assume that $p \neq 2$. Then the general member of $|C|$ is smooth.

PROOF. If $C^2 = 0$ this will be explained in Chapter 5.

Assume $C^2 = 2$. Then φ_{2C} defines a double covering to a Del Pezzo quartic surface and $|C|$ is the inverse transform of a pencil of conics. Obviously, not all of them are tangent to the branch locus everywhere ($p \neq 2$).

If $|C|$ is hyperelliptic, equivalently $\Phi(C) = 1$; this is Corollary 4.5.1.

If $C^2 = 4$ and $\Phi(C) = 2$, we have a map of degree 4 onto $\mathbb{P}^2$ and $|C|$ are inverse transforms of lines. Since one can find a line transversal to the branch locus, the assertion is true.

If $|C|$ is superelliptic, equivalently $C^2 = 6$ or 8 and the $\Phi(C) = 2$, the general member of $|C|$ is a separable double cover of a general hyperplane section of a Del Pezzo surface hence smooth.

If $|C|$ is not superelliptic and $\Phi(C) = 2$ the map φ_C is a birational map onto a non-normal surface . The one-dimensional part of its singular locus consists of the lines which are the images of curves of arithmetic genus E with $E\cdot C = 2$ (Theorem 4.6.2). Since a divisor 2p (p is a point) does not move on an elliptic curve E in characteristic $\neq 2$, the general member of $|C|$ intersects E transversally at 2 points. Thus it is nonsingular at E hence nonsingular everywhere.

If $\Phi(C) \geq 3$, φ_C is a birational morphism onto a normal surface . Hence the inverse image of a general hyperplane section is smooth.

Appendix. A theorem of Igor Reider.

In this appendix we deduce some of the results of this chapter by using a quite different approach from **[Rei]** It is based on Bogomolov's theorem on stable vector bundles on algebraic surfaces **[Bog,Ray 5]**. However, as the latter uses the assumption of the zero characteristic of the ground field, we have to assume here that char(k) = 0.

Recall that a divisor D on a smooth projective surface X is said to be **nef** (numerically effective) if $D \cdot C \geq 0$ for every curve C on X.

Theorem. Let X be a smooth projective surface over k and D be a nef divisor on X.

(i) If $D^2 \geq 5$ and p is a base point of $|K_X+D|$, then there exists an effective divisor E passing through p such that either $D \cdot E = 0, E^2 = -1$ or $D \cdot E = 1$, $E^2 = 0$.

(ii) If $D^2 \geq 10$ and points p,q are not separated by $|K_X+D|$ (p,q can be infinitely near), then there exists an effective divisor E on X passing through p and q such that either $D \cdot E = 0, E^2 = -1$ or -2, or $D \cdot E = 1$, $E^2 = 0$ or -1, or $D \cdot E = 2$, $E^2 = 0$.

PROOF.(i) If p is a base point of $|K_X+D|$, then the restriction map

$$H^0(X, \mathcal{O}_X(K_X+D)) \to H^0(X, \mathcal{O}_p(K_X+D))$$

is not surjective. In the terminology of **[G-H 2]** this means that the 0-cycle p has the Cayley-Bacharach property relative to $|K_X+D|$. By Proposition 1.33 from [**loc.cit.**], there exists a pair $(\mathcal{E},e)$, where $\mathcal{E}$ is a rank 2 vector bundle on X and e is its section such that $\Lambda^2(\mathcal{E}) \cong \mathcal{O}_X(D)$ and the zero cycle of e is equal to p. By the assumption of the theorem

$$c_1(\mathcal{E})^2 = D^2 > 4\deg c_2(\mathcal{E}) = 4.$$

By Bogomolov's theorem **[Re 1]**, $\mathcal{E}$ is unstable. This means that there exists an exact sequence

$$0 \to \mathcal{O}_X(D_1) \to \mathcal{E} \to \mathcal{I}_Z(D_2) \to 0,$$

where D_1, D_2 are divisors on X and $\mathcal{I}_Z$ is the ideal of an effective 0-cycle Z on X. We have (see **[Re 1, Ray 5]**)

a) $D = D_1 + D_2 = c_1(\mathcal{E})$,

b) $D_1 \cdot D_2 + \deg(Z) = c_2(\mathcal{E}) = 1$,

c) the class of $D_1 - D_2$ in NS(X) is effective.

On the other hand, the Koszul exact sequence for the pair $(\mathcal{E}, e)$

$$0 \to \mathcal{O}_S \xrightarrow{e} \mathcal{E} \to \mathcal{I}_p(D) \to 0$$

implies that $H^0(X, \mathcal{I}_p(D_2)) \neq 0$ (by tensoring the both exact sequences with $\mathcal{O}_S(-D_1)$). Let $E \sim D_2$ be the zero divisor of a non-zero section of $\mathcal{I}_p(D_2)$. We will check that E satisfies the assertion (i) of the theorem.

From b) above we obtain

$$D_1 \cdot E = 1 - \deg(Z) \leq 1.$$

From a) and c) we deduce, using that D is nef,

$$D_1 \cdot E \geq 0 \ , \ D \cdot (D_1 - D_2) = D_1^2 - E^2 > 0.$$

Assume that $E^2 > 0$. The equality

$$E \cdot ((D_1 \cdot E)/E^2)E - D_1) = 0$$

together with the Hodge Index theorem implies

$$0 \geq ((D_1 \cdot E)/E^2)E - D_1)^2 = D_1^2 - (D_1 \cdot E)^2/E^2.$$

This shows that

$$(E^2)^2 < E^2 D_1^2 \leq (D_1 \cdot E)^2 \leq 1$$

which contradicts the assumption on E^2. Thus $E^2 \le 0$, and

$$D\bullet E = D_1\bullet E + E^2 \le D_1\bullet E \le 1.$$

Since D is nef, this implies that either $D\bullet E = 1$ and $E^2 = 0$, or $D\bullet E = 0$ and $E^2 = -1$. This proves the assertion (i).

The proof of (ii) is similar: we replace the 0-cycle p by the cycle p+q or 2p and repeat the argument. We leave the details to the reader or refer him to **[Rei]**.

Corollary 1 (cf.Theorem 4.4.1). Let F be an Enriques surface and D be a nef divisor on F with $D^2 \ge 6$. Then $|D|$ has a base-point if and only if $\Phi([D]) = 1$.

PROOF. Assume $|D| = |D+K_F+K_F|$ has a base point, we apply the Theorem to a divisor $D+K_F \equiv D$ to obtain a an effective divisor E such that $D\bullet E = 1$, $E^2 = 0$. Since [E] is an isotropic vector in Num(F), we get $\Phi([D]) = 1$. The converse is easy (see the proof of Theorem 4.4.1). Since $\varphi([D]) = 1$, there exists an elliptic pencil $|2E'|$ with $D\bullet E' = 1$. The point $D \cap E'$ is a base-point of $|D|$.

Corollary 2 (cf. Theorem 4.6.1). Let F be an Enriques surface and D be a nef divisor on F with $D^2 \ge 10$. Then $\varphi_{|D|}$ is an isomorphism outside a finitely many nodal curves which are blown down to rational double points if and only if $\Phi([D]) \ge 3$.

PROOF. By Lemma 4.6.1, the condition $\Phi([D]) \ge 3$ is necessary. Let us prove its sufficiency. We apply the Theorem to a nef divisor $D+K_F$. If $\varphi_{|D|}$ does not separate points p and q, then there exists an effective divisor E passing through p and q and such that $E^2 = -2$ and $D\bullet E = 0$ (the other two cases do not occur by the assumption $\Phi([D]) \ge 3$). Since D is nef, every irreducible component of E is a nodal curve R with $D\bullet R = 0$. This shows that $|D|$ blows down E (hence p and q) to a rational double point.

Bibliographical notes to Chapter IV.

A systematic study of linear systems on Enriques surfaces was initiated by F.Cossec **[Cos 2]** as a natural generalization of the corresponding results for K3 surfaces obtained by B.Saint-Donat **[SD]**. §1-7 of this chapter give an exposition of these results with some ameliorations which take into acount the case of characteristic 2.

The superelliptic maps and their branch loci were known to many people as a version of the classical double plane construction (see **[B-P, Do 5, Cos 2]** for the precise statements in characteristic $\neq 2$). The case of characteristic 2 is considered here for the first time.

Most of the projective models described in §9 are classical. If $p\neq 2$ a sextic model was first model of an Enriques surface considered by Enriques **[En 1]**. Castelnuovo observed that the thetraedron of the double curves could degenerate. Later Enriques found a double plane model **[En 2]** and obseved that it could also degenerate. M. Artin **[Art 1]** and B. Averbukh **[AS,Av]** were first who put Enriques's study of these models on the firm ground. They showed that degenerate models exist only if the surface F is special in the sence of existence a degenerate U-pair. We will show in Part 2 that this condition is equivalent to the condition that F is nodal, that is, contains a smooth rational curve. The generalization of Enriques sextic models to characteristic 2 was considered in **[La 3]**. The existence of a non-degenerate sextic and double plane models for non-exra-special surfaces (e.g. classical Enriques surfaces) is a new result even in the case $p\neq 2$. The double quadric models defined by hyperelliptic maps were used by E.Horikawa **[Ho 2]** for the study of the periods of Enriques surfaces and by J.Shah **[Sha]** for the study of stable degenerations of Enriques surfaces. The fact that the K3-cover of a generic Enriques surface is isomorphic to the intersection of three quadrics is observed in **[En 3]**. Theorem 4.9.2 asserting that this is true for every Enriques surface over a field of characteritic $\neq 2$ was proven by different methods independently by F. Cossec **[Cos 1]** and A.Verra **[Ver 2]**. There are many special constructions of Enriques surfaces obtained by non-complete linear systems (cf. **[Ca,Sta]**).

The results of §10 are taken from **[Cos 2]**.

Chapter V

GENUS ONE FIBRATIONS

§1. Genus 1 fibrations: generalities.

Let S be a regular integral scheme of dimension 1, η be its generic point and $K = \kappa(\eta)$ be its residue field. A projective morphism $f: X \to S$ is said to be **a genus 1 fibration** if X is regular and irreducible, and the general fibre X_η is a geometrically integral regular algebraic curve of arithmetic genus 1.

A genus 1 fibration is called an **elliptic** (resp. **quasi-elliptic**) **fibration** if its general fibre is smooth (resp. non-smooth). It is called **minimal** if it cannot be birationally factored through another genus 1 fibration. It follows from the theory of minimal models of two-dimensional schemes (**[Sh 2,Li,Ch]**) that the latter condition is equivalent to the absence of exceptional curves of the first kind in the fibres of f.

We will be concerned with the following two cases:

Global case: S is a complete smooth algebraic curve over an algebraically closed field k. In this case X is a smooth algebraic surface.

Local case: S = Spec A, where A is the local ring of a global S at its closed point or its henselization, or its completion. In the latter cases we call S **strictly local**. More generally S may be the spectrum of any excellent equicharacteristic discrete valuation ring.

Proposition 5.1.1. Let $f: X \to S$ be a genus 1 fibration. Then

(i) f is flat;

(ii) all geometric fibres of f are connected;

(iii) $f_*\mathcal{O}_X = \mathcal{O}_S$.

PROOF. This follows from the known properties of morphisms of schemes. We refer for this to **[Har 2]**, Chapter III (Proposition 9.7 for (i), Corollary 11.3 and 11.5 for (ii) and (iii)).

Proposition 5.1.2. Let $f: X \to S$ be a quasi-elliptic fibration. Then the geometric general fibre $X_{\bar{\eta}} = X_\eta \otimes_K \bar{K}$ is a rational curve with a cusp. Moreover, char(K) = 2 or 3.

PROOF.This is a well-known fact due to J.Tate (**[Ta 1]**). We will prove it in §5.

Let $f:X \to S$ be an elliptic fibration. According to the general properties of morphisms of schemes there exists a finite subset D of closed points on S such that a fibre $X_s = f^{-1}(s)$ is non-smooth if and only if $s \in S$ (**[EGA]**).

Let $f:X \to S$ be a quasi-elliptic fibration. Then there exists a finite subset D of closed points on S such that the geometric fibre X_s is reducible if and only if $s \in D$ (**[EGA]**).

We will call the set D defined above the **degeneracy set**. The fibres X_s, $s \in D$ (resp. $s \notin D$), are called **degenerate** (resp. **non-degenerate**).

Let X_s be a fibre over a closed point $s \in S$. Then it is a positive divisor on X such that

$$\mathcal{O}_X(-X_s) \cong f^*(\mathcal{O}_S(-s)).$$

Since $f_*(\mathcal{O}_X) = \mathcal{O}_S$, each fibre X_s is connected (**[Har 2]** Chapter 3, Corollary 11.3). Let

$$X_s = \sum_{i \in I} n_i C_i$$

be its decomposition into irreducible components, the number n_i is called the

multiplicity of the component C_i. Let

$$m(s) = \text{g.c.d.}((n_i)_{i\in I}).$$

This number is called the **multiplicity** of X_s, the fibre X_s is called **multiple** if $m(s) > 1$ and **non-multiple** otherwise. For every fibre X_s we denote by $\bar{X}_s$ the divisor $(1/m(s))X_s$. We have the equality of divisors:

$$X_s = m(s)\bar{X}_s.$$

From now on we will assume that f is a minimal fibration (if not stated otherwise).

Proposition 5.1.3. Let $f:X \to S$ be a genus 1 fibration. Then

$$\omega_X = f^*(\mathcal{L}^{-1}\otimes\omega_S)\otimes\mathcal{O}_X(\sum_{s\in S} a_s X_s),$$

where $\mathcal{L}$ is an invertible sheaf on S such that

$$R^1f_*\mathcal{O}_X = \mathcal{L} \oplus T,\ T = \mathrm{Tors}(R^1f_*\mathcal{O}_X).$$

$$\deg(\mathcal{L}^{-1}\otimes\omega_S) = \chi(\mathcal{O}_X) - 2\chi(\mathcal{O}_S) + h^0(T).$$

Moreover, $0 \le a_s < m(s)$ and $a_s = m(s)-1$ if and only if $T_s = 0$.

PROOF. See Theorem 2 of **[B-M 2]**.

The fibres X_s, where $T_s \neq 0$ are called **wild**. Clearly, they are multiple. Since $R^2f_*\mathcal{O}_X = 0$, we have (**[Har 2]**, Chapter III, §7)

$$(R^1f_*\mathcal{O}_X)\otimes\kappa(s) = T_s \oplus(\mathcal{L}\otimes\kappa(s)) \cong H^1(X_s, \mathcal{O}_{X_s}).$$

This shows that

$$\dim H^1(X_s, \mathcal{O}_{X_s}) = 1+\dim T_s,$$

and by the invariance of the Euler characteristic of fibres of a flat map, we get

$$\dim T_s = \dim H^0(X_s, \mathcal{O}_{X_s})-1.$$

This shows that the fibre X_s is not wild if and only if f is **cohomologically**

flat over s. Recall (**[EGA]**) that the latter means that f is flat and $f_*\mathcal{O}_X \cong \mathcal{O}_S$ universally over a neighborhood of s.

Corollary 5.1.1. Every fibre X_s of a genus 1 fibration over a closed point s is a connected divisor of canonical type. The divisor $\bar{X}_s$ is an indecomposable divisor of canonical type. In particular,

$$H^0(X_s, \mathcal{O}_{X_s}) \cong K$$

PROOF. For every irreducible component C_i of X_s we have $C_i \cdot X_s = C_i \cdot X_{s'} = 0$, where $s' \neq s$. The formula for ω_X shows that $C_i \cdot K_X = 0$ also. The last assertion follows from Lemma 3.1.1.

We recall the following fact noted in Chapter 3, **S**1.

Proposition 5.1.4. Let $X_s = \Sigma n_i C_i$ be a degenerate fibre. Denote by $Num_s(X)$ the sublattice of the Neron-Severi lattice Num(X) spanned by the irreducible components of X_s. Then $Num_s(X)$ is negative semi-definite and its radical is spanned by the divisor $\bar{X}_s = (1/m(s))X_s$. If X_s is reducible,then each $C_i \cong \mathbb{P}^1$ and their classes $[C_i]$ form a canonical root basis in $Num_s(X)$ of type $\tilde{A}_n$, $\tilde{D}_n$ or $\tilde{E}_n$, where $n = \#I - 1$.

We will use freely the classification of indecomposable curves of canonical types (Chapter 3, Proposition 3.1.1). We have the following pictures describing the **types** of fibres, by which will be meant the type of the corresponding indecomposable divisor of canonical type $\bar{X}_s$.

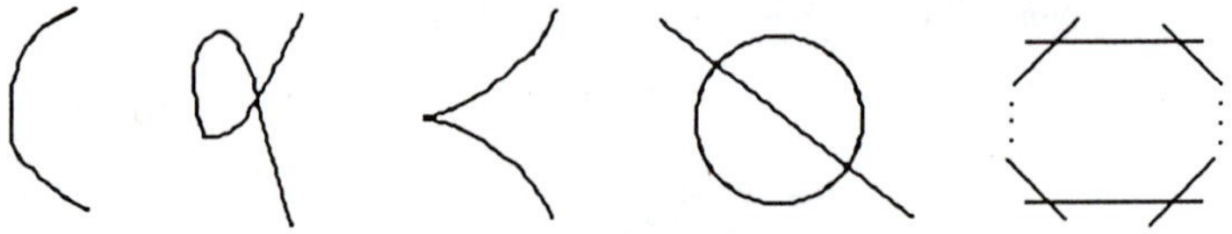

$\tilde{A}_0$ $\tilde{A}_0{}^*$ $\tilde{A}_0{}^{**}$ $\tilde{A}_1$ $\tilde{A}_n$

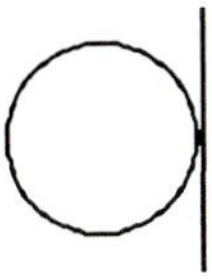

A_2^*

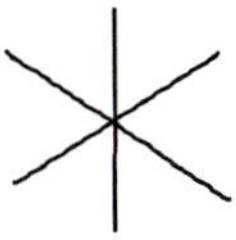

$\tilde{A}_3$

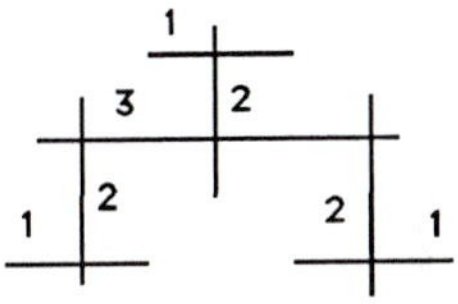

$\tilde{E}_6$

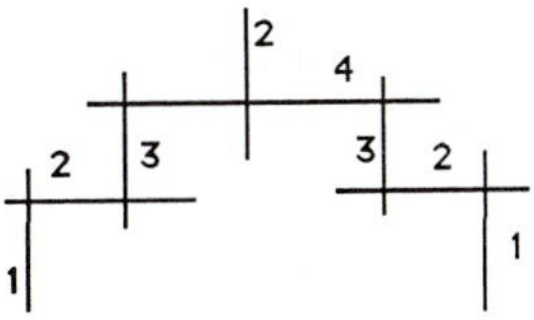

$\tilde{E}_7$

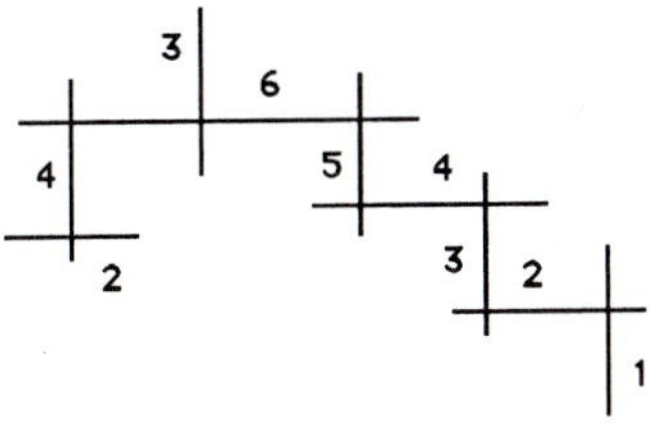

$\tilde{E}_8$

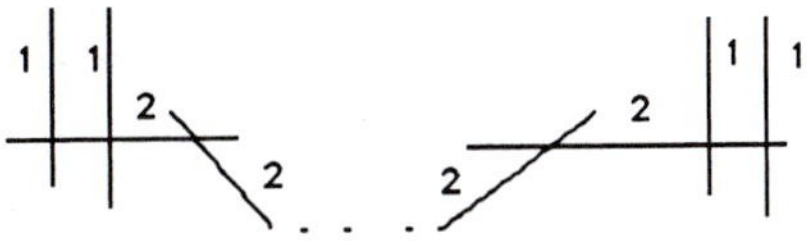

$\tilde{D}_n$

Here the numbers indicate the multiplicities of the irreducible components of $\bar{X}_s$

Proposition 5.1.5. Let $\mathfrak{N}_{\bar{X}_s} = \mathcal{O}_{\bar{X}_s}(\bar{X}_s)$ be the normal sheaf of $\bar{X}_s$, where X_s is a multiple fibre. Let ν_s be the order of $\mathfrak{N}_{\bar{X}_s}$ in $\mathrm{Pic}(\bar{X}_s)$. Then

(i) ν_s divides $m(s)$ and a_s+1;

(ii) $\nu_s = m(s)$ if and only if X_s is not wild;

(iii) if $h^1(\mathcal{O}_X) = 1$, then $m(s)=a_s+1$ or $\nu_s = m(s)-a_s-1$;

(iv) if $h^1(\mathcal{O}_X) = 0$, then $T = 0$;

(v) $m(s)/\nu_s$ is a power of $\mathrm{char}(k)$;

In particular, X_s is not wild if $\mathrm{char}(k) \nmid m(s)$ or if $h^1(\mathcal{O}_X) = 0$.

PROOF. For (i) and (ii)(resp.(iii)) see Proposition 4 (resp. its Corollary) of **[B-M 2]** (iv) follows immediately from the existence of a surjective map

$$H^1(X,\mathcal{O}_X) \to H^0(S,R^1f_*\mathcal{O}_X) \supset H^0(S,T)$$

(a boundary map in the Leray spectral sequence).(v) is proven in **[Ray 3]**, Proposition 6.3.5.

Let $e(Z)$ denote the l-adic (or topological if $\kappa = \mathbb{C}$) Euler-Poincaré characteristic $\Sigma(-1)^i b_i(Z)$, where $b_i(Z) = \dim H^1_{et}(Z,\mathbb{Q}_\ell))$ of an algebraic variety Z over an algebraically closed field . Recall that $e(Z) = c_2(Z)$, the second Chern number of X . An easy calculation shows that

$$\begin{aligned} e(X_s) &= 0 \text{ if } s \notin D, \text{ f is an elliptic fibration;} \\ &= 2 \text{ if } s \notin D, \text{ f is a quasi-elliptic fibration;} \\ &= 1+b_2(X_s) \text{ if } s \in D, X_s \text{ is not of type } \tilde{A}_n; \\ &= b_2(X) = n \text{ if } s \in D, X_s \text{ is of type } \tilde{A}_n. \end{aligned}$$

Proposition 5.1.6. Assume S is global. Then

$$e(X) = c_2(X) = e(X_{\bar{\eta}})e(S) + \Sigma_{s\in S}(e(X_s) - e(X_{\bar{\eta}}) + \delta_s),$$

where $X_{\bar{\eta}} = X_\eta \otimes_K \bar{K}$, $\delta_s \geq 0, \delta_s = 0$ if char(κ) $\neq$ 2,3, or f is a quasi-elliptic fibration, or X_s is of type $\tilde{A}_n$.

PROOF. This is well-known if $\kappa = \mathbb{C}$ (see, for example, **[G-H 1]**, p.510) For the case char(κ) > 0, see **[Del 1,Do 2]**

Let f: X → S be a quasi-elliptic fibration. It follows from Remark 5.1.1 that the degeneracy set D consists of all points $s \in D$ such that X_s is not a rational curve with a cusp. Let Σ be the closure in X of the set of cusps of all fibres X_s, $s \notin D$. Equivalently, Σ is the closure of the point of X_η which becomes the cusp of $X_{\bar{\eta}}$. We will call Σ the **curve of cusps**.

Proposition 5.1.7. The curve of cusps Σ is a smooth curve on X such that $\Sigma \cdot X_s = p$ for every $s \in S$. The restriction of f to Σ is a purely inseparable morphism of degree p.

PROOF. See **[B-M 2]** or Remark 5.5.3.

Corollary 5.1.2. Let $f: X \to S$ be a quasi-elliptic fibration. Then every multiple fibre has multiplicity equal to p ($= 2$ or 3).

Proposition 5.1.8. Suppose that X_s is a multiple fibre of multiplicity m. Assume that char(κ)∤m. Then X_s is of type $\tilde{A}_n$ for some $n \geq 0$.

PROOF. Let $\mathfrak{N}_{\bar{X}_s}$ be the normal sheaf of $\bar{X}_s$. By Proposition 5.1.5, $\mathfrak{N}_{\bar{X}_s}$ is an element of order $m > 1$ in $\mathrm{Pic}(X_s)$. Since char(κ)∤m, we can apply the Kummer exact sequence for étale topology (see **[Mi 2]**) to obtain an isomorphism

$$_m\mathrm{Pic}(\bar{X}_s) \cong H^1_{et}(\bar{X}_s, \mu_m).$$

An elementary computation shows that this group is not trivial ($\cong \mathbf{Z}/m$ or $(\mathbf{Z}/m)^2$) only in the case $\tilde{A}_n$.

Finally, we explain the process of **elimination of multiple fibres** due to Kodaira.

Proposition 5.1.9. Let $f:X \to S$ be a genus 1 fibration and X_s be a multiple fibre of multiplicity m. There exists a finite separable cover $\varphi:S' \to S$, a genus 1 fibration $f':X' \to S'$ and a morphism $\pi:X' \to X$ such $\varphi\circ f' = f\circ\pi$ and for every point $s' \in S'$ lying over s the fibre $X'_{s'}$ is non-mutiple.

PROOF. Let $\hat{K}_s$ be the fraction field of the formal completion of the local ring $\mathcal{O}_{S,s}$. There exists a finite separable extension $L/\hat{K}_s$ such that the curve $X_\eta \otimes_K \hat{K}_s$ has a rational smooth point over L. Let $\varphi:S' \to S$ be a Galois cover of S such that for every point s' over s, the canonical homomorphism $\hat{\mathcal{O}}_{S,s} \to \hat{\mathcal{O}}_{S',s'}$ induces an extension of the fraction fields isomorphic to the extension $L/\hat{K}_s$. Let X' be a nonsingular model of the base change $X \times_S S'$ which does not contain exceptional curves in the fibres of its canonical projection $f':X' \to S'$. For every point $s' \in S'$ lying over s, the curve $X'_{\eta'} \otimes_{K'} \hat{K}_{s'}$ has a rational point over $\hat{K}_{s'}$. The closure of this point in the scheme $X' \otimes_{S'} \hat{\mathcal{O}}_{S',s'}$ intersects a component of the closed fibre $X'_{s'}$ transversally at one point. This shows that $X'_{s'}$ contains a reduced component, in particular it is not multiple. We define π as the natural projection $X' \to X$ and this concludes the proof.

Remark 5.1.1. A much less trivial argument shows that one can reduce any fibre of an elliptic fibration to a fibre of type $\tilde{A}_n$ by a process similar to one described in the previous proposition. This is the so-called **semi-stable reduction** theorem (see **[Gro 5]**).

§2. The Picard group.

We keep the notation of the previous section and of §7 of Chapter 0. Let $f: X \to S$ be a genus 1 fibration. We denote by $\mathcal{P}$ the relative Picard functor $\mathcal{P}ic_{X/S}$. Since $Br(\eta) = 0$ and $Br(S) \hookrightarrow Br(\eta)$, we have

$$\mathcal{P}(S) = Pic(X)/f^*Pic(S).$$

For every complete curve Z over a field we denote by

$$d: Pic(Z) \to \mathbf{Z}^K, \text{ (resp. } d^t: Pic(Z) \to \mathbf{Z})$$

the degree (resp. the total degree) homomorphism.

Let

$$Pic(Z)^0 = Ker(d), \quad Pic(Z)_0 = Ker(d^t).$$

If Z is integral, $Pic(Z)^0 = Pic(Z)_0$

For every point $s \in S$ we denote by

$$r_s: Pic(X) \to Pic(X_s)$$

the homomorphism obtained by restriction of invertible sheaves on X to the fibre X_s.

We set

$$Pic(X)_f = Ker(r_\eta),$$

$$E(S) = Pic(X)_f/f^*(Pic(S)) \subset \mathcal{P}(S),$$

$$Pic(X)_0 = Ker(d \circ r) = r_\eta^{-1}(Pic(X_\eta)^0),$$

$$\mathcal{P}(S)' = Pic(X)_0/f^*(Pic(S)) \subset \mathcal{P}(S).$$

Lemma 5.2.1. Let S be strictly local and s be its closed point. Then the restriction homomorphism

$$Pic(X) \to Pic(X_s)$$

is surjective and its kernel is uniquely divisible by any integer n prime to char(k).

PROOF. The surjectivity follows from **[Art 5]**. By the proper base change theorem (**[Mi 2]**, Chapter 6, Corollary 2.7), the canonical reduction homomorphism

$$H^1_{et}(X,\mu_n) \to H^1_{et}(X_s,\mu_n)$$

is bijective. Now the assertion immediately follows from the Kummer theory which gives isomorphisms

$$Pic(X)_n \cong H^1_{et}(X,\mu_n),\ Pic(X_s)_n \cong H^1_{et}(X_s,\mu_n),$$

$$Pic(X)^{(n)} \cong H^2_{et}(X,\mu_n),\ Pic(X_s)^{(n)} \cong H^2_{et}(X_s,\mu_n)$$

(one uses here also the fact that $H^2(X,\mathbf{G}_m) = H^2(X_s,\mathbf{G}_m) = 0$ **[Gro 3]**).

Proposition 5.2.1.
(i) $E(S) = \bigoplus_{s\in S}(Num_s(X)' \oplus \mathbf{Z}/m(s)\mathbf{Z})$, where $Num_s(X)' = Num_s(X)/Radical$;
(ii) $E(S)\cap\mathcal{P}^0(S) = \bigoplus_{s\in S}(\mathbf{Z}/m(s)\mathbf{Z})$;
(iii) $\mathcal{P}(S)'/\mathcal{P}^0(S)+E(S) \cong \bigoplus_{s\in S} Discr(Num_s(X)')$;
(iv) if S is strictly local and s is its closed point, Pic(S) = 0 and the homomorphisms

$$r_s':\mathcal{P}(S)'/E(S) \to Pic(X_s)_0,$$

$$r_s':\mathcal{P}(S)'/E(S)\cap\mathcal{P}(S)^0 \to Pic(X_s)^0,$$

induced by the restriction homomorphism r_s are surjective;
(v) $Pic(X)_0/Pic(X)_f \cong J(X_\eta)(K)$. If S is global and f is elliptic, $J(X_\eta)(K)$ is a

finitely generated abelian group unless there exists an elliptic curve E over k such that $J(X_\eta) \cong E\otimes_k K$.

(vi) $J(X_\eta)(K)$ is an elementary abelian p-group if f is quasi-elliptic.

PROOF. (i) Clearly $\mathrm{Pic}(X)_f$ is generated by the classes of irreducible components of fibres X_s over closed points $s\in S$. Every $D\in f^*(\mathrm{Pic}(S))$ is equal to a linear combination of some fibres $X_s = m(s)\bar{X}_s$. The assertion follows from the fact that the class of $\bar{X}_s$ spans the radical of $\mathrm{Num}_s X$ and the irreducible components of X_s are linearly independent modulo $\bar{X}_s$.

(ii) Obviously we may assume that S is local and isomorphic to the spectrum of $\mathcal{O}_{S,s}$. Every element of $\mathcal{P}(S)^0\cap E(S)_s$ is represented by the class of a divisor $D = \sum_{i\in I} n_iE_i$, where the E_i's are irreducible components of the fibre X_s and

$$(\sum_{i\in I} n_iE_i)\cdot E_j = \sum_{i\in I} n_i(E_i\cdot E_j) = 0 \text{ for every } j\in I.$$

This shows that $[D] = m[\bar{X}_s]$ for some $m>0$. It remains to note that D is a principal divisor if and only if $m = m(s)$.

(iii) Again we may assume that S is local. Let

$$\alpha: \mathbf{Z}^k \to \mathbf{Z}^k,$$

be the homomorphism given by the Cartan matrix $(E_i\cdot E_j)$ of the root basis in $\mathrm{Num}_s(X)$ formed by the classes of all irreducible components of X_s. Let

$$\beta: \mathbf{Z}^k \to \mathbf{Z}$$

be the homomorphism $(a_1,\dots,a_k) \to \Sigma m_ia_i$, where m_i is the multiplcity of E_i. By definition

$$\mathcal{P}(S)' = \{\mathcal{L}\in\mathrm{Pic}(X): d\tau(r_s(\mathcal{L})) = 0\},$$

$$\mathcal{P}(S)^0 = \{\mathcal{L}\in\mathrm{Pic}(X): d(r_s(\mathcal{L})) = 0\}.$$

Hence

$$\mathcal{P}(S)'/E(S)+\mathcal{P}(S)^0 \cong \mathrm{Ker}(\alpha)/\mathrm{Im}(\beta).$$

On the other hand, we have obviously

$$\mathrm{Ker}(\alpha)/\mathrm{Im}(\beta) \cong \mathrm{Discr}(\mathrm{Num}_s(X)).$$

(iv) Since S is strictly local, $\mathrm{Pic}(S) = 0$ and the restriction homomorphism r_s is surjective by Lemma 5.2.1. This immediately proves the assertion.
(v) The definition of $J(X_j)$ and the surjectivity of r_j shows that

$$\mathrm{Pic}(X)_0/\mathrm{Pic}(X)_f \cong J(X_j).$$

Assume S is global. Then $\mathbf{Pic}_{X/K}$ and $\mathbf{Pic}_{S/K}$ are defined and the morphism of group schemes $f^*\colon \mathbf{Pic}_{X/K} \to \mathbf{Pic}_{S/K}$ induces a homomorphism of abelian varieties:

$$f^*\colon \mathbf{Pic}_{X/K}{}^0 \to \mathbf{Pic}_{S/K}{}^0.$$

By Poincaré's reducibility theorem (**[Mu 3]**, Ch. IV, §19), there exists an abelian k-variety $\mathbf{A}$ and an isogeny $\mathbf{A}\times_k\mathbf{Pic}_{X/K}{}^0 \to \mathbf{Pic}_{S/K}{}^0$. Assume $\mathbf{A} = 0$. Since

$$\mathrm{Pic}(X)/f^*\mathrm{Pic}(S) = \mathbf{Pic}_{X/K}(k)/\mathbf{Pic}_{S/K}(k),$$

$$\mathbf{Pic}_{X/K}(k)/\mathbf{Pic}_{X/K}{}^0(k) \cong \mathrm{NS}(X),$$

$$\mathbf{Pic}_{S/K}(k)/\mathbf{Pic}_{S/K}{}^0(k) \cong \mathbf{Z},$$

the group $\mathrm{Pic}(X)_0/f^*\mathrm{Pic}(S)$ is finitely generated of rank equal to $rk(\mathrm{NS}(X))-1$. Hence $J(X_\eta)$ is finitely generated and

$$rk(J(X_\eta)) = rk(\mathrm{NS}(X))-1- rk(\mathrm{Pic}(X)_f/f^*\mathrm{Pic}(S) =$$
$$= rk(\mathrm{NS}(X))-2 - \sum_{i\in I}(rk(\mathrm{Num}_s X)-1).$$

Thus it suffices to investigate the case $\mathbf{A} \neq 0$. Obviously, the restriction homomorphism $\mathrm{Pic}(X)^0 \to J(X_\eta)(K)$ factors through $\mathbf{A}(k)$ and defines a morphism of group K-varieties:

$$\bar{r}\colon \mathbf{A}\otimes_k K \to J(X_\eta).$$

Since $\mathrm{Pic}(X)_f$ is a discrete group, $\mathrm{Ker}(\bar{r})$ is finite and $\bar{r}$ is an isogeny. In particular $J(X_\eta) \cong \mathbf{A}'\otimes_k K$ for some elliptic curve $\mathbf{A}'$ isogeneous to $\mathbf{A}$.
(vi) By (v) $J(X_\eta)(K)$ is finitely generated. Since $X_\eta\otimes_K\bar{K}$ is a rational curve with a cusp if f is quasi-elliptic, $J(X_\eta)(K)$ is a finitely generated subgroup of the additive group $\bar{K}$ hence is killed by multiplication by $p = \mathrm{char}(k)$.

An algebraic variety V over K is said to be **constant** if $V \cong V'\otimes_k K$ for some variety V' over k.

The following results follow immediately from the proof of the previous proposition.

Corollary 5.2.1. Assume S is global and $J(X_\eta)$ is not constant if f is elliptic. Let $r = \mathrm{rk}(J(X_\eta)(K))$, $\rho = \mathrm{rk}(NS(X))$, $\rho_f = \mathrm{rk}(\mathrm{Pic}(X)_f)$ and κ_s be the number of irreducible components in a fibre X_s. Then

$$\rho = r+2+\sum_{s\in S}(\kappa_s-1) = r+1+\rho_f.$$

In particular,

$$\rho = 2+\sum_{s\in S}(\kappa_s-1) = 1+\rho_f$$

if f is quasi-elliptic.

Corollary 5.2.2. Assume S is global. Then

$$b_1(X) = b_1(S)$$

unless f is elliptic and $J(X_\eta)$ is constant. In the latter case $b_1(X) = b_1(S)+2$.

Proposition 5.2.2. Let $f:X \to S$ be a genus 1 fibration. There exists a maximal subsheaf $\mathcal{E}$ of the sheaf $\mathcal{P} = \mathcal{P}ic_{X/S} = R^1f_*\mathbf{G}_m$ satisfying the following properties:

(i) for every closed $s\in S$, $\mathcal{E}_s$ is representable by an affine group scheme $\mathbf{E}_s$ of finite type over k;

(ii) $(\mathbf{E}_s^0)_{red}$ is a unipotent algebraic group of dimension $h^0(\mathcal{O}_{X_s})-1$;

(iii) $(\mathbf{E}_s/\mathbf{E}_s^0)(k) \cong \mathrm{Num}_s(X)'$;

(iv) $\mathcal{Q} = \mathcal{P}/\mathcal{E}$ is representable by a separated group S-scheme of locally finite type and $\mathcal{Q}^0$ is smooth;

(v) $\mathcal{E}_s^0 = 0$ if X_s is not wild and $\mathcal{E}_s = 0$ if it is moreover irreducible. $\mathcal{P}$ is representable by an algebraic space if and only if $\mathcal{E}^0 = 0$.

PROOF. See **[Ray 3]**.

Recall **[Ray 1,3]** that a separated group S-scheme G of locally finite type is said to be **néronian** if G^0 is smooth and for every smooth S-scheme X and every η-morphism φ: $X_\eta \to G_\eta$ there exists an S-morphism $X \to G$ which exends φ.

Lemma 5.2.2. G is néronian if and only if its base change $G \otimes \hat{\mathcal{O}}_{S,s}$ is néronian for every closed $s \in S$. If S is strictly local, G is néronian if and only if the canonical map:

$$G(S) \to G(\eta)$$

is bijective.

PROOF. See **[Ray 1, Art 8]**.

Proposition 5.2.3. The group scheme $\mathcal{a}$ is néronian.

PROOF. By the previous lemma we may assume that S is the spectrum of a complete local ring. Since $\mathcal{a}$ is represented by a separated S-scheme, the canonical map

$$\mathcal{a}(S) \to \mathcal{a}(\eta)$$

is injective. The surjectivity of this map follows from the surjectivity of the similar map

$$\mathcal{P}(S) \to \mathcal{P}(\eta).$$

Since $Br(\eta) = Br(S) = 0$, this map is equal to the restriction map

$$Pic(X) = Pic(X)/f^*Pic(S) \to Pic(X_\eta).$$

The latter is obviously surjective because X is a regular scheme.

Let $i: \eta \hookrightarrow S$ be the canonical inclusion morphism. As always we identify the Picard functor $\mathcal{P}$ or its subfunctor, both considered as abelian sheaves in the étale or flat topology of S, with the group scheme which represents it. We know that

$$\mathcal{P}_\eta = i^*(p) = i^*(q) = \mathcal{a}_\eta = \mathbf{Pic}_{X_\eta/\eta}.$$

Since $\mathcal{A}$ is néronian,

$$\mathcal{A} = i_*(i^*(\mathcal{A})).$$

as sheaves in the étale topology. Extending the degree homomorphism on η

$$d_\eta : i^*(\mathcal{A}) \to \mathbf{Z}$$

to a degree homomorphism on S

$$d_S : i_*(i^*(\mathcal{A})) = \mathcal{A} \to i_*(\mathbf{Z}) = \mathbf{Z}_S,$$

we set

$$\mathfrak{A} = \mathrm{Ker}(d_S).$$

Clearly

$$\mathfrak{A} = i_*(i^*(\mathcal{P}^0)) = i_*(\mathcal{A}_\eta{}^0) = i_*(J(X_\eta)).$$

We will call this sheaf the **Néron sheaf** of f.

Proposition 5.2.4. The Néron sheaf $\mathfrak{A}$ is represented by a smooth néronian group scheme **A**. We have

(i) $\mathfrak{A}^0 = \mathcal{A}^0$;

(ii) $\mathfrak{A}^0 = \mathcal{P}^0$ if f does not have multiple fibres;

(iii)

$$\mathbf{A}_s(k)/\mathbf{A}_s^0(k) = (\mathfrak{A}/\mathfrak{A}^0)_s \cong \mathrm{Discr}(\mathrm{Num}_s(X)')/B$$

where $m(s)b = 0$ for every $b \in B$ and $B = 0$ if X_s is not wild.

PROOF. Since $\mathfrak{A}$ is a closed subscheme of $\mathcal{A}$ with discrete factor group it is obviously néronian. Since $\mathcal{A}^0 \subset \mathfrak{A}$ and evidently $\mathfrak{A}^0 \subset \mathcal{A}^0$, we obtain $\mathfrak{A}^0 = \mathcal{A}^0$. We know that $\mathcal{A}^0$ is represented by a smooth group scheme. So the last assertion implies that $\mathfrak{A}$ is of finite type over S and hence is represented by a smooth scheme.

Let $\mathcal{P}'$ be the inverse image of $\mathfrak{A}$ under the natural surjection $\mathcal{P} \to \mathcal{A}$. Then in the notation of the beginning of this section

$$\mathcal{P}'(S) = \mathcal{P}(S)'.$$

By Proposition 5.2.1

$$\mathcal{P}(S)'/E(S)+\mathcal{P}(S)^0 \cong \mathrm{Discr}(\mathrm{Num}_s(X)'),$$

which defines a surjection

$$\gamma\colon \mathrm{Discr}(\mathrm{Num}_s(X)') \to \mathcal{Q}(S)/\mathcal{Q}^0(S).$$

If X_s is not wild, $\mathcal{E}_s$ is a constant group scheme (Proposition 5.2.2) and the morphism $\mathcal{P}^0 \to \mathcal{Q}^0$ is an étale morphism of algebraic spaces. It is an isomorphism if X_s is not multiple (Proposition 5.2.1 (ii)). Hence $\mathcal{P}^0(S) \to \mathcal{Q}^0(S)$ is surjective and γ is injective. If X_s is wild, we apply Proposiion 7.2.4 of **[Ray 3]** where it is shown that the image of the morphism $\mathcal{Q}^0 \to \mathcal{Q}^0$ of multiplication by m(s) lies in the image of the canonical map $\mathcal{P}^0 \to \mathcal{Q}^0$. Note that in the notation of this lemma the equalities $\mathcal{P}^0 = \bar{P}^0$ and $\mathcal{Q}^0 = \bar{Q}^0$ are verified in the proof of Proposition 4.2.1 of loc. cit. (the needed fact that $\mathcal{P}^0$ is formally smooth follows from Corollary 2.3.2 of loc. cit.).

For every strictly local S, we have

$$\mathcal{Q}(S) = \mathcal{Q}_s(s) = \mathcal{Q}_s(k),\ \mathcal{Q}(S) = \mathcal{Q}_s(s) = \mathcal{Q}_s(k)$$

because $\mathcal{Q}^0$ and $\mathcal{Q}^0$ are smooth. This implies our last assertion

$$(\mathbf{A}_s/\mathbf{A}_s^0)(k) = \mathbf{A}_s(k)/\mathbf{A}_s^0(k) \cong \mathrm{Discr}(\mathrm{Num}_s(X)').$$

For any scheme Z of finite type over a scheme T we denote by $Z^{\#}$ the maximal open subscheme of Z such that the restriction of f is smooth.

Proposition 5.2.5. There exists a unique (up to S-isomorphisms) minimal genus 1 fibration $j\colon J \to S$ satisfying the following properties:

(i) $J(S) = \mathbf{J}(X_\eta)(K) \neq \emptyset$;

(ii) $(J_\eta)^{\#} = \mathbf{J}(X_\eta)$;

(iii) $J^{\#} = \mathbf{A}$, where J is considered as a scheme over S.

PROOF. This follows from the theory of minimal models of two-dimensional schemes. We consider a minimal projective regular curve over K, $C \subset \mathbb{P}^n_K$ containing $J(X_\eta)$ as an open dense subset (equal to $\mathbf{J}(X_\eta)$ itself if f is elliptic). Then we take the schematic closure J' of C in $\mathbb{P}^n_S$ if S is local or find a smooth projective model J' of the field K(C). The morphism $\mathbf{J}(X_\eta) \to \eta$ defines a rational morphism $J' \to S$ which can be replaced by a projective morphism $J'' \to S$ with the generic fibre isomorphic to $J(X_\eta)$. Resolving singularities of J'' and blowing down exceptional curves of the first kind in closed fibres, we obtain the needed genus 1 fibration $J \to S$. The uniqueness of J follows from the theory of minimal models.

Since $J(X_\eta)$ is a group variety, $J(X_\eta)(K) \neq \emptyset$. The points of $J(X_\eta)$ can be considered as rational maps $S \to J$. Since J is projective over S and S is one-dimensional and regular, every such rational map extends to a morphism $S \to J$, i.e. an element of J(S). In this way, we obtain a bijection

$$J(X_\eta)(K) \to J(S).$$

The group law

$$J_\eta^\# \times_K J_\eta^\# \to J_\eta^\#$$

defines a rational map

$$\varphi : J \times_S J \dashrightarrow J.$$

It can be shown that the restriction of φ to $J^\# \times_S J^\#$ defines a morphism:

$$\varphi^\# : J^\# \times_S J^\# \to J^\#.$$

which is a group law (see **[Art 81]**).

Since the restriction map

$$J(S) = J^\#(S) \to J(X_\eta)(K)$$

is surjective and this property is preserved under base changes, we obtain that $J^\#$ is néronian hence coincides with **A**.

The genus 1 fibration $j:J \to S$ constructed in the previous proposition is said to be the **Jacobian fibration** of $f:X \to S$. The group scheme $\mathbf{A} = J^\#$ is said to be the **Néron model** of $J(X_\eta)$.

Lemma 5.2.3. Let $f : X \to S$ be a quasi-elliptic fibration. For every geometric fibre X_s

$$b_1(X_{\bar{s}}) = 0.$$

PROOF. The assertion is true for the geometric generic fibre $X_{\bar{\eta}}$. Obviously we may assume that S is strictly local. Let $a: X_{\bar{\eta}} \to X$ be the natural morphism which is composed of the projection $X_{\bar{\eta}} \to X_{\eta}$ and the inclusion morphism $X_{\eta} \to X$. It induces an injective homomorphism of cohomology groups

$$H^1(X, a_* \mathbb{Q}_l) \hookrightarrow H^1(X_{\bar{\eta}}, \mathbb{Q}_l) = 0.$$

Since $a_* \mathbb{Q}_l = \mathbb{Q}_l$ (**[Mi 2]**), we obtain that $H^1(X, \mathbb{Q}_l) = 0$. It remains to use that

$$H^1(X, \mathbb{Q}_l) \cong H^1(X_s, \mathbb{Q}_l)$$

for the closed point s of S **[Mi 2]**.

Corollary 5.2.3. Let $j: J \to S$ be the Jacobian fibration of a genus 1 fibration $f: X \to S$ and $\mathbf{A} = J^{\#}$ be the Neron model of $J(X_{\eta})$. For every closed point s the algebraic group $\mathbf{A}_s^0$ is an abelian variety if J_s is non-degenerate, the multiplicative group $\mathbf{G}_m$ if J_s is of type $\tilde{A}_n$, and the additive group $\mathbf{G}_a$ otherwise. The latter case occurs always if f is quasi-elliptic. Moreover, the group of connected components $\mathbf{A}_s/\mathbf{A}_s^0$ is isomorphic to the following group:

$\mathbf{Z}/(n+1)\mathbf{Z}$ if J_s is of type $\tilde{A}_n$, $n \neq 0$,

$\mathbf{Z}/4\mathbf{Z}$ if J_s is of type $\tilde{D}_{2K+1}$,

$(\mathbf{Z}/2\mathbf{Z})^2$ if J_s is of type $\tilde{D}_{2K}$,

$\mathbf{Z}/3\mathbf{Z}$ if J_s is of type $\tilde{A}_2^*$, $\tilde{E}_6$,

$\mathbf{Z}/2\mathbf{Z}$ if J_s is of type $\tilde{A}_1^*$, $\tilde{E}_7$,

{1} otherwise.

PROOF. This follows directly from the known classification of fibres.

From now on degenerate fibres of type $\tilde{A}_n$, $n \neq 0$, will be called fibres of **multiplicative type**. All other degenerate fibres are said to be of **additive type**.

Corollary 5.2.4. Let $f:X \to S$ be a quasi-elliptic fibration and $s \in S$ be a closed point. Then

(i) X_s is of type $\tilde{A}_0^{**}$, $\tilde{A}_1^{*}$, $\tilde{D}_{2K}$, $\tilde{E}_7$ or $\tilde{E}_8$ (resp. $\tilde{A}_0^{**}$, $\tilde{A}_2^{*}$, $\tilde{E}_6$ or $\tilde{E}_8$) if $p = 2$ (resp. if $p = 3$) and X_s is not wild;

(ii) X_s is of type $\tilde{A}_0^{**}, \tilde{A}_1^{*}, \tilde{D}_n, \tilde{E}_7$ or $\tilde{E}_8$ (resp. $\tilde{A}_0^{**}$, $\tilde{A}_2^{*}$, $\tilde{E}_6$ or $\tilde{E}_8$) if $p = 2$ (resp. $p = 3$) and X_s is wild of multiplicity p.

PROOF. Since $\mathbf{A}_\eta = (J_\eta)^{\#}$ is a connected unipotent group, the Néron sheaf $\mathcal{A} = i_* \mathbf{A}_\eta$ is annihilated by multiplication by p. This implies that $\mathcal{A}_s/\mathcal{A}_s{}^0$ is an elementary abelian p-group for every closed point $s \in S$. By Proposition 5.2.4, $\mathrm{Discr}(\mathrm{Num}_s(X)')$ is an elementary p-group if X_s is not wild and an extension of an elementary p-group by an elementary p-group if X_s is wild of multiplicity p. After this remark the assertion follows from the classification of degenerate fibres of genus 1 fibrations.

S3. Jacobian fibrations.

A genus 1 fibration $f:X \to S$ is said to be **jacobian** if $X(S) \neq \emptyset$. We will identify a section $q: S \to X$ from $X(S)$ with its image in X. This is a curve C such that $f|C$ is an isomorphism,or equivalently $C \cdot X_s = 1$ for every closed point $s \in S$. We can also identify C or q with its generic point $\eta' \in X_\eta(K)$. Obviously, a jacobian fibration does not have multiple fibres. We call f **trivial** (resp. **isotrivial**) if $X \cong E \times S$ for some elliptic curve over k and f is equal to the projection $E \times S \to S$ (resp. this is so after an étale base change $S' \to S$).

Proposition 5.3.1. Let $f: X \to S$ be a genus 1 fibration over a strictly local S. Assume the fibre over the closed point is not multiple. Then f is jacobian.

PROOF. It follows from the classification of degenerate fibres that X_s has a smooth point. By the Hensel lemma, there exists a section of f passing through this point.

Proposition 5.3.2. Every jacobian fibration $f:X \to S$ is isomorphic over S to its Jacobian fibration $j: J \to S$. In particular, $X^{\#} \cong J^{\#}$ and after fixing a section of f $X^{\#}$ can be identified with the Néron model **A** of f.

PROOF. By Riemann-Roch, $\mathrm{Pic}(X_\eta)^1 \cong X_\eta^{\#} \cong J(X_\eta)$, thus $X_\eta \cong J_\eta$. After this, the assertion follows from the minimality of f and j.

From now on we will always assume that X(S) has a fixed section $C_0 \in X(S)$ and is identified with the group $J(X_j)(K)$ the zero element of which corresponds to the fixed section. In particular, $X^{\#}$ will be identified with the Néron scheme **A** . We denote by $(X^{\#})^0$ the open subset of $X^{\#}$ obtained by deleting the irreducible components of $X^{\#}$ which do not intersect C_0. It represents the sheaf $\mathfrak{a}^0$ and is identified with the connected group scheme $\mathbf{A}^0$.

Corollary 5.3.1. A jacobian fibration $f: X \to S$ is trivial if and only if $J(X_\eta)$ is constant.

The next result is a special case of Proposition 5.2.1:

Corollary 5.3.2. Let $f: X \to S$ be a jacobian fibration. If S is global, X(S) is a finitely generated abelian group unless f is smooth. If f is a quasi-elliptic fibration, X(S) is an elementary abelian p-group.

Since jacobian fibrations have no multiple fibres, we deduce from Proposition 5.1.2 .

Proposition 5.3.3. Let $f: X \to S$ be a jacobian fibration over a global S. Then

$$\omega_X \cong f^*(\mathcal{L}^{-1} \otimes \omega_S),$$

where $\mathcal{L} = R^1 f_* \mathcal{O}_X$ is an invertible sheaf on S of degree equal to $-\chi(\mathcal{O}_X)$.

Corollary 5.3.3. For every section C of a jacobian fibration $f: X \to S$

$$C^2 = -\chi(\mathcal{O}_X),\ C \cong S.$$

Proposition 5.3.4. Let $f:X \to S$ be a non-trivial jacobian elliptic fibration and $X(S)(l)$ be the l-primary component of the torsion subgroup $X(S)_{tors}$ of the group $X(S)$.

(i) For every prime $l \neq p$ and every closed point $s \in S$ the restriction homomorphism

$$X(S)(l) \to X_s^{\#}$$

is injective.

(ii) For every non-zero section $C \in X(S)(l)$, $l \neq p$, $C \not\subset (X^{\#})^0$ unless f is trivial or X is a hyperelliptic surface, S is an elliptic curve and f is isotrivial.

(iii) Assume that $\rho(J) = \rho_f(J)+1$. Then the group $X(S) = X(S)_{tors}$ is finite of order d which is determined by the equality

$$d^2 \# \mathrm{Disc}(\mathrm{Num}(X)) = \prod_{s \in S^{\#}} \mathrm{Disc}(\mathrm{Num}_s(X)/\mathrm{Rad}).$$

PROOF. (i) Let $f_s:X(s) \to \tilde{S}_s$ be the morphism obtained by the base change $\tilde{S}_s = \mathrm{Spec}(\tilde{\mathcal{O}}_{S,s}) \to S$, where $\tilde{\mathcal{O}}_{S,s}$ denotes the henselization of the local ring $\mathcal{O}_{S,s}$. Obviously,

$$X(S) \subset X(s)(\tilde{S}_s)\ .$$

(since K is included in the quotient field of $\tilde{\mathcal{O}}_{S,s}$). Under the identification of $X(S)$ with a subgroup of $X(s)(\tilde{S}_s)$, the restriction homomorphism $X(S) \to X_s^{\#}(\kappa)$ is induced by the reduction homomorphism

$$X(s)(\tilde{S}_s) \to X_s^{\#}(\kappa).$$

Thus it remains to prove the assertion assuming that $S = \tilde{S}_s$ is strictly local. We know from S2 that $\mathfrak{A} \cong \mathcal{P}/\mathcal{E}$, where $\mathcal{P}$ is the relative Picard functor of f and $\mathcal{E}$ is represented by a discrete group scheme concentrated at the closed point s. By Lemma 5.2.1 the reduction homomorphism $\mathcal{P}(S) = \mathrm{Pic}(X) \to \mathcal{P}(s) = \mathrm{Pic}(X_s)$ is surjective and its kernel is a group uniquely divisible by any number prime to p. The reduction homomorphism

$\mathfrak{B}(S) \to \mathfrak{B}(s)$ is obviously bijective. Together this implies the assertion.

(ii) Let C be a nontrivial element of X(S) of order n prime to p which lies in $(X^{\#})^0$ of j. The sections C and C_0 meet the same irreducible component X_s° of each fibre X_s. By (i) C and C_0 are disjoint. Since $n(C-C_0)|X_\eta$ is linearly equivalent to zero, we obtain

$$[n(C-C_0)] \in \mathrm{Pic}_f(X).$$

Since $(nC-nC_0)\cdot R = 0$ for every irreducible curve contained in fibres, $n(C-C_0)$ is a linear combination of the divisors X_s. Hence for some m and any $s \in S$

$$n(C-nC_0)^2 = 0 .$$

By Corollary 5.3.2

$$C\cdot C = -\chi(\mathcal{O}_X)$$

for every section C of f: X → S. Thus

$$0 = (C-C_0)^2 = C^2 + C_0^2 = -2\chi(\mathcal{O}_X)$$

and we obtain $\chi(\mathcal{O}_X) = C^2 = 0$. By Noether's formula e(J) = 0 and hence by Proposition 5.1.6 all fibres of j are smooth elliptic curves. Also we know that

$$\omega_X = f^*(\mathfrak{L})$$

for some invertible sheaf of degree $\chi(\mathcal{O}_S) - 2\chi(\mathcal{O}_X) = \chi(\mathcal{O}_S)$. By Riemann-Roch

$$0 = \chi(\mathcal{O}_X) = \chi(\omega_X) = \chi(\mathfrak{L}) = 2\chi(\mathcal{O}_S) .$$

This shows that S is an elliptic curve. By Corollary 5.2.2

$$q(X) = 1,\ p_g(X) = 0$$

if f is not trivial. It follows from the classification of algebraic surfaces (or from S5) that X is a hyperelliptic surface and there exists an étale finite cover S' → S such that $X \times_S S'$ is a trivial elliptic fibration over S'.

(iii) By Proposition 5.2.1 $X_\eta(K)$ is finite and is isomorphic to the factorgroup $\mathrm{Num}(X)_0/\mathrm{Num}_f(X)$, where $\mathrm{Num}(X)_0$ is the image of $\mathrm{Pic}(X)_0$ in Num(X) and $\mathrm{Num}_f(X)$ is its sublattice spanned by the lattices $\mathrm{Num}_s(X)$, $s \in S$. Let C be a section. Since $\rho(X) = \rho_f(X)+1$, for every divisor D there exists an integer n such that

$$[D] \quad -n[C] \in \mathrm{Num}(X)_0,$$

This shows that

$$\mathrm{Num}(X)\ /L_1 \perp L_2 \cong \mathrm{Num}(X)_0/\mathrm{Num}_f(X),$$

where $L_1 = \mathbf{Z}[X_s]+\mathbf{Z}[C]$ is a unimodular sublattice of Num(X), L_2 is spanned by the components of fibres which do not intersect C. Note that the latter is isomorphic to the direct sum of the root lattices of finite type corresponding to the type of the fibre. The formula now follows from the known relationships between the discriminant of a lattice and its sublattice of finite index.

Next we will study the relationship between the geometry of a genus 1 fibration and its Jacobian fibration.

Lemma 5.3.1. Let $f:X \to S$ be a genus 1 fibration. There is a canonical isomorphism

$$\mathrm{Br}(X) \cong H^1_{et}(S, R^1 f_* \mathbf{G}_m).$$

PROOF. Since the fibres of f over closed points are curves over algebraically closed field,

$$R^i f_* \mathbf{G}_m = 0,\ i > 1$$

(**[Gro 3, Mi 2]**). We also know that $f_*\mathbf{G}_m = \mathbf{G}_m$ because $f_*\mathcal{O}_X = \mathcal{O}_S$ and $\mathrm{Br}(S)=0$. Thus the Leray spectral sequence

$$E_2^{p,q} = H^q(S, R^p f_* \mathbf{G}_m) \Rightarrow H^{p+q}(X, \mathbf{G}_m)$$

degenerates and gives us the needed canonical isomorphism

$$H^2(X, \mathbf{G}_m) = \mathrm{Br}(X) \to E_2^{1,1} = H^1_{et}(S, R^1 f_* \mathbf{G}_m).$$

Let $f:X \to S$ be a genus 1 fibration, its **index** ind(f) is defined to be the minimal degree of a separable extension L/K such that $X_\eta(L) \neq \emptyset$. Equivalently, ind(f) is the minimal degree of an element in $\mathrm{Pic}(X_\eta)$. It is easy to see that

$$m(s) | ind(f)$$

for every $s \in S$.

Proposition 5.3.5. There is an exact sequence of abelian groups:

$$0 \to Br(J) \to Br(X) \xrightarrow{\varphi} (\sum_{s \in S} \mathbf{Z}/m(s)\mathbf{Z})/\varphi(\mathbf{Z}/ind(f)\mathbf{Z}) \to 0,$$

where $\varphi = \oplus_{s \in S} \varphi_s$, and φ_s: $\mathbf{Z}/ind(f)\mathbf{Z} \to \mathbf{Z}/m(s)\mathbf{Z}$ is induced by the identity map $\mathbf{Z} \to \mathbf{Z}$.

PROOF. By the previous lemma

$$Br(X) \cong H^1_{et}(S, R^1 f_* \mathbf{G}_m), \; Br(J) \cong H^1_{et}(S, R^1 j_* \mathbf{G}_m).$$

Let $\mathcal{A}(f)$ and $\mathcal{A}(j)$ be the respective representable factors of $R^1 f_* \mathbf{G}_m$ and $R^1 j_* \mathbf{G}_m$ defined in **S2**. We have exact sequences of abelian sheaves in the étale topology:

$$0 \to \mathcal{E} \to R^1 f_* \mathbf{G}_m \to \mathcal{A}(f) \to 0$$

$$0 \to \mathcal{E}' \to R^1 j_* \mathbf{G}_m \to \mathcal{A}(j) \to 0.$$

Since $\mathcal{E}$ and $\mathcal{E}'$ are concentrated at a finite set of closed points, the cohomology sequence defines isomorphisms:

$$Br(X) \cong H^1(S, \mathcal{A}(f)), \; Br(J) \cong H^1(S, \mathcal{A}(j))$$

We have observed already that $\mathbf{J}(X_\eta) = \mathbf{J}(J_\eta)$ hence the Néron sheaf $\mathfrak{a}(f)$ of f and the Néron sheaf $\mathfrak{a}(j)$ of j coincide and may be denoted by $\mathfrak{a}$. Recall that they were defined as the kernels of the corresponding total degree maps:

$$\mathfrak{a}(f) = Ker(\mathcal{A}(f) \to \mathbf{Z}_S), \; \mathfrak{a}(j) = Ker(\mathcal{A}(j) \to \mathbf{Z}_S).$$

Since $J(S) \neq \emptyset$, the second map is surjective . Using the vanishing of all cohomology of non-zero degree of the constant sheaf $\mathbf{Z}_S$ we obtain an isomorphism

$$Br(J) \cong H^1(S, \mathcal{A}(j)) \cong H^1(S, \mathfrak{a})$$

The first degree map is not surjective and we can write two exact sequences:

$$0 \to \mathcal{A} \to \mathcal{A}(f) \to \mathcal{F} \to 0,$$

$$0 \to \mathcal{F} \to \mathbf{Z}_S \to \bigoplus_{s\in S}(i_s)_*(\mathbf{Z}/m(s)\mathbf{Z}) \to 0,$$

where i_s: $s \hookrightarrow S$ is the natural inclusion morphism and f is the image of the total degree map.

Passing to the cohomology groups, we obtain an exact sequence

$$0 \to H^0(S,\mathcal{F}) \to \mathbf{Z} \to \bigoplus_{s\in S}(\mathbf{Z}/m(s)\mathbf{Z}) \to H^1(S,\mathcal{F}) \to 0.$$

Obviously the cokernel of the map $H^0(S,\mathcal{F}) \to \mathbf{Z}$ is isomorphic to $\mathbf{Z}/\mathrm{ind}(f)\mathbf{Z}$. Thus

$$H^1(S,\mathcal{F}) \cong \bigoplus_{s\in S}(\mathbf{Z}/m(s)\mathbf{Z})/(\mathbf{Z}/\mathrm{ind}(f)\mathbf{Z})$$

and the exact sequence

$$0 \to H^1(S,\mathcal{A}) \to H^1(S,\mathcal{A}(f)) \to H^1(S,\mathcal{F}) \to 0$$

can be rewritten as the exact sequence from the assertion of the proposition.

Proposition 5.3.6. Assume S is global. Then

$$\chi(\mathcal{O}_X) = \chi(\mathcal{O}_J)\ ,\ e(X) = e(J).$$

PROOF. It follows from Proposition 5.1.3 that $c_1^2 = 0$ for every minimal elliptc fibration. Applying the Noether formula, we see that it suffices to prove one of the equalities. The canonical surjection q: $R^1f_*\mathbf{G}_m = \mathcal{P}(f) \to \mathcal{A}(f)$ defines a morphism of the corresponding Lie algebras:

$$\mathrm{Lie}(q)\colon\ \mathcal{L}ie(R^1f_*\mathbf{G}_m) = R^1f_*\mathcal{O}_X \to \mathcal{L}ie(\mathcal{A}(f))$$

By a result of Raynaud (**[Ray 6]**, Theorem 3.5.5), the kernel and cokernel of $\mathcal{L}ie(q)$ have the same length at each point $s\in S$ which is equal to the length of the sheaf $T = \mathrm{Tors}(R^1f_*\mathcal{O}_X)$. Applying this to j for which T = 0, we get

$$R^1j_*\mathcal{O}_X \cong \mathcal{L}ie(\mathcal{A}(j)).$$

Since $\mathcal{A}(f)$ and $\mathcal{A}(j)$ have a common subsheaf $\mathcal{A}$ with a discrete factor,

$$\mathcal{L}ie(\mathcal{A}(f)) \cong \mathcal{L}ie(\mathcal{A}(j))$$

and we obtain a map

$$R^1f_*\mathcal{O}_X \to R^1j_*\mathcal{O}_X$$

with kernel and cokernel of equal length at each point. In particular, we have

$$\chi(R^1f_*\mathcal{O}_X) = \chi(R^1j_*\mathcal{O}_X)$$

hence

$$\chi(\mathcal{O}_X) = \chi(\mathcal{O}_S) - \chi(R^1f_*\mathcal{O}_X) = \chi(\mathcal{O}_S) - \chi(R^1j_*\mathcal{O}_X) = \chi(\mathcal{O}_J),$$

as follows immediately from the Leray spectral sequences for f and j.

Proposition 5.3.7. There is an isomorphism of sheaves in the étale cohomology:

$$R^if_*\mu_n \cong R^ij_*\mu_n\ ,\ i \geq 0$$

for all n prime to p and to all m(s), s∈S.

PROOF. We assume that n satisfies the restrictions from above. The Kummer exact sequence

$$0 \to \mu_n \to \mathbf{G}_m \to \mathbf{G}_m \to 0$$

immediately implies the existence of an isomorphism

$$R^1f_{**}\mu_n \cong {}_n(R^1f_*\mathbf{G}_m),$$

where ${}_n(\ )$ denotes the kernel of the homomorphism of multiplication by n. Using the relationships between the sheaves $R^1f_*\mathbf{G}_m$ and $R^1j_*\mathbf{G}_m$ explained in the proof of the previous proposition, we immediately see that

$$R^1f_*\mu_n \cong R^1j_*\mu_n\ .$$

It remains to prove the assertion about $R^2f_*\mu_n$. Note that these sheaves contain the constant sheaf μ_n and the quotient sheaves are concentrated at the points where the fibres are reducible. To show that these sheaves are isomorphic, it suffices to check that

$$b_2(X_s) = rk_{\mathbf{Z}/n\mathbf{Z}}(R^2f_*\mu_n)_s = b_2(J_s) = rk_{\mathbf{Z}/n\mathbf{Z}}(R^2j_*\mu_n)_s\ .$$

for every closed point s of S. So the assertion is local and we may assume for a moment that S is strictly local. By Propositions 5.3.1 and 5.3.2, f = j if m(s) = 1. Going back to the general situation, we obtain that

$$(R^2f_*\mu_n)_s \cong (R^2j_*\mu_n)_s$$

if m(s) = 1. Applying Artin's Approximation Theorem we may assume that S is global. Assume that all multiple fibres X_{s_i}, i = 1,...,k , of f are of the same type. and the same is true for the fibres J_{s_i}. By Proposition 5.1.6

$$e(X) = e(X_{\bar\eta})e(S)+\sum_{s\neq s_i}(e(X_s)-e(X_{\bar\eta}))+k(e(X_{s_1})-e(X_{\bar\eta}))+\sum_{s\in S}\delta_s(f),$$
$$e(J) = e(J_{\bar\eta})e(S)+\sum_{s\neq s_i}(e(J_s)-e(J_{\bar\eta}))+k(e(J_{s_1})-e(J_{\bar\eta}))+\sum_{s\in S}\delta_s(j),$$

We know already that $c_2(X) = c_2(J)$. Also the definition of the invariants depends only on the sheaves $R^1f_*\mu_n$ which are the same for f and j. Comparing the two formulas we obtain that

$$b_2(X_{s_i}) = rk_{\mathbf{Z}/n\mathbf{Z}}(R^2f_*\mu_n)_{s_i} = b_2(J_{s_i}) = rk_{\mathbf{Z}/n\mathbf{Z}}(R^2j_*\mu_n)_{s_i}.$$

This proves the assertion in this case.

Assume now that f has multiple fibres of different types, one of them being X_s. We choose a finite separable cover φ: S' → S which is unramified over s and eliminates the other multiple fibres in sense of Proposition 5.1.9. Let f': X' → S' be the corresponding genus 1 fibration and j': J'→ S' its Jacobian fibration. Obviously J' is birationally isomorphic to the base change $J\times S' \to S'$. Both f' and j' have $k = \#\varphi^{-1}(s)$ fibres over points $s'\in\varphi^{-1}(s)$ which are of the same type isomorphic to the type of X_s and J_s respectively. Thus we are in the situation of the previous special case which shows that $b_2(X_s) = b_2(J_s)$.

Corollary 5.3.4. For every closed $s\in S$

$$b_i(X_s) = b_i(J_s), \quad i \geq 0$$

Corollary 5.3.5. Assume S is global .Then

$$b_i(X) = b_i(J), \quad i \geq 0,$$

$$\rho(X) = \rho(J).$$

PROOF. Clearly we have to verify only that $b_i(X) = b_i(J)$ for $i = 1$ and 2. Since $J(X_\eta) \cong J(J_\eta)$, Corollary 5.2.1 implies $\rho(X) = \rho(J)$ unless j is constant. In the latter case the equality $b_1(X) = b_1(J)$ follows from Corollary 5.2.2 and the equality $b_2(X) = b_2(J)$ is implied by the equality $e(X) = e(J)$. Hence we may assume that $\rho(X) = \rho(J)$. By Proposition 1.2.5 of Chapter 1 for every smooth projective surface Z we have the equality

$$b_2(Z) = \rho(Z)+t(Z),$$

where $T_1(Br(X)) \cong \mathbf{Z}_1^{t(X)}$. It remains to apply Proposition 5.3.4 which shows that $t(X) = t(J)$.

Theorem 5.3.1. Let $f:X \to S$ be a genus 1 fibration and $j: J \to S$ be its Jacobian fibration. Let $s \in S$ be a closed point. Then

(i) $X_s \cong J_s$ if X_s is not multiple;

(ii X_s and J_s are of the same type if X_s is not wild;

(iii) if X_s is wild, then X_s and J_s are of the same type unless J_s is of type $\tilde{E}_7$ (resp. $\tilde{E}_8$). In the latter case X_s is of the same type or of type $\tilde{D}_7$ (resp. $\tilde{D}_8$) and $p = 2$;

(iv) if f is quasi-elliptic and X_s is wild, then X_s and J_s are of the same type unless J_s is of type $\tilde{E}_8$. In the latter case X_s is of the same type or of type $\tilde{D}_8$ and $p = 2$.

PROOF. By Artin's Approximation Theorem **[Art 4]**, we may assume that S is global. The assertion (i) follows immediately from Proposition 5.3.1 by localization at the point s. By Corollary 5.3.4, X_s and J_s have the same Betti numbers. The classification of degenerate fibres shows that X_s and J_s must be of the same type unless J_s is of type $\tilde{E}_n$ (resp. $\tilde{D}_n$) and X_s is of type $\tilde{D}_n$ (resp. $\tilde{E}_n$). By Proposition 5.2.4, there exists a surjection

$$\mathrm{Disc}(\mathrm{Num}_s(J)') \cong \mathrm{Discr}(\mathrm{Num}_s(X)')/B$$

where $m(s)B = 0$ and $B = 0$ if X_s is not wild. Comparing the discriminant groups of the root lattices of type $\tilde{D}_n$ and $\tilde{E}_n$, we obtain (ii) and (iii). To prove (iv) we use Corollary 5.2.4 which shows that X_s cannot be of type $\tilde{D}_7$.

Remark 5.3.1. We do not know whether the exceptional cases in (iii) or (iv) do occur. Note that X_s and J_s are of the same type if $p \neq 2$. This was proven in **[K-U]** by a different method.

§4. Ogg-Shafarevich theory.

Our next goal is to describe the set of all genus 1 fibrations with isomorphic Jacobian fibration.

Let $j:X \to S$ be a genus 1 fibration , we know that $X_\eta^\#$ can be identified with the component of its Picard scheme corresponding to invertible sheaves of degree 1. Thus it has a natural structure of a torseur (= principal homogeneous space) over its Jacobian variety $J(X_\eta)$. It is the trivial torseur (i.e. isomorphic to $J(X_\eta)$) if and only if $X_\eta(K) \neq 0$, i.e. f is a jacobian fibration.

Fix a jacobian fibration $j: J \to S$ and let Elf(j) denote the set of isomorphism classes of minimal genus 1 fibrations the Jacobian fibration of which is isomorphic to j.

Proposition 5.4.1. The map

$$(f:X \to S) \to X_\eta^\#$$

defines a bijection between the set Elf(j) and the group of isomorphism classes of torseurs over $J_\eta^\#$

PROOF. This follows immediately from the theory of minimal models of two-dimensional schemes (cf. the proof of Proposition 5.2.5).

Corollary 5.4.1. The set Elf(j) has a natural structure of an abelian group isomorphic to the cohomology group:

$$H^1(Gal(\bar{K}/K), J_\eta^\#(\bar{K})) \cong H^1_{et}(\eta, J_\eta^\#),$$

where $Gal(\bar{K}/K)$ is the Galois group of the separable closure $\bar{K}$ of K.

PROOF. This is well-known. We refer for the cohomological interpretation of the group of isomorphism classes of torseurs to **[Ser 3]**.

We will use the following notations:

$\tilde{S}_s = Spec(\tilde{\mathcal{O}}_{S,s})$, the spectrum of the henselization $\tilde{\mathcal{O}}_{S,s}$ of $\mathcal{O}_{S,s}$ at a closed point;

$\tilde{K}_s$ = the field of fractions of $\tilde{\mathcal{O}}_{S,s}$;

$\tilde{f}_s: \tilde{X}(s) = X \times_S \tilde{S}_s \to \tilde{S}_s$ the base change corresponding to the natural map $\tilde{S}_s \to S$.

Let f: X → S be a genus 1 fibration the isomorphism class [f] of which belongs to the group Elf(j). It is clear that the map $\tilde{f}_s$: $\tilde{X}(s) \to \tilde{S}_s$ is a genus 1 fibration the isomorphism class $[\tilde{j}_s]$ of which is an element of the group $Elf(\tilde{j}_s)$. This defines a map:

$$\varphi_s: Elf(j) \to Elf(\tilde{j}_s),$$

which is a homomorphism of groups.

We will call the image of [f] under φ_s the **local invariant** of f at s.

Proposition 5.4.2. Let f: X → S be a genus 1 over a strictly local S, j:J → S be its Jacobian fibration, s be the closed point of S. Then

$$ord([f])) = ind(f) = m(s)$$

PROOF. Note that the assertion is true if $m(s) = 1$ (Proposition 5.3.1). Since $\bar{X}_s$ has a smooth point x, one can find 2 local parameters t_1, t_2 at x such that the local equation of $\bar{X}_s$ at x is $t_1 = 0$. Let V be a closed subscheme defined by $t_2 = 0$ at x. Since

$$H^0_{et}(V, \mu_n) \cong H^0_{et}(V \times_S k(s), \mu_n), \quad p \nmid n,$$

we can replace V by an irreducible component if needed to assume that $V \cdot \bar{X}_s = 1$. The generic fibre of V defines a divisor on X_η of degree m(s). Clearly this shows that m(s) = ind(f) and is equal to the minimal degree of divisors on X_η. It is easy to see from the definition that n[f] is equal to the isomorphism class of the torseur $\mathcal{P}^n_\eta$ of the Picard scheme $\mathcal{P}_\eta$ classifying divisor classes of degree n. The latter is trivial if and only if X_η contains a divisor of degree n. This proves the assertion.

Corollary 5.4.2. Let $[f] \in Elf(j)$. For almost all closed points $s \in S$

$$\varphi_s([f]) = 0.$$

We denote by

$$\varphi: Elf(j) \to \sum_{s \in S} Elf(j_s)$$

the homomorphism $\sum_{s \in S} \varphi_s$.

Let $\eta = Spec(K) \hookrightarrow S$ be the canonical inclusion of the generic point of S. Recall from §2 that we can identify $J^\#_\eta$ with the generic fibre of the Néron sheaf

$$\mathcal{A} = i_*(J^\#_\eta)$$

represented by the group scheme $\mathbf{A} = J^\#$.

Proposition 5.4.3. There is an exact sequence

$$0 \to H^1(S, \mathcal{A}) \to Elf(j) \xrightarrow{\varphi} \sum_{s \in S} Elf(j_s) \to H^2(S, \mathcal{A})$$

PROOF. The Leray spectral sequence for the morphism i gives an exact sequence of the étale cohomology groups:

$$0 \to H^1(S,\mathcal{A}) \to H^1(\eta,\mathcal{A}_\eta) \to H^0(S,R^1i_*\mathcal{A}_\eta) \to H^2(S,\mathcal{A}).$$

For any closed point $s\in S$, we have

$$(R^1i_*\mathcal{A}_\eta)_s \cong H^1(\bar{K}_s,\mathcal{A}) \cong \mathrm{Elf}(\tilde{j}_s),$$

It is easy to see that the composition of the map $H^1(\eta,\mathcal{A}_\eta) \to H^0(S,R^1i_*\mathcal{A}_\eta)$ and the canonical map $H^0(S,R^1i_*\mathcal{A}_\eta) \to (R^1i_*\mathcal{A}_\eta)_s$ is equal to the map φ_s from above. This shows that we may replace $H^0(S,R^1i_*\mathcal{A}_\eta)$ by the group $\sum_{s\in S}\mathrm{Elf}(\tilde{j}_s)$ and the map $H^1(\eta,\mathcal{A}_\eta) \to H^0(S,R^1i_*\mathcal{A}_\eta)$ by the map φ preserving the exactness of the sequence.

To describe Elf(j) we have to compute the local parts $\mathrm{Elf}(j_s)$ and the groups $H^i(S,\mathcal{A})$, i= 1,2. Note that the group $H^1(S,\mathcal{A})$ classifies "locally trivial torseurs". It is often denoted by $Ш(\mathcal{A})$ (or $Ш(J_\eta^\#)$) and is said to be the **Shafarevich** (or **Shafarevich-Tate**) **group** of $J_\eta^\#$.

We begin with the groups $\mathrm{Elf}(j_s)$, that is we will compute Elf(f) in the case where **S is strictly local**. Note that in this case the exact sequence from Proposition 5.4.3 is of no help. The groups $H^i(S,\mathcal{A})$ are zero for $i > 0$.

For every abelian group A and a prime l we denote by $A(\neq l)$ (resp. $A(l)$) the group

$$\lim_{l\nmid n}\mathrm{ind}\ {}_nA \quad (\text{resp. } \lim_{l\mid n}\mathrm{ind}\ {}_nA)$$

where, as before, ${}_nA$ denotes the kernel of the homomomorphism [n] of multiplication by n.

Theorem 5.4.1. Let $j:J \to B$ be a jacobian elliptic fibration over a strictly local S. There exists a natural isomorphism

$$\mathrm{Elf}(j)(\neq p) \cong (\mathbb{Q}/\mathbb{Z})(\neq p)^{b_1(J_s)}$$

PROOF. Let **A** be the Néron model of j representing the sheaf $\mathcal{A}$. For every integer n prime to p, the multiplication by n defines a surjective homomorphism of abelian varieties $\mathbf{A}_\eta \to \mathbf{A}_\eta$. Let

$$0 \to {}_n(\mathbf{A}_\eta) \to \mathbf{A}_\eta \xrightarrow{[n]} \mathbf{A}_\eta \to 0,$$

be the corresponding Kummer exact sequence. Passing to cohomology, we obtain an exact sequence

$$0 \to \mathbf{A}_\eta(K)^{(n)} \to H^1(K,{}_n(\mathbf{A}_\eta)) \xrightarrow{[n]} {}_nH^1(K,\mathbf{A}_\eta) \to 0.$$

Let

$$\mathbf{A}_\eta(K) = \mathfrak{a}(S) \to \mathfrak{a}(k) = J^{\#}{}_s = \mathbf{A}_s$$

be the reduction homomorphism. By the previous lemma

$$\mathbf{A}_\eta(K)^{(n)} \cong \mathbf{A}_s(k)^{(n)}.$$

Since the connected component $\mathbf{A}_s^\circ$ is a connected commutative group, and the factor group $\mathbf{A}_s/\mathbf{A}_s^\circ$ is finite we have

$$\lim_{p\nmid n}.\text{ind}\,\mathbf{A}_\eta(K)^{(n)} = \lim_{p\nmid n}.\text{ind}\,\mathbf{A}_s(k)^{(n)} = \lim_{p\nmid n}.\text{ind}\,\mathbf{A}_s{}^0(k)^{(n)} = 0$$

This shows that

$$\lim_{p\nmid n}.\text{ind}\; H^1(K,{}_n\mathbf{A}_\eta) = \lim_{p\nmid n}.\text{ind}\; ({}_nH^1(K,\mathbf{A}_\eta)) = H^1(K,\mathbf{A}_\eta)(\neq p)$$

Thus we are reduced to the computation of the groups $H^1(K,{}_n\mathbf{A}_\eta)$, $p\nmid n$. Let

$${}_n\mathbf{A}_\eta \times {}_n\mathbf{A}_\eta \to \mu_n$$

be the Weil pairing on the subgroups of points of finite order of the elliptic curve $\mathbf{A}_\eta$ (**[Mu 3]**, Ch. 4, §20). The corresponding cup product in the cohomology produces a pairing

$$H^1(K,{}_n\mathbf{A}_\eta) \times H^0(K,{}_n\mathbf{A}_\eta) \to H^1(K,\mu_n).$$

By Kummer theory and Hensel's lemma,

$$H^1(K,\mu_n) \cong K^*/K^{*n} \cong \mathbf{Z}/n\mathbf{Z},$$

and, hence we have a homomorphism

$$H^1(K,{}_n\mathbf{A}_\eta) \to \text{Hom}({}_n\mathbf{A}_\eta(K),\mathbf{Z}/n\mathbf{Z}) = \text{Hom}({}_n\mathbf{A}_s(k),\mathbf{Z}/n\mathbf{Z}).$$

It follows from the local duality theory **[Mi 3]** that this homomorphism is bijective. Passing to the limit when n is a power of a prime $l \neq p$, we obtain

$$\lim_{\substack{\to \\ r}}.\mathrm{ind}\ H^1(K,\mathbf{A}_J) = \lim_{\substack{\to \\ r}}.\mathrm{ind}\ \mathrm{Hom}({}_{l^r}\mathbf{A}_s(k),\mathbf{Z}/l^r\mathbf{Z}) =$$
$$= \mathrm{Hom}(\lim.\mathrm{proj}_r\, {}_{l^r}\mathbf{A}_s(k),\mathbb{Q}_l/\mathbf{Z}_l) \cong \mathrm{Hom}(\lim.\mathrm{proj}_r\, {}_{l^r}\mathbf{A}_s(k),\mathbb{Q}_l/\mathbf{Z}_l)$$
$$\cong \mathrm{Hom}(\mathbf{Z}_l)^{b_1(J_s)},\mathbb{Q}_l/\mathbf{Z}_l) \cong (\mathbb{Q}_l/\mathbf{Z}_l)^{b_1(J_s)}.$$

Remark 5.4.1. One can give the following interpretaion of the local invariant of an elliptic fibration $f: X \to S$ over a local base S with m(s) prime to p. We fix an isomorphism $\mu_n \cong \mathbf{Z}/n\mathbf{Z}$, where $n = m(s)$. By Proposition 5.1.5, the normal bundle of $\bar{X}_s$ in X is of order n in Pic(X). In fact $n\bar{X}_s \sim X_s = (f^*(\pi))$, where π is a local parameter in S. By standard construction (see Chapter 0, §1), this defines a cyclic cover $X' \to X$ ramified along $\bar{X}_s$. It is easy to see that it comes from the base change $S' \to S$ which is obtained by the normalization of S in the field $K(\pi^{1/n})$. A regular minimal model of the projection $f':X' \to S'$ is an elliptic fibration with the fibre X'_s of multiplicity 1. It follows from Proposition 5.3.1 that f' is a jacobian fibration; hence its generic fibre is isomorphic to the generic fibre J'_η of the projection $J' = J\times_S S' \to S'$. The other projection $J' \to J$ defines a cyclic cover $J'_\eta \to J_\eta$ of degree n which in its turn defines an element $\mathfrak{X} \in {}_n\mathrm{Pic}(J_\eta) = {}_n\mathbf{A}_\eta(K)$. Using the selfduality of the Jacobian, we get a homomorphism

$${}_n\mathbf{A}_\eta(K) \cong \mathrm{Hom}({}_n\mathbf{A}_\eta(K),\mu_n) \cong \mathrm{Hom}({}_n\mathbf{A}_\eta(K),\mathbf{Z}/n\mathbf{Z}) \cong$$
$$H^1(K,{}_n\mathbf{A}_\eta) \to {}_nH^1(K,\mathbf{A}_\eta) \cong (\mathbf{Z}/n\mathbf{Z})^{b_1(J_s)}$$

The image of $\mathfrak{X}$ is the local invariant.

Note also that we can view $\mathfrak{X}$ as an element of

$${}_n\mathbf{A}_\eta(K) \cong {}_n\mathbf{A}(S) \cong {}_n\mathbf{A}_s(k) \cong {}_nJ_s^{\#}.$$

If n is prime to $\#(\mathbf{A}_s/\mathbf{A}_s^0)(k) = b_2(J_s)-1$ (Propositions 5.1.8 and 5.2.4), then

$${}_nJ_s^{\#} \cong {}_n\mathrm{Pic}(J_s) \cong H^1(J_s,\mu_n) \cong (\mathbf{Z}/n\mathbf{Z})^{b_1(J_s)}.$$

Remark 5.4.2. Let $\hat{S}$ = Spec $\hat{A}$, where $\hat{A}$ is the formal completion of A, and $\hat{j}_S$ be the jacobian genus 1 fibration obtained from j by the base change. We have a canonical injective map:

$$\mathrm{Elf}(j) \rightarrow \mathrm{Elf}(\hat{j}).$$

which induces an isomorphism:

$$\mathrm{Elf}(j) \cong \mathrm{Elf}(\hat{j})$$

(see **[Ray 2]** and **[Mi 3]**, p.59). In the original Ogg-Shafarevich theory, Proposition 5.4.3 is stated in the case S = $\hat{S}$.

The computation of the p-part Elf(j)(p) is a much more complicated matter. To state a corresponding result we have to recall some definitions from **[Ser 2]**.

Let G be a smooth group scheme over S. We assume that S ≅ Spec(R) where R is complete. Let $R_i = R/\mathfrak{m}^{i+1}$ and S_i = Spec(R_i). The projective system $\mathcal{G}(G) = (G(S_i))_{i\geq 0}$ has a natural structure of a pro-algebraic group over κ. In particular, its homotopy groups $\pi_i(\mathcal{G}(G))$ are defined and are denoted by $\pi_i(G)$. They are topological compact totally discontinuous groups. For example,

$$\pi_0(G) \cong (G_s/G_s{}^0)(\kappa),$$

$$\pi_1(\mathbf{G}_m) \cong \mathrm{Gal}(K_{ab}/K),$$

where K_{ab} is the maximal abelian extension of K.

If **A** is an elliptic curve over R and A_s is an ordinary (resp. supersingular) elliptic curve over κ, then $\pi_1(\mathbf{A})$ is determined by the following exact sequence:

$$0 \rightarrow \pi_1(\mathbf{G}_m) \rightarrow \pi_1(\mathbf{A}) \rightarrow \mathbf{Z}_p \rightarrow 0$$

(resp.

$$0 \rightarrow \pi_1(\mathbf{G}_a) \rightarrow \pi_1(\mathbf{A}) \rightarrow \pi_1(\mathbf{G}_a) \rightarrow 0\ )$$

Theorem 5.4.2.

$$\mathrm{Elf}(j) \cong \mathrm{Hom}_{\kappa}(\pi_1(\mathbf{A}), \mathbf{Q}/\mathbf{Z}),$$

where **A** is the Neron model of j.

PROOF. Use **[Mi 3]**, p.378 and the semi-stable reduction theorem.

Corollary 5.4.3. Let j be an elliptic fibration over local complete S. Then (i) if J_s is a nonsingular ordinary elliptic curve , then there exists an extension

$$0 \rightarrow Elf(j)(p)^{rad} \rightarrow Elf(j)(p) \rightarrow Elf(j)^{et} \rightarrow 0,$$

where $Elf(j)(p)^{rad}$ is isomorphic to $\mathbb{Q}_p/\mathbb{Z}_p$, and $Elf(j)(p)^{et}$ is isomorphic to $Hom(Gal(K_{ab}/K),\mathbb{Q}/\mathbb{Z})$.

(ii) Suppose J_s is of type $\tilde{A}_n$, $n \neq 0$, then Elf(j)(p) is isomorphic to $Hom(Gal(K_{ab}/K),\mathbb{Q}/\mathbb{Z})$.

Moreover, if J_s is of type $\tilde{A}_n$ or a supersingular elliptic curve (resp. an ordinary elliptic curve), then $f:X \rightarrow S$ defines a non-zero element in Elf(j)(p) (resp. in $Elf(j)(p)^{rad}$) if and only if X_s is wild.

Next we consider the case where j is a quasi-elliptic fibration.

Lemma 5.4.1. Let G be a connected unipotent group of dimension 1 over a field K of characteristic $p > 0$. Then G is isomorphic to the subgroup of $\mathbf{G}_a^2$ defined by an equation

$$y^{p^n}+a_0x+a_1x^p+ \ldots +a_rx^{p^r} = 0,$$

where $a_i \in K$, $a_0 \neq 0$.

PROOF. See **[Ru]**.

Proposition 5.4.4. Let A be given in the above lemma and denote by P(t) the polynomial $a_0+a_1t+\ldots+a_rt^r$. Let **F** be the Frobenius map of the field K and $P(\mathbf{F})\colon \mathbf{G}_a \rightarrow \mathbf{G}_a$ be the corresponding homomorphism of the additive group of K obtained by plugging **F** in P(t). Then

$$H^1(K,A) \cong K^+/F^{p^n}(K^+)+P(F)(K^+)$$

PROOF. We have an exact sequence of groups

$$0 \to A \to \mathbf{G}_a^2 \to \mathbf{G}_a \to 0,$$

where $\alpha(y,x) = \mathbf{F}^n(y)+P(\mathbf{F})(x)$. Applying the cohomology exact sequence, we obtain

$$H^0(K,\mathbf{G}_a^2) \to H^0(K,\mathbf{G}_a) \to H^1(K,A) \to H^1(K,\mathbf{G}_a).$$

It is known, that $H^1(K,\mathbf{G}_a) = 0$ (**[Ser 3]**). This gives an isomorphism

$$H^1(K,A) \cong K^+/\mathbf{F}^n(K^+)+P(\mathbf{F})(K^+).$$

Corollary 5.4.4. Let $j: J \to \dot{S}$ be a quasi-elliptic fibration. Then j is isomorphic to a subgroup of $\mathbf{G}_a^2$ given by an equation

$$y^2+a_0x+a_1x^2+a_2x^4 \quad \text{if } p = 2,$$

$$y^3+a_0x+a_1x^3 = 0 \quad \text{if } p = 3,$$

where $a_0,a_2,a_1 \neq 0$. If $P(t) = a_0+a_1t^2+a_2t^4$ $(p=2)$ or $a_0+a_1t^3$, then

$$\mathrm{Elf}(j) \cong K^+/\mathbf{F}(K^+)+P(\mathbf{F})(K^+).$$

PROOF. Applying Lemma 5.2.1, we easily check the first assertion by computing the arithmetic genus of a regular projective completion of A.

Corollary 5.4.5. Let j be a quasi-elliptic fibration. Then

$$\mathrm{Elf}(j) = \mathrm{Elf}(j)(p).$$

Remark 5.4.3. Explicitly, if $a \in K_s^+$ is a representative of the group $K_s^+/\mathbf{F}^n(K_s^+)+P(\mathbf{F})(K_s^+)$, then the corresponding torseur is isomorphic to the subvariety of $\mathbf{A}_K^2$ given by an equation

$$y^2+a+a_0x+a_1x^2+a_2x^4 \quad \text{if } p = 2,$$

$$y^3+a+a_0x+a_1x^3 = 0 \quad \text{if } p = 3.$$

If $f:X \to S$ is a quasi-elliptic fibration over complete local S, one should be able to compute the multiplicity m(s) and also decide whether X_s is wild from the Laurent expansions of the corresponding coefficients a_i. We refer to **[La 1]** for a conjecture in this respect in the case $p = 3$.

Next we assume that we are in the **global situation**. We have to compute the groups $H^1(S,\mathfrak{a})$ and $H^2(S,\mathfrak{a})$.

Theorem 5.4.3. Assume j is a jacobian genus 1 fibration. Then

$$H^1(S,\mathfrak{a}) \cong Br(J) = H^2(J,\mathbf{G}_m).$$

In particular,

$$H^1(S,\mathfrak{a})(l) \cong (\mathbb{Q}_l/\mathbf{Z}_l)^{b_2(J)-t(F)} \oplus Tors(H^2_{et}(J,\mathbf{Z}_l[1])) \text{ if } l \neq p,$$

$$H^1(S,\mathfrak{a})(p) \cong (\mathbb{Q}_p/\mathbf{Z}_p)^{b_2(J)-t_p(F)} \oplus Tors(H^2_{fl}(J,\mathbf{Z}_p[1])).$$

PROOF. It follows from the proof of Proposition 5.3.4 that

$$H^1(B,\mathfrak{a}(j)) \cong Br(J).$$

Now, the first assertion follows from the existence of an exact sequence

$$0 \to \mathfrak{a} \to \mathfrak{a}(j) \to \mathbf{Z}_B \to 0.$$

The second one was explained in Chapter I,**S**2.

Corollary 5.4.6. Assume that J is a rational surface. Then φ is injective. Since the Brauer group is a birational invariant, we get $Br(J) = Br(\mathbb{P}^2) = 0$ (see **[Gro 3]** or cf. **[Mi 2]**,Chapter V,§,3).

Remark 5.4.4. If j is quasi-elliptic, we obtain from the previous proposition and Corollary 5.4.4 that

$$Br(J) = Br(J)(p).$$

It also follows from the fact that there exists an inseparable cover $J' \to J$ of degree p, where J' is a ruled surface.

For the proof of the next theorem we have to recall the existence of the theory of cohomology with compact support in the étale or flat topology of a scheme **[Mi 2]**. If $H_c(Z,F)$ denotes the corresponding cohomology groups with coefficients in a sheaf F, we have an exact sequence:

$$\dots \to H_c(V,F) \to H^r(V,F) \to \bigoplus_{z\in Z} \backslash_U H^r(K_z,F_z) \to H_c^{r+1}(Z,F) \to \dots$$

where V is any open subset of Z and K_z is the fraction field of the henselization of the local ring of Z at a point z.

Theorem 5.4.4. Let j be a jacobian non-trivial elliptic fibration . Then

$$\mathrm{Tors}(H^2(S,\mathcal{A})) = 0$$

PROOF. Let V be an open subset such that the restriction $\mathcal{A}_U$ of $\mathcal{A}$ to V is an abelian scheme. Since κ is algebraically closed, the fraction field of the henselization of the local ring of every closed point of S is of cohomological dimension 1. The exact sequence of cohomology with compact support from above shows that

$$H^2_c(V,\mathcal{A}) \cong H^2(X,\mathcal{A})$$

For every integer n we have the following exact sequence of sheaves in the flat topology of V:

$$0 \to {}_n\mathcal{A}_U \to \mathcal{A}_U \to \mathcal{A}_U \to 0$$

corresponding to the homomorphism of multiplication by n. Passing to the cohomology, we obtain the following exact sequence:

$$0 \to H^1_c(V,\mathcal{A})^{(n)} \to H^2_c(V,{}_n\mathcal{A}) \to H^2_c(V,\mathcal{A})_n \to 0$$

If n is coprime to p, ${}_n\mathcal{A}$ is a smooth group scheme hence the flat cohomology coincide with the étale cohomology and we have by the duality in étale cohomology of constructible sheaves **[Mi 2]**:

$$H^2_c(V,\mathcal{A}_n) \cong \mathrm{Hom}(H^0(V,({}_n\mathcal{A})^D),\mathbb{Q}/\mathbf{Z}),$$

where $(\ )^D$ denotes the Cartier dual of a finite group scheme. Since the Weil pairing on the sheaf $\mathcal{A}_n$ is a selfduality, we can drop $(\ )^D$. The sheaf ${}_n\mathcal{A}$ is néronian hence

$$H^0(V,{}_n\mathcal{A}) \cong {}_n\mathcal{A}(K) = {}_nJ_\eta(K).$$

It follows from Proposition 5.2.1 that $J_\eta(K)$ is finitely generated. Thus taking n to be a power of a prime $l \neq p$, we obtain

$$\lim_{\substack{\\ r}}.\mathrm{proj}\ {}_{1}{}^{r}\mathfrak{a}(K) = J_\eta(K)\otimes \mathbb{Q}_1/\mathbf{Z}_1 = 0,$$

$$H_c^2(V,\mathfrak{a})(l) = \lim_{\substack{\\ r}}.\mathrm{ind}\ \mathrm{Hom}({}_{1}{}^{r}\mathfrak{a}(K),\mathbb{Q}/\mathbf{Z}) =$$

$$= \mathrm{Hom}(\lim_{\substack{\\ r}}.\mathrm{proj}\ {}_{1}{}^{r}\mathfrak{a}(K),\mathbb{Q}/\mathbf{Z}) = 0.$$

If $p|n$ we argue in the same way by replacing the étale duality by the flat duality. The cohomology $H_c^1(V,{}_n\mathfrak{a})$ has a natural structure of an extension of a discrete group $D_c^1(V,\mathfrak{a}_n)$ by a connected quasi-algebraic group $U_c^1(V,\mathfrak{a}_n)$. By the flat duality **[Ber,Mi 1]**:

$$D_c^2(V,\mathfrak{a}_n) \cong \mathrm{Hom}(D^0(V,({}_n\mathfrak{a})^D),\mathbb{Q}/\mathbf{Z}) \cong \mathrm{Hom}(D^0(V,{}_n\mathfrak{a}),\mathbb{Q}/\mathbf{Z}),$$

$$U_c^2(V,\mathfrak{a}_n) \cong \mathrm{Hom}(U^1(V,({}_n\mathfrak{a})^D),\mathbb{Q}/\mathbf{Z}) \cong \mathrm{Hom}(U^1(V,{}_n\mathfrak{a}),\mathbb{Q}/\mathbf{Z}),$$

As above

$$D^0(V,{}_n\mathfrak{a}) = {}_nJ_\eta(K).$$

and we conclude that

$$\lim_{\substack{p|n}}.\mathrm{ind} D_c^2(V,\mathfrak{a}_n) = 0\ .$$

Repeating the proof of Theorem 5.4.3, we obtain

$$H^1(V,{}_p\mathfrak{a}) \cong {}_p\mathrm{Br}(J_V)$$

and is an extension of the group $H_{fl}^2(J_V,\mu_{p^r})$ by a finite group $NS(J_V)^{(p^r)}$. This shows that

$$U^1(V,\mathfrak{a}_{p^r}) \cong U^2(J_V,\mu_{p^r}) \cong \mathrm{Hom}(U_c{}^3(J_V,\mu_{p^r}),\mathbb{Q}/\mathbf{Z})\ .$$

This implies

$$U^2(V,\mathfrak{a}_{p^r}) \cong U_c^2(V,{}_p\mathfrak{a}_r) \cong U_c^2(J_V,\mu_{p^r}) \cong U^3(J_V,\mu_{p^r}).$$

Also we have

$$U^1(V,\mathfrak{a})^{(p^r)} \cong (\mathrm{Br}(J_V)^{cont})^{(n)}$$

and the homomorphism

$$H^1(V,\mathfrak{a})^{(p^r)} \to H_c^2(V,{}_p\mathfrak{a}_r)$$

induces on its continuous parts the homomorphism

$$(Br(J_V)^{cont})^{(p^r)} \to U^3(J_V,\mu_{p^r}).$$

The latter arises from the Kummer exact sequence :

$$0 \to \mu_{p^r} \to \mathbf{G}_m \to \mathbf{G}_m \to 0.$$

The group $Br(J_V)^{cont}$ is a quasi-unipotent algebraic group over k. Thus

$$(Br(J_V)^{cont})^{(p^r)} \cong Br(J_V)^{cont} \cong {}_{p^r}(Br(J_V)^{cont}) \cong U^2(J_V,\mu_{p^r}).$$

The homomorphism

$$H^1(V,\mathfrak{a})^{(p^r)} \to H^2_c(V,\mathfrak{a}_{p^r})$$

induces on its continuous part the homomorphism

$$\delta: U^2(J_V,\mu_{p^r}) \to U^3(J_V,\mu_{p^r})$$

which is induced by the Bockstein map $H^2(J_V,\mu_{p^r}) \to H^3(J_V,\mu_{p^r})$. The latter is defined by the exact sequence

$$0 \to \mu_{p^r} \to \mu_{p^{2r}} \to \mu_{p^r} \to 0$$

the corresponding cohomology sequence of which is

$$H^2(J_V,\mu_{p^r}) \to H^2(J_V,\mu_{p^{2r}}) \to H^2(J_V,\mu_{p^r}) \to H^3(J_V,\mu_{p^r}).$$

The first map is an isomorphism on U^2. Hence the last map must be an isogeny on its continuous parts. This shows that δ is surjective and in the exact sequence

$$0 \to H^1_c(V,\mathfrak{a})^{(n)} \to H^2_c(V,{}_n\mathfrak{a}) \to {}_nH^2_c(V,\mathfrak{a}) \to 0$$

the map $H^2_c(V,{}_n\mathfrak{a}) \to {}_nH^2_c(V,\mathfrak{a})$ factors through the discrete parts.
So

$$H^2(S,\mathfrak{a})(p) = 0.$$

as wanted.

Corollary 5.4.6. Assume j is elliptic and not trivial. Then the map

$$\varphi: \mathrm{Elf}(j) \to \bigoplus_{s\in S} \mathrm{Elf}(j_s)$$

is surjective.

PROOF. Since both groups are torsion groups, this immediately follows from the previous theorem and the exact sequence from Proposition 5.4.3.

Corollary 5.4.7. Assume j is elliptic and not trivial. For every finite set of closed point $s_1,\dots,s_r$ such that $b_1(J_s) \neq 0$ and a set of integers $m_1,\dots,m_r$ prime to p there exists an elliptic fibration $f\colon X \to S$ with multiple fibres at s_i of multiplicity m_i the isomorphism class of which belongs to Elf(j).

Corollary 5.4.8. Let $j{:}J \to S$ be the Jacobian fibration of an elliptic fibration $f\colon X \to S$. Suppose that J is a rational surface. Then the isomorphism class of f in Elf(j) is uniquely determined by the local invariants of f and every collection of local invariants can be realized in Elf(j).

One can explicitly reconstruct a genus 1 fibration $f\colon X \to B$ corresponding to an element $x \in \mathrm{Elf}(j)$. This is done as follows.

We may assume that $x \in H^1(G, J_\eta^{\#}(K'))$ for some finite Galois extension K'/K with the Galois group G. Let

$$\varphi_x : G \to J_\eta^{\#}(K')$$

be the corresponding 1-cocycle. Let $j'{:}J' \to S'$ be a minimal jacobian genus 1 fibration with the general fibre $J_{\eta'}' \cong J_\eta \otimes_K K'$ (here $\kappa(S') = K'$ and η' is the generic point of X'). For every $g \in G$, the corresponding automorphism of K' over K acts on $J_{\eta'}'$. Also $\varphi_x(g) \in J_\eta(K') = J_{\eta'}'(K')$ acts on $J_{\eta'}'$ by translation $t_{\varphi_x(g)}\colon a \to a + \varphi_x(g)$, $a \in J_{\eta'}'(K')$.

Lemma 5.4.2. The map

$$\rho_{\varphi_x}{:}G \to \mathrm{Aut}(J_{\eta'}'),\ g \to t_{\varphi_x}(g)\circ g,$$

is a homomorphism of groups. If φ_x' is another cocycle representing x, then

$$\rho_{\varphi_x'}(g) = t_{-a_0}\circ\rho_{\varphi_x}(g)\circ t_{-a_0}, \qquad \forall g \in G,$$

where a_0 is a fixed element of $J_\eta(K') = J_{\eta'}'(K')$.

PROOF. This trivially follows from the definition of a cocycle.

Proposition 5.4.5.

$$X_\eta \cong J'_{\eta'}/\rho_{\varphi_X}(G).$$

PROOF. This is a standard construction of the principal homogeneous space corresponding to a given cocycle in the Galois cohomology group (see **[Ser 3]**).

In the next section we will compute the groups $H^1(S,\mathfrak{a})$ and $H^2(S,\mathfrak{a})$ for a quasi-elliptic fibration j: J → S and deduce the following result (see Corollary 5.5.4):

Theorem 5.4.5. Let $\mathfrak{L} = R^1 j_* \mathcal{O}_J$. Assume $H^1(S,\mathfrak{L}^{-n}) = 0$ for $1 \le n \le 4$. Then

$$H^1(S,\mathfrak{a}) = H^2(S,\mathfrak{a}) = 0.$$

Corollary 5.4.9. Let j:J → S be the Jacobian fibration of a quasi-elliptic fibration f: X → S. Suppose that J is a rational surface. Then the isomorphism class of f in Elf(j) is uniquely determined by the local invariants of f. For every finite set of elements $a_i \in \mathrm{Elf}(j_s)$ there exists a unique element from Elf(j) whose local invariant at b_i is equal to a_i.

PROOF. If J is a rational surface, then $S = \mathbb{P}^1$ and

$$\deg(R^1 j_* \mathcal{O}_J) = \chi(\mathcal{O}_J) - 2\chi(\mathcal{O}_J) = -1.$$

Thus, the condition of the previous theorem is satisfied.

S5. Weierstrass models.

Let j:J → S be a jacobian genus 1 fibration. We keep the notation from the previous sections.

Proposition 5.5.1. Let X be a geometrically irreducible complete regular curve of arithmetic genus 1 over a field K. Suppose that $X(K) \neq \emptyset$. Then X is isomorphic over K to the curve in $\mathbb{P}^2_K$ given in projective coordinates (x,y,t) by one of the following equations:

(i) $y^2t+x^3+a_1xt^2+a_2t^3 = 0$, $\Delta = 4a_1^3+27a_2^2 \neq 0$ (X is smooth, $p \neq 2,3$),

(ii) $y^2t+a_1xyt+a_2yt^2+x^3+a_3xt^2+a_4t^3 = 0$, $\Delta = a_1^3a_2^3+a_2^4+a_1^4(a_1a_2a_3+a_3^2+a_1^2a_4) \neq 0$ (X is smooth, $p = 2$),

(iii) $y^2t+x^3+a_1x^2t+a_2xt^2+a_3t^3 = 0$, $\Delta=a_2^3+(a_2^2+2a_1a_3)a_1^2 \neq 0$ (X is smooth, $p = 3$),

(iv) $y^2t+x^3+a_1xt^2+a_2t^3 = 0$, $a_1 \notin K^2$ or $a_1 \in K^2$, $a_2 \notin K^2$ (X is not smooth, $p = 2$),

(v) $y^2t+x^3+at^3 = 0$, $a \notin K^3$ (X is not smooth, $p = 3$).

PROOF. In the case X is smooth, this is well-known (see **[Ta 3]**. We give a proof for the sake of completeness.

Let $P \in X(K)$. It defines a divisor of degree 1 on X. By a standard argument, the linear system $|3P|$ defines an isomorphism of X onto a plane curve given by an equation

$$F = y^2t+b_1yt^2+b_2xyt+x^3+b_3x^2t+b_4xt^2+b_5t^3 = 0.$$

Assume that $p \neq 2,3$. Then, replacing y by $y+\frac{1}{2}b_1t+b_2x$ and x by $x+(1/3)b_3t$, we may assume that $b_1 = b_2 = b_3 = 0$. Hence X is isomorphic to a plane curve

$$y^2t+x^3+a_1xt^2+a_2t^3 = 0.$$

Clearly, $\Delta = 4a_1^3+27b^2 = 0$ if and only if the polynomial $x^3+a_1xt^2+a_2t^3$ is reducible. One of its irreducible factors must be a linear polynomial $x+\alpha t$. The point $Q = (-\alpha,0,1)$ is a singular K-point of X. Since X is regular, this is impossible. Hence $\Delta \neq 0$. It is immediately checked that X is smooth.

Let $p = 2$. Replacing x by $x+(1/3)b_3t$, we may assume that $b_3=0$. Computing the partial derivatives, we obtain that a singular point $Q = (x_0,y_0,t_0) \in X(\bar{K})$ satisfies the equations:

$$F_x' = x^2+b_4t^2+b_2yt = 0, \quad F_y' = b_1t^2+b_2xt = 0,$$

$$F_t' = y^2+b_2xy+b_3x^2+b_5t^2 = 0.$$

It is easy to see that $t_0 \neq 0$. Then $b_1t_0+b_2x_0 = 0$, $y_0 = (x_0^2+b_4t_0^2+b_2)/t_0$. If $b_1t+b_2x \neq 0$, $Q \in X(K)$. This is impossible. Thus, either X is smooth or $b_1 = b_2 = 0$, and X is isomorphic to a curve

$$F = y^2t+a_1yt^2+a_2xyt+x^3+a_3x^2t+a_4xt^2+a_5t^3 = 0.$$

$$F = y^2t+x^3+a_1xt^2+a_2t^3 = 0.$$

It can be proven that X is smooth if and only if $\Delta \neq 0$. Clearly, X is regular and non-smooth if and only if $a_1 \notin K^2$ or $a_1 \in K^2$ but $a_2 \notin K^2$.

Let p = 3. Replacing y by $y + \frac{1}{2}(b_1t+b_2x)$, we may assume that $b_1=b_2 = 0$. Computing the partial derivatives, we obtain that a singular point $Q = (x_0,y_0,t_0) \in X(\bar{K})$ satisfies the equations:

$$F_x' = b_3tx+b_4t^2 = 0 \ , \ F_y' = yt = 0 \ , \ F_t' = y^2+b_3x^2+b_4xt = 0.$$

If $t_0 \neq 0$, then $y_0 = 0$ and $b_3x_0+b_4t_0 = 0$. Thus, $b_3 = b_4 = 0$, otherwise $Q \in X(K)$ and X is not regular. This shows that either X is smooth and isomorphic to a curve,

$$y^2t+x^3+a_1x^2t+a_2xt^2+a_3t^3 = 0$$

or X is regular non-smooth and is isomorphic to a curve

$$y^2t+x^3+at^3 = 0.$$

Obviously, in the latter case $a \notin K^3$. It is easy to check that X is smooth, if and only if $\Delta \neq 0$.

Corollary 5.5.1. A quasi-elliptic fibration exists only in char(K) = 2 or 3.

PROOF. We have checked it in the case when the general fibre has a rational point. The general case can be reduced to this case by extending the field K.

Recall that Δ defined above is called the **discriminant** of the elliptic curve X. Let

$$c = a_1 \quad (\text{ if } p \neq 2,3),$$

$$= a_1^4 \ (\text{ if } p = 2),$$

$$= a_1^2 \ (\text{if } p = 3).$$

Then

$$j = c^3/\Delta$$

is called the **absolute invariant** of X.

Note that if $K = \mathbb{C}$, $j = (6912)^{-1}j^{cl}$, where j^{cl} is the classical invariant defined by modular forms.

A smooth (resp. non-smooth but regular) geometrically irreducible complete curve E of arithmetic genus 1 over K with $E(K) \neq 0$ will be called an **elliptic curve** (resp. a **quasi-elliptic curve]**).

Proposition 5.5.2. Two isomorphic elliptic curves over a field K have the same absolute invariant. Conversely if their absolute invariants are equal, they become isomorphic over a separable quadratic extension of K.

PROOF. This is a well-known fact (cf. **[Ta 3]**). We recall its proof in the case $p \neq 2,3$. Let E and E' be two ellipic curves over K and φ: $E \to E'$ be an isomorphism. Let f: $E \hookrightarrow \mathbb{P}^2$ and $f'{:}E' \hookrightarrow \mathbb{P}^2$ be the corresponding embedding with the images given by the equations

$$y^2t+x^3+a_1xt^2+a_2t^3 = 0,$$

$$y'^2t'+x'^3+a_1'x't'^2+a_2't'^3 = 0$$

respectively. Let $P = f^{-1}((0,1,0)$, $P' = f'^{-1}((0,1,0))$. By composing φ with a translation, we may assume that $\varphi(P) = P'$. It is easy to see that $\{1,x/t\}$ (resp. $\{1,x/t,y/t\}$) is a basis of the space $\Gamma(E,\mathcal{O}_E(2P))$ (resp. $\Gamma(E,\mathcal{O}_E(3P))$), and a similar fact is true for the curve E'. Thus

$$\varphi^*(x'/t') = a+bx/t, \quad \varphi^*(y'/t') = c+dx/t+ey/t$$

for some constants a,b,c,d, and e with $b \neq 0$, $e \neq 0$. Substituting into the equations, we immediately obtain

$$a = c = d = 0, \; b^3 = e^2,$$

$$a_1 = (b/e)^4a_1', \quad a_2 = (b/e)^6a_2',$$

Recalling the definition of the absolute invariant, we see that they are equal.

Conversely, assume that the absolute invariants are equal. In the above notation this implies:

$$\frac{a_1^3}{4a_1^3+27a_2^2} = \frac{a_1'^3}{4a_1'^3+27a_2'^2}$$

or, equivalently,

$$\frac{a_2^2}{a_2'^2} = \frac{a_1^3}{a_1'^3} .$$

Let L/K be an extension such that

$$\frac{a_1'a_2}{a_1a_2'} = \alpha^2$$

for some $\alpha \epsilon L$. Then we immediately verify that

$$a_1 = \alpha^4 a_1', \; a_2 = \alpha^6 a_2'.$$

Hence we can define an isomorphism $\varphi : E_L \rightarrow E_L'$ by a projective transformation:

$$x = \alpha^2 x', \; y = \alpha^3 y', \; t = t'.$$

This checks the assertion.

Proposition 5.5.3. Let

$$X: \; y^2t+x^3+a_1xt^2+a_2t^3 = 0 \text{ and } X': \; y^2t+x^3+a_1'xt^2+a_2't^3 = 0$$

$$(\text{resp. } X : \; y^2t+x^3+at^3 = 0 \text{ and } X': \; y^2t+x^3+a't^3 = 0)$$

be two quasi-elliptic curves over a field K of characteristic 2 (resp. 3). Then $X \cong X'$ if and only if there exist $c,b \epsilon K$, $\varepsilon \epsilon K^*$ such that

$$a_1' = \varepsilon^4(a_1+c^4), \; a_2' = \varepsilon^6(a_2+a_1c^2+b^2)$$

(resp. there exist $c \epsilon K, g \epsilon K^*$ such that $a' = \varepsilon^6(a+c^3)$).

PROOF. We do only the case p = 2 and leave the other case to the reader. Similarly to the proof of the previous proposition we find that $X \cong X'$ if and only if the equation of X' is obtained from the equation of X by the change of coordinates:

$$y \rightarrow g^3y + kg^2xt^2 + st^3, \; x \rightarrow g^2x+ht^2, \; t \rightarrow t$$

for some $g \in K^*$ and $h, \kappa, s \in K$. Plugging in, we obtain the equation of X' in the form

$$(g^6y^2+\kappa^2g^4x^2t^4+s^2t^6)t+(g^6x^3+g^4x^2ht^2+g^2h^2t^4x+h^3t^6)+a_1(g^2x+ht^2)t^4+a_2t^3.$$

Dividing by g^6, we must have

$$\kappa^2 = h,\ a_1' = g^{-4}(h^2+a_1) = g^{-4}(\kappa^4+a_1),$$

$$a_2' = g^{-6}(a_2+s^2+h^3+a_1h) = g^{-6}(a_2+s^2+\kappa^6+a_1\kappa^2).$$

It remains to set

$$c = \kappa,\ b = s+\kappa^3 \text{ and } \varepsilon = g^{-1}.$$

We can apply Proposition 5.5.1 to obtain an embedding $\varphi_\eta : J_\eta \to \mathbb{P}^2_K$, where $K = \kappa(S)$. Next we will extend this embedding to a birational morphism

$$\varphi: J \to W \subset \mathbb{P},$$

where $\mathbb{P}$ is a certain $\mathbb{P}^2$-bundle over S.

Let C be a fixed section of j. Applying R^1j_* to the exact sequence

$$0 \to \mathcal{O}_J((n-1)C) \to \mathcal{O}_J(nC) \to \mathcal{O}_C(nC) \to 0,$$

we find (see details in **[M-S]**).

Lemma 5.5.1.

(i) $j_*(\mathcal{O}_J(nC))$ is locally free of rank n;

(ii) $R^1j_*\mathcal{O}_J(nC) = 0$ for $n>0$ and invertible for $n = 0$;

(iii) $R^1j_*\mathcal{O}_J(nC) = 0$ for $i > 0$ and any $n \in \mathbf{Z}$;

(iv) the natural inclusion $\mathcal{O}_C \to j_*\mathcal{O}_J(C)$ is an isomorphism;

(v) $j_*(\mathcal{O}_J(nC)/\mathcal{O}_J((n-1)C)) \cong \mathfrak{N}_C^n \cong (R^1j_*\mathcal{O}_J)^n$, for $n>1$.

Let $\mathfrak{L} = R^1j_*\mathcal{O}_J$, $V = j_*\mathcal{O}_J(2C)$, $V' = j_*\mathcal{O}_J(3C)$. Choose an affine cover $\{U_i\}$ of S such that $\mathfrak{L}|U_\alpha$ is free with basis t_i. Then t_i^n is a basis of $\mathfrak{L}^n|U_i$ for all n. Let $x_i \in \Gamma(U_i, V)$ projects to t_i^2 in the exact sequence

$$0 \to \mathcal{O}_J \to V \to \mathfrak{X}^2 \to 0.$$

Similarly, we choose $y_i \in \Gamma(U_i, j_*\mathcal{O}_J(3C))$ which projects to t_i^3 in the exact sequence

$$0 \to V \to j_*\mathcal{O}_J(3C) \to \mathfrak{X}^3 \to 0.$$

Then $1, x_i$ and y_i form a basis for $j_*\mathcal{O}_J(3C)|U_i$ and

$$1, x_i, x_i^2, x_i^3, y_i, x_i, y_i, y_i^2 \in \Gamma(U_i, j_*\mathcal{O}_J(6C)).$$

Since $\Gamma(U_i, j_*\mathcal{O}_J(6C)) \cong \mathcal{O}_S(U_i)^6$, these sections are linearly dependent over $\mathcal{O}_S(U_i)$. Restricting to the general fibre of j, we may assume by applying Proposition 5.5.1, that one of the following holds:

(i) $y_i^2 + x_i^3 + a_1^{(i)} x_i + a_2(i) = 0$, $\Delta = 4a_1^{(i)2} + 27a_2^{(i)3} \neq 0$ (j is elliptic , $p \neq 2,3$),

(ii) $y_i^2 + a_1^{(i)} x_i y_i + a_2^{(i)} y_i + x_i^3 + a_3^{(i)} x_i + a_4^{(i)} t = 0$, (j is elliptic, $p = 2$),

(iii) $y_i^2 + x_i^3 + a_1^{(i)} x_i^2 + a_2^{(i)} x_i + a_3^{(i)} = 0$, (j is elliptic, $p = 3$),

(iv) $y_i^2 + x_i^3 + a_1^{(i)} x_i + a_1^{(i)} = 0$, (j is quasi-elliptic, $p = 2$),

(v) $y_i^2 + x_i^3 + a_1^{(i)} = 0$, (j is quasi-elliptic, $p = 3$).

Since $1, x_i, y_i$ generate $\mathcal{O}_J(3C)|j^{-1}(U_i)$, they define a morphism

$$\varphi_i: j^{-1}(U_i) \to \mathbb{P}^2 \times U_i.$$

Since j is proper, φ is proper. Also, by the previous proposition, its restriction to the general fibre is a birational isomorphism onto its image. Thus, φ defines a birational morphism onto the subscheme of $\mathbb{P}^2 \times U_i$ given by the homogenized equations above.

Replacing U_i by U_j , we easily see (cf.the proof of Proposition 5.5.2) that

$$t_i = g_{ij} t_j,\ x_i = g_{ij}^4 x_j + h_{ij} g_{ij}^6 t_j^2,\ y_i = g_{ij}^2 y_j + k_{ij} g_{ij}^4 x_j t_j^2 + s_{ij} g_{ij}^3 t_j^3,$$

where $g_{ij} \in \Gamma(U_i \cap U_j, \mathcal{O}_S^*)$, $h_{ij}, k_{ij}, s_{ij} \in \Gamma(U_i \cap U_j, \mathcal{O}_S)$. Plugging these coordinate changes into (i)-(v), we find in the respective cases

(i) $h_{ij} = k_{ij} = s_{ij} = 0$, $a_1^{(i)} = g_{ij}^4 a_1^{(j)}$, $a_2^{(i)} = g_{ij}^6 a_1^{(j)}$;

(ii) $k_{ij}^2 + a_1^{(i)} k_{ij} + h_{ij} = 0$, $a_1^{(i)} = g_{ij} a_1^{(j)}$, $a_1^{(i)} h_{ij} + a_2^{(i)} = g_{ij}^3 a_2^{(j)}$, $a_3^{(i)} + a_2^{(i)} k_{ij} + a_1^{(i)}(k_{ij} h_{ij} + s_{ij}) + h_{ij}^2 = g_{ij}^4 a_3^{(j)}$, $a_4^{(i)} + a_3^{(i)} h_{ij} + a_2^{(i)} s_{ij} + a_1^{(i)} h_{ij} s_{ij} + s_{ij}^2 + h_{ij}^3 = g_{ij}^6 a_4^{(j)}$;

(iii) $\kappa_{ij}=s_{ij}=0$, $a_1^{(i)} = g_{ij}^2 a_1^{(j)}$, $a_2^{(i)}+2h_{ij}a_1^{(i)} = g_{ij}^4 a_2^{(j)}$ $a_3^{(i)}+h_{ij}^3+a_1^{(i)}h_{ij}^2+a_2^{(i)}h_{ij} = g_{ij}^6 a_3^{(j)}$;

(iv) $\kappa_{ij}^2=h_{ij}$, $a_1^{(i)}+h_{ij}^2 = g_{ij}^4 a_1^{(j)}$, $a_2^{(i)}+a_1^{(i)}h_{ij}+s_{ij}^2+h_{ij}^3 = g_{ij}^6 a_2^{(j)}$;

(v) $\kappa_{ij}=s_{ij}=0$, $a^{(i)}+h_{ij}^3=g_{ij}^6 a^{(j)}$.

This shows that in the respective cases:

(i) $\{a_1^{(i)}\}$ defines a section $a_1\epsilon\Gamma(\mathcal{I}^{-4})$, $\{a_2^{(i)}\}$ defines a section $a_2\epsilon\Gamma(\mathcal{I}^{-6})$, $(a_0\equiv 1,a_1,a_2)$ defines a section of the vector bundle $E \cong \mathcal{O}_S \oplus \mathcal{I}^{-4} \oplus \mathcal{I}^{-6}$ of rank 3;

(ii) $\{a_1^{(i)}\}$ defines a section $a_1\epsilon\Gamma(\mathcal{I})$, if $a_0^{(i)} \equiv 1$, then $(\{a_0^{(i)}\},\{a_1^{(i)}\},\{a_2^{(i)}\},\{a_3^{(i)}\},\{a_4^{(i)}\})$ defines a section (a_0,a_1,a_2,a_3,a_4) of a certain vector bundle of rank 5 whose transition matrices are given by

$$\begin{bmatrix} a_0^{(i)} \\ a_1^{(i)} \\ a_2^{(i)} \\ a_3^{(i)} \\ a_4^{(i)} \end{bmatrix} = \begin{bmatrix} 1 & 0 & 0 & 0 & 0 \\ 0 & g_{ij} & 0 & 0 & 0 \\ 0 & g_{ij}h_{ij} & g_{ij}^3 & 0 & 0 \\ h_{ij}^2 & g_{ij}s_{ij} & g_{ij}^3\kappa_{ij} & g_{ij}^4 & 0 \\ s_{ij}^2+h_{ij}^3 & g_{ij}h_{ij}s_{ij} & g_{ij}^3(s_{ij}+\kappa_{ij}h_{ij}) & g_{ij}^4 h_{ij} & g_{ij}^6 \end{bmatrix} \begin{bmatrix} a_0^{(j)} \\ a_1^{(j)} \\ a_2^{(j)} \\ a_3^{(j)} \\ a_4^{(j)} \end{bmatrix}$$

(iii) $\{a_1^{(i)}\}$ defines a section $a_1\epsilon\Gamma(\mathcal{I}^{-2})$, $(\{a_0^{(i)}\}\equiv 1,\{a_1^{(i)}\},\{a_2^{(i)}\},\{a_3^{(i)}\})$ defines a (a_0,a_1,a_2,a_3,a_4) of certain vector bundle E of rank 4 whose transition matrices are given by

$$\begin{bmatrix} a_0^{(i)} \\ a_1^{(i)} \\ a_2^{(i)} \\ a_3^{(i)} \end{bmatrix} = \begin{bmatrix} 1 & 0 & 0 & 0 \\ 0 & g_{ij}^2 & 0 & 0 \\ 0 & g_{ij}^2 h_{ij} & g_{ij}^4 & 0 \\ -h_{ij}^3 & g_{ij}^2 h_{ij}^2 & -g_{ij}^4 h_{ij} & g_{ij}^6 \end{bmatrix} \begin{bmatrix} a_0^{(j)} \\ a_1^{(j)} \\ a_2^{(j)} \\ a_3^{(j)} \end{bmatrix}$$

(iv) $(\{a_0^{(i)}\}\equiv 1,\{a_1^{(i)}\},\{a_2^{(i)}\})$ defines a section (a_0,a_1,a_2) of a certain vector bundle of rank 3 whose transition matrices are given by

$$\begin{bmatrix} a_0^{(i)} \\ a_1^{(i)} \\ a_2^{(i)} \end{bmatrix} = \begin{bmatrix} 1 & 0 & 0 \\ \kappa_{ij}^4 & g_{ij}^4 & 0 \\ s_{ij}^2+\kappa_{ij}^6 & g_{ij}^4\kappa_{ij}^2 & \kappa_{ij}^6 \end{bmatrix} \begin{bmatrix} a_0^{(j)} \\ a_1^{(j)} \\ a_2^{(j)} \end{bmatrix}$$

(v) $\{\{a_0^{(i)}\}\equiv 1, a^{(i)}\}$ define a section (a_0,a) of a certain vector bundle of rank 2 whose transition functions are given by

$$\begin{bmatrix} a_0^{(i)} \\ a_1^{(i)} \end{bmatrix} = \begin{bmatrix} 1 & 0 \\ h_{ij}^3 & g_{ij}^6 \end{bmatrix} \begin{bmatrix} a_0^{(j)} \\ a_1^{(j)} \end{bmatrix}$$

It is easy to compute the transition matrices for the bundles V and V'. They are

$$\begin{bmatrix} g_{ij}^2 & h_{ij}^2 \\ 0 & 1 \end{bmatrix}^{-1} = \begin{bmatrix} g_{ij}^{-2} & -h_{ij}^{2}g_{ij}^{-2} \\ 0 & 1 \end{bmatrix}$$

$$\begin{bmatrix} g_{ij}^3 & g_{ij}^2 k_{ij} & s_{ij} \\ 0 & g_{ij}^2 & h_{ij} \\ 0 & 0 & 1 \end{bmatrix}^{-1} = \begin{bmatrix} g_{ij}^{-3} & -k_{ij}g_{ij}^{-3} & (k_{ij}h_{ij}-s_{ij})g_{ij}^{-3} \\ 0 & g_{ij}^{-2} & -h_{ij}g_{ij}^{-2} \\ 0 & 0 & 1 \end{bmatrix}$$

respectively. Also we see that V and V' can be obtained as extensions:

$$0 \to \mathcal{O}_S \to V \to \mathfrak{L}^2 \to 0\,, \quad 0 \to V \to V' \to \mathfrak{L}^3 \to 0$$

and the vector bundle E is determined from the extensions

$E \cong \mathcal{O}_S \oplus \mathfrak{L}^{-4} \oplus \mathfrak{L}^{-6}$ in case (i),

$0 \to V'\otimes\mathfrak{L}^{-6} \to E \to \mathfrak{L}^{-1}\oplus\mathcal{O}_S \to 0$ in case (ii),

$0 \to \mathrm{Sym}^3(V)\otimes\mathfrak{L}^{-6} \to E \to \mathcal{O}_S \to 0$ in case (iii),

$0 \to \mathfrak{L}^{-6}\otimes V \to E \to \mathcal{O}_S \to 0$ in case (iv),

$0 \to \mathfrak{L}^{-6} \to E \to \mathcal{O}_S \to 0$ in case (v).

Summarizing, we have proven

Theorem 5.5.1. Let $j:J \to S$ be a jacobian elliptic surface, C a section of j, $\mathbb{P} = \mathbb{P}(j_*\mathcal{O}_J(3C))$, $\pi: \mathbb{P} \to S$ the natural projection. There exists a birational morphism $\varphi: J \to W \cong V(s) \subset \mathbb{P}$, where

$$s\in\Gamma(\mathbb{P},\mathcal{O}_{\mathbb{P}}(3)\otimes\pi^*E) = \Gamma(S,\mathrm{Sym}^3(j_*\mathcal{O}_S(3C))\otimes E)$$

is given by

(i) $a_0y^2z + a_0x^3 + a_1xz^2 + a_2z^3 = 0$, j is elliptic, $p \neq 2,3$,

(ii) $a_0y^2z + a_1xyz + a_2yz^2 + x^3 + a_3xz^2 + a_4z^3 = 0$, j is elliptic, $p = 2$,

(iii) $a_0y^2z + a_0x^3 + a_1x^2z + a_2xz^2 + a_3z^3 = 0$, j is elliptic, $p = 3$,

(iv) $a_0y^2z + a_0x^3 + a_1xz^2 + a_2z^3 = 0$, j is quasi-elliptic, $p = 2$,

(v) $a_0y^2z + a_0x^3 + a_1z^3 = 0$, (j is quasi-elliptic, $p = 3$).

Moreover, $(a_0, a_1, \dots, a_r) \to a_0 = 1$ under the map $\Gamma(E) \to \Gamma(\mathcal{O}_S)$ corresponding to the extension $E \to \mathcal{O}_S \to 0$.

We will call the surface W, defined as above, the **Weierstrass model** of the jacobian fibration $j:J \to S$. We denote by $j:W \to S$ the restriction to W of the natural projection $\mathbb{P}(V') \to S$. Notice that the birational morphism $\varphi: J \to W$ is a morphism over S (i.e. $j = j \circ \varphi$).

Remark 5.5.1. Note that W is not smooth in general, thus $j:W \to S$ is not a genus 1 fibration by our definition.

Remark 5.5.2. If $p \neq 2$ or 3, the bundles V and V' split and we have

$$V \cong \mathcal{L}^2 \oplus \mathcal{O}_S,\ V' \cong \mathcal{L}^3 \oplus \mathcal{L}^2 \oplus \mathcal{O}_S.$$

This follows from inspection of the transition matrix and the fact that $s_{ij} = h_{ij} = k_{ij} = 0$ in this case.

If $p = 2, 3$, the same is true, if

$$H^1(S, \mathcal{L}^{-2}) = 0 \text{ (for V)}$$

$$H^1(S, \mathcal{L}^{-2}) = H^1(S, \mathcal{L}^{-1}) = 0 \text{ (for V').}$$

For example, it is true if $\deg(\mathcal{L}^{-1}) = \chi(\mathcal{O}_J) > -2\chi(\mathcal{O}_S)$.

Proposition 5.5.4. Let W be the Weierstrass model of $j: J \to S$ and $\mathbf{A} = J^{\#}$ be the Néron scheme of j. Then φ defines an isomorphism of group schemes over S

$$W^{\#} \cong \mathbf{A}^0,$$

where the zero section of $\mathbf{A}$ is defined by the section C.

PROOF. Obvious.

Corollary 5.5.2. Assume that $j:J \to S$ is a quasi-elliptic fibration. Then the following exact sequences of groups schemes are defined:

$$0 \to J^{\#} \to \mathbf{V}(V \otimes \mathfrak{L}^{-3}) \to \mathbf{V}(\mathfrak{L}^{-3}) \to 0 \quad (p = 3),$$

$$0 \to J^{\#} \to \mathbf{V}(V'' \otimes \mathfrak{L}^{-4}) \to \mathbf{V}(\mathfrak{L}^{-4}) \to 0 \quad (p = 2),$$

where $\mathbf{V}(\mathcal{E})$ is the additive group scheme associated to a vector bundle $\mathcal{E}$, V'' is a subbundle of V' spanned by 1 and y.

PROOF. Assume first that p = 3. Then, in the notation of Weierstrass models,

$$J^{\#} \cong W^{\#} \cong D(y) = \mathbb{P}(V') \setminus V(y),$$

where $V(y)$ is the zero subscheme of the section $y \in \Gamma(\mathbb{P}, \mathcal{O}_{\mathbb{P}}(1)) = \Gamma(S, V')$. Under the natural map $V' \to \mathfrak{L}^3$, y goes to a generator of $\mathfrak{L}^3$ (locally). This shows that

$$D(y) \cong \mathbf{V}(\mathrm{Ker}(V' \otimes \mathfrak{L}^{-3} \to \mathfrak{L}^3 \otimes \mathfrak{L}^{-3} = \mathcal{O}_S) \cong \mathbf{V}(V \otimes \mathfrak{L}^{-3}).$$

It is immediately verified that $W^{\#} = W \cap D(y)$ is given locally by the equation

$$v_i + u_i^3 + a_1^{(i)} v_i^3 = 0,$$

where $v_i = z_i/y_i$, $u = x_i/y_i$. One checks that the homomorphisms of the additive groups:

$$\mathcal{O}_{U_i}^3 \to \mathcal{O}_{U_i}, \quad (v_i, u_i) \to v_i + u_i^3 + a_1^{(i)} v_i^3$$

are glued together to give a global surjective homomorphism

$$\mathbf{V}(V \otimes \mathfrak{L}^{-3}) \to \mathbf{V}(\mathfrak{L}^{-3})$$

whose kernel is isomorphic to $W^{\#}$.

Assume p = 2. We immediately check that

$$J^{\#} = W^{\#} = D(x^2 + a_1) = \mathbb{P}(V') \setminus Z(x^2 + a_1),$$

where $Z(x^2+a_1)$ is the zero subscheme of the section

$$x^2 + a_1 \in \Gamma(S, \mathrm{Sym}^2(V)) \subset \Gamma(S, \mathrm{Sym}^2(V')) = \Gamma(\mathbb{P}(V'), \mathcal{O}_{\mathbb{P}(V')}(2)).$$

It is easy to verify that $W^{\#} = W \cap D(x^2+a_1)$ is given locally by the equation

$$u_i^4 + v_i + a_1^{(i)} v_i^2 + a_2^{(i)2} v_i^4 = 0,$$

where

$$v_i = \frac{1}{x_i^2 + a_1^{(i)}}, \quad u_i = \frac{y_i}{x_i^2 + a_1^{(i)}}$$

One checks that (v_i, u_i) are related to (v_j, u_j) by

$$v_i = g_{ij}^{-4} v_j, \quad u_i = g_{ij}^{-1}(u_j + s_{ij} v_j)$$

Thus the homomorphisms of the additive groups:

$$\mathcal{O}_{U_i}^2 \to \mathcal{O}_{U_i}, \quad (v_i, u_i) \to u_i^4 + v_i + a_1^{(i)} v_i^2 + a_2^{(i)2} v_i$$

are glued together to define a global surjective homomorphism of unipotent group schemes

$$\mathbf{V}(V'' \otimes \mathcal{L}^{-4}) \to \mathbf{V}(\mathcal{L}^{-4}),$$

where V'' is a subbundle of V' spanned by 1 and y. Clearly the kernel of κ is isomorphic to $J^{\#}$.

The previous corollary allows one to compute the cohomology of the Néron sheaf $\mathfrak{a}$ represented by the scheme $J^{\#}$. We state only a special case which will be applied to Enriques surfaces.

Corollary 5.5.3. Assume $H^1(S, \mathcal{L}^{-i}) = 0$ for $i = 1,2,3,4$. Then

$$H^i(S, \mathfrak{a}) = 0, \ i > 0.$$

Remark 5.5.3. It follows from the previous proof that the image $\mathfrak{C}'$ of the curve of cusps $\mathfrak{C}$ in W is isomorphic to the subscheme

$$\mathfrak{C}' = Z(x^3 + a_1) \cap Z(y) \subset \mathbb{P}(V'), \qquad p = 3,$$

$$\mathfrak{C}' = Z(x^2 + a_1) \cap Z(y^2 + a_2) \subset \mathbb{P}(V'), \qquad p = 2.$$

Local computations show that this curve is smooth at a regular point of W and intersects the fibre passing through this point with multiplicity p. The induced cover $\mathfrak{C}' \to S$ is purely inseparable of degree p. One can also show that a minimal resolution of W resolves the singularities of $\mathfrak{C}'$, hence $\mathfrak{C}$ is smooth.

Corollary 5.5.4. Assume j is an elliptic fibration. Then

$$\Delta_i = 4a_1^{(i)3} + 27a_2^{(i)2} a_0^{(i)} \quad (p \neq 2,3),$$

$$\Delta_i = a_1^{(i)3} a_2^{(i)3} a_0^{(i)} + a_2^{(i)4} a_0^{(i)3} + a_1^{(i)4}(a_1^{(i)} a_2^{(i)} a_3^{(i)} + a_3^{(i)2} + a_1^{(i)2} a_4^{(i)}) \quad (p = 2),$$

$$\Delta_i = 2a_2^{(i)\,3}a_0^{(i)} + (a_0^{(i)\,2} + a_1^{(i)}a_3^{(i)})a_1^{(i)\,2} \quad (p = 3).$$

can be glued together to define a section

$$\Delta = 4a_1^3 + 27a_2^2a_0 \in \Gamma(\mathfrak{L}^{-12})\setminus\{0\} \subset \Gamma(\mathrm{Sym}^3(E))\setminus\{0\} \quad (p \neq 2,3),$$

$$\Delta = a_1^3a_2^3a_0 + a_2^4a_0^3 + a_1^4(a_1a_2a_3 + a_3^2 + a_1^2a_4) \in \Gamma(\mathfrak{L}^{-12})\setminus\{0\} \subset \Gamma(\mathrm{Sym}^7(E)) \quad (p=2)$$

$$\Delta = 2a_2^3a_0 + (a_0^2 + a_1a_3)a_1^2 \in \Gamma(\mathfrak{L}^{-12})\setminus\{0\} \subset \Gamma(\mathrm{Sym}^4(E)) \quad (p = 3).$$

We will call both Δ and the corresponding divisor $\mathrm{div}(\Delta)$ on S the **discriminant** of j. The following result follows from Proposition 5.5.1.

Proposition 5.5.5. Let j be an elliptic jacobian fibration $\mathfrak{L} = R^1j_*\mathcal{O}_J$, Δ is the discriminant. Then

(i) $s \in \Delta$ if and only if $W_s = j^{-1}(b)$ is not smooth;

(ii) $\deg(\Delta) = c_2(J) = e(J) \geq 0$;

(iii) the following properties are equivalent:

(a) $\deg(\mathfrak{L}) = 0$;

(b) $\mathfrak{L} \cong \mathcal{O}_S$;

(c) $\mathrm{div}(\Delta) = 0$;

(d) j is smooth;

(e) there exists a finite étale cover $S' \to S$ such that $J \times_S S' \cong S' \times E$, where E is a smooth elliptic curve.

PROOF. (i) Follows immediately from Proposition 5.5.1. We know from §1 that

$$\deg(\mathfrak{L}) = C^2 = -\chi(\mathcal{O}_J) = \frac{1}{12}c_2(F) .$$

Since $\Delta \in \Gamma(\mathfrak{L}^{-12})$, we immediately obtain (ii) and that the properties (a)-(d) from (iii) are equivalent. The implication (e) $\Rightarrow$ (d) is obvious. Assume (d) holds. Let us give first a simple proof in the case $p \neq 2,3$. Since $\Delta = 0$, $\mathfrak{L}^{-12} \cong \mathcal{O}_S$. We take the cover $S' \to S$ which trivializes $\mathfrak{L}$ and obtain a fibration j' over S' with $\mathfrak{L}' = \mathcal{O}_{S'}$. Thus the sections a_1' and a_2' for the fibration j' are constants and the Weierstrass equation of J_η' is defined over K. This obviously proves the assertion. In the general case we use the theory of a fine moduli space of elliptic curves with a level structure (see **[D-R]**. Let $S' \to S$ be an étale cover of S which trivializes the sheaf $R^1j_*\mu_n$ for some $n > 2$ prime to char(κ). The base change $f': J \to S'$ defines a family of elliptic curves with level n. Thus it is a pull-back of the universal family with respect to a morphism $S' \to \mathfrak{M}(n)$, where $\mathfrak{M}(n)$ is the fine moduli scheme. Since $\mathfrak{M}(n)$ is affine and S' is complete this morphism is constant. This proves our assertion.

Corollary 5.5.5. Assume $S \cong \mathbb{P}^1$. Then every elliptic fibration $j:F \to S$ has a degenerate fibre unless $J \cong S\times E$ for some elliptic curve and j is the second projection.

PROOF. Since $\pi_1(\mathbb{P}^1) = 0$ this is clear if f is jacobian. Assume f is not jacobian and smooth. By above its Jacobian fibration is trivial. The Brauer group $Br(E\times\mathbb{P}^1)$ is obviously trivial. Since all local invariants of f are trivial, we apply Theorem 5.4.3 to obtain that $[f] = 0$ in Elf(j).

Corollary 5.5.6. Let $f: J \to S$ be a smooth jacobian elliptic fibration over a global base S. Then all fibres over closed points are isomorphic elliptic curves.

Now let $j:J \to S$ be a jacobian quasi-elliptic fibration. In this case, we can also define the discriminant.

Assume p = 2. Then

$$a_1^{(i)}(da_1^{(i)})^2+(da_2^{(i)})^2 \epsilon \Omega^1(U_i)^2$$

are glued to a section

$$\Delta = a_1(da_1)^2+(da_2)^2 \ \epsilon \ \Gamma(\mathfrak{L}^{-12}\otimes(\Omega_S^1)^2) = \Gamma(\mathfrak{L}^{-12}\otimes\omega_S^2).$$

Similarly, if p = 3, then

$$(da_1^{(i)})^2$$

are glued together to form a section

$$\Delta = (da_1)^2 \ \epsilon \ \Gamma(\mathfrak{L}^{-12}\otimes\omega_S^2).$$

We will call this section, identified with its divisor of zeroes, the

discriminant of the quasi-elliptic fibration j.

Proposition 5.5.6. Let Δ be the discriminant of a quasi-elliptic fibration. Then

(i) $s\in\Delta$ if and only if W is regular over s;

(ii) $\deg(\Delta) = c_2(F)-4\chi(\mathcal{O}_S)$;

(iii) the following properties are equivalent

(a) $\deg(\Delta) = 0$;

(b) $\mathfrak{L}^{12} \cong \omega_S^2$;

(c) W is regular.

PROOF. Similar to the previous proposition and left to the reader.

Corollary 5.5.7. Assume $3\nmid\chi(\mathcal{O}_S)$, e.g. $S \cong \mathbb{P}^1$, then every quasi-ellptic fibration $f:X \to S$ has a degenerate fibre.

PROOF. By Theorem 5.3.3, we may assume that f = j is jacobian. Under the above assumptions the Weierstrass model is not regular. Thus, φ: J → W resolves a singular point x of W by blowing up an exceptional curve over x. The corresponding fibre $\varphi^{-1}(\bar{f}^{-1}(\bar{f}(x)))$ of j is reducible.

Proposition 5.5.7. Let W be a Weierstrass model of a genus 1 fibration j: J → S. Then W is normal. Let $w\in W$ and $s = j(w)$. Then w is nonsingular if and only if J_s is irreducible. If J_s is reducible, then w is a rational double point of type A_n (resp.D_n, resp. E_n) if and only if $\mathrm{Num}_s(J)/\mathrm{Rad}$ is a root lattice of type A_n (resp.D_n, resp.E_n).

PROOF. The restriction of φ: J → W to a fibre J_s blows down all irreducible components except for a component J_s° which intersects the section C. Clearly the multiplicity of J_s° is 1. By inspection of the list of reducible degenerate fibres, we see that $(J_s\setminus J_s^\circ)_{red}$ is a Dynkin curve of the type determined by the type of $NS_s(J)$. Since the general fibre of W → S is regular, W has only isolated singularities. Moreover, W is a locally hypersurface in a regular scheme. Thus, it is normal (even, Gorenstein). It remains to end by observing that w is a normal singularity which is resolved by nodal curves and by applying Proposition 0.2.4.

The next propositions describe the type of the degenerate fibre by analyzing the coefficients of the corresponding Weierstrass model W. All of them are verified by applying the classification of rational double points and resolving explicitly the singular point of W. We keep the earlier notations.

Proposition 5.5.8. Let $f : X \to S$ be a jacobian elliptic fibration and j be the absolute invariant of its generic fibre. Assume that $p \neq 2,3$. Let $n_s = ord_s(\Delta)$, $\alpha_s = ord_s(a_1)$, $\beta_s = ord_s(a_2)$, $\gamma_s = ord_s(j)$. Then $\alpha_s \leq 4$, $\beta_s \leq 5$ and the type of the fibre J_s is determined by the following table:

Type	n_s	α_s	β_s	γ_s
$I_0(\tilde{A}_0)$	0	-	-	≥ 0
$\tilde{A}_0^*$	1	0	0	-1
$\tilde{A}_{n-1}(n>1)$	n	0	-	-n
$\tilde{A}_0^{**}$	2	≥ 1	1	≥ 1
$\tilde{A}_1^*$	3	1	≥ 2	0
$\tilde{A}_2$	4	2	≥ 2	2
$\tilde{D}_4$	6	2	≥ 3	0
$\tilde{D}_{n+4}(n\neq 0)$	6+n	2	3	n
$\tilde{E}_6$	8	≥ 3	4	≥ 1
$\tilde{E}_7$	9	3	≥ 5	0
$\tilde{E}_8$	10	≥ 4	5	≥ 2

If p = 2 or 3 the description of the degenerate fibres similar to above is less explicit but can be done. We refer to **[Ogg 2]** or **[Ta 3]** for the corresponding computations.

Proposition 5.5.9. Let j: J → S be a quasi-elliptic jacobian fibration over a complete local S ≅ Speck[[t]] of characteristic 3 and $y^2t+x^3+at^3 = 0$ be its Weierstrass model over S. Write $a = \alpha^3+\varepsilon t^m$, where ε is invertible, $3\nmid m$. Let n = ord(Δ). Then m ≤ 5 and the possible type of J_s is determined by the following table:

Type	n	m
$\tilde{A}_0^{**}$	0	1
$\tilde{A}_2^{*}$	2	2
$\tilde{E}_6$	6	4
$\tilde{E}_8$	8	5

Lemma 5.5.2. Let j: J → S be a quasi-elliptic jacobian fibration over a complete local S ≅ Speck[[t]] of characteristic 2 and $y^2t+x^3+a_1xt^2+a_2t^3 = 0$ be its Weierstrass model W over S. Then the equation of W can be put in the form:

$$y^2+x^3+(t^{2s}\alpha^4+t^m\beta^2)x+t^\kappa\gamma^2,$$

where each s,m or κ is an odd positive integer(resp. 0), κ ≤ m and the respective coefficent α, β or γ is a unit in k[[t]] (resp.zero).

PROOF. We can write

$$a_1 = \varepsilon_1^4+\alpha^4t^{2s}+t^m\beta^2,\ a_2 = t^\kappa\gamma^2+\delta^2,$$

where m,s,κ > 0 are odd, α, β, γ are units or zero. By Proposition 5.5.3, we can add ε_1^4 to a_1 and $\varepsilon_1^2a_1$ to a_2 to transform the equation to the form:

$$y^2+x^3+(t^{2s}\alpha^4+t^m\beta^2)x+t^\kappa\gamma'^2+\delta^2+(\varepsilon_1^4+\alpha^4t^{2s}+t^m\beta^2)\varepsilon_1^2.$$

Adding $\delta+\varepsilon_1^3+\alpha^2\varepsilon_1t^s$ to y, we may assume that $a_2 = t^\kappa\gamma'^2+t^m\beta^2\varepsilon_1^2$. It remains to factor out t^l from a_2, where $l = \min(\kappa,m)$ and take for γ^2 the corresponding coefficient.

Proposition 5.5.10 Let j: J → S be a quasi-elliptic jacobian fibration over a complete local S ≅ Speck[[t]] of characteristic 2 and $y^2t+x^3+a_1xt^2+a_2t^3 = 0$ be its Weierstrass model over S taken in the form from Lemma 5.5.2. Let $n = \mathrm{ord}_s(\Delta)$. Then the possible type of J_s is determined by the following table:

(I)$\gamma \neq 0$

Type	n	k	s
$\tilde{A}_0{}^{**}$	0	1	≥ 0
$\tilde{D}_4$	4	3	≥ 0
$\tilde{E}_8$	8	5	$\neq 1$
D_{2k-2}	k-2	≥ 5	1

(II)$\gamma = 0$

Type	n	m	s
$\tilde{A}_1{}^{*}$	1	1	≥ 0
$\tilde{E}_7$	7	3	$\neq 1$
$\tilde{D}_{2k}$	2k	≥ 3	1

Corollary 5.5.8. Let $j:J \to S$ be a genus 1 fibration over a global S. Assume that $p \neq 2,3$ or j is a quasi-elliptic. Then

$$\mathrm{ord}_s(\Delta) = e(J_s)-e(J_\eta).$$

PROOF. This is obtained by inspecting the tables.

Remark 5.5.4. If j is elliptic, then it was shown in **[Ogg 2]**, that

$$\mathrm{ord}_s(\Delta) = \chi(J_s) + \delta_s,$$

where δ_s is the invariant of wild ramification (see §1). Let $w \in W$ and $\mu(w)$ be its Milnor number (see **[Del 2]**). Then, this result implies that

$$\mathrm{ord}_s(\Delta) = \mu(w)$$

for the singular point w of W with $j(w)=s$.

Also note, that applying Corollary 5.3.4, we obtain the formula

$$\deg \Delta = 12\chi(\mathcal{O}_J) = e(J) = \sum_{s \in S} \mathrm{ord}_s(\Delta) =$$

$$= e(X_{\bar\eta})e(S) + \sum_{s\in S} (e(X_s)-e(X_{\bar\eta})+\delta_s)$$

from Proposition 5.1.9.

Similarly, if j is quasi-elliptic, we obtain from Corollary 5.5.8:

$$\deg(\Delta) = 12\chi(\mathcal{O}_J)+2\chi(\mathcal{O}_S) = e(J)-e(X_{\bar\eta})e(S = \sum_{s\in S} \mathrm{ord}_s(\Delta)$$

$$= \sum_{s\in S} (e(X_s)-e(X_{\bar\eta}).$$

which is the same formula for a quasi-elliptic fibration.

Let C be a section of a jacobian fibration $j{:}J \to S$, $V' = j_*\mathcal{O}_J(3C)$ and $W \subset \mathbb{P}(V')$ be the corresponding Weierstrass model of j. The surjection

$$V' \to \mathfrak{L}^3$$

defines a section $c{:}S \to \mathbb{P}(V')$. It follows from our construction of W that this section factors through W and its image in W is isomorphic to the section C. The inclusion

$$V = j_*\mathcal{O}_J(2C) \to V'$$

defines a projection (from C)

$$pr_C{:}\ \mathbb{P}(V') \to \mathbb{P}(V)$$

which induces a morphism

$$\bar\varphi{:}\ W \to \mathbb{P}(V)$$

whose composition with $\varphi : J \to W$ defines a morphism

$$\psi{:}\ J \to \mathbb{P}(V).$$

Proposition 5.5.11

(i) $\bar\varphi$ is a finite map of degree 2;

(ii) $\bar\varphi$ is a separable cover if and only j is an elliptic fibration;

(iii) the surjection $V \to \mathfrak{L}^2$ defines a section of $\mathbb{P}(V)$ which is the image of C under $\bar\varphi$;

(iv) if j is an elliptic fibration, then the ramification locus of $\bar\varphi$ is the union of $\bar\varphi(C)$ and the subscheme $R = V(s)$ of $\mathbb{P}(V)$, where

$$s = a_0x^3+a_1xz^2+a_2z^3\in\Gamma(\mathbb{P}(V),\mathcal{O}_{\mathbb{P}(V)}(3)\otimes\pi^*(E)) \quad (p \neq 2, 3),$$

$$s = a_1x+a_2z\in\Gamma(\mathbb{P}(V),\mathcal{O}_{\mathbb{P}(V)}(1)\otimes\pi^*(V\otimes\mathfrak{L}^{-3})) \quad (p = 2),$$

$$s = a_0x^3+a_1x^2z+a_2xz^2+a_3z^3\in\Gamma(\mathbb{P}(V),\mathcal{O}_{\mathbb{P}(V)}(3)\otimes\pi^*(E)) \quad (p = 3)$$

(here the a_i's are the coefficients in the equation of the Weierstrass model W, $\pi:\mathbb{P}(V) \to S$ is the natural projection).

(v) if j is an elliptic fibration, then $\bar{\varphi}$ coincides with the projection map

$$W \to W/(\tau),$$

where τ is an involution of W which induces the inversion map $x \to -x$ on the group scheme $W^\# \cong J^\#$. Its set of fixed points consists of CUR and isolated singular points of fibres of $j:W \to S$.

PROOF. This is checked immediately.

Remark 5.5.5. We see from the above proposition that almost all smooth fibres of $j:F \to S$ are ordinary elliptic curves if $p \neq 2$. If $p = 2$, then all smooth fibres are supersingular if and only if $a_1 = 0$. In this case they are isomorphic to each other.

§ 6. Genus 1 fibrations on rational surfaces.

We will see in the next section that the jacobian fibration of every genus 1 fibration on an Enriques surface is a genus 1 fibration on a rational surface. In this section we study genus 1 fibrations on rational surfaces.

Proposition 5.6.1. Let $f:X\to S$ be a genus 1 fibration of a rational surface X and $j: J \to S$ be its Jacobian fibration. Then

(i) $S \cong \mathbb{P}^1$; $b_2(X) = 10$;

(ii) J is a rational surface;

(iii) f has at most one multiple fibre, which is not wild;

(iv) $\omega_X \cong \mathcal{O}_X(-\bar{X}_s)$, where X_s is the unique multiple fibre or any fibre if f does not have multiple fibres.

(v) f is jacobian if f does not have multiple fibres ;

(vi) ind(f) = m(s), where X_s is the unique multiple fibre or any fibre if f is jacobian.

Conversely, if $f:X \to S$ is a quasi-elliptic fibration satisfying (ii) and (iii),then X is a rational surface.

PROOF. (i) Obviously $S \cong \mathbb{P}^1$. Since $K_X^2 = 0$, $e(X) = 12$ and $b_2(X) = 10$;

(ii) By the formula for the canonical class of a jacobian genus 1 fibration, we have $K_J \sim mJ_s$ for any fibre of j and some integer m. Applying Proposition 5.3.6 and Corollary 5.3.5, we obtain $p_g(J) = q(J) = 0$. This implies that $m < 0$ and hence all pluri-genera of J are zero. Hence J is rational.

(iii) and (iv) We know that $H^1(X,\mathcal{O}_X) = 0$. By Proposition 5.1.5 $\mathrm{Tors}(R^1f_*\mathcal{O}_X) = 0$ and every multiple fibre of f is non-wild. The formula for ω_X given in Proposition 5.1.3 shows that

$$\omega_X = f^*(\mathfrak{X}) + \Sigma a_s\bar{X}_s$$

where $\deg(\mathfrak{X}) = \chi(X)-2\chi(S) = -1$ and $a_s = m(s)-1$ (again Proposition 5.1.5). Let X_0 be some fibre of f. For every $n > 0$ we can write

$$nK_X \sim -n\bar{X}_0+\sum_{s\in S} \epsilon_s\bar{n}(m(s)-1)\bar{X}_s = n(-1+\sum_{s\in S}(m(s)-1)/m(s))X_0$$

The number

$$-1+\sum_{s\in S}(m(s)-1)/m(s)) < 0,$$

otherwise some multiple of K_X is effective. It is easy to see that this can happen only if $m(s) = 1$ except for at most one s. This implies that

$$K_X \sim -m(s)\bar{F}_s+(m(s)-1)\bar{F}_s = -\bar{F}_s.$$

This proves (iii) and (iv).

(v) follows from Corollary 5.4.5.

(vi) Since $K_X^2 = 0$, X is not a minimal rational surface. Thus, there exists an exceptional curve of the first kind R on X. By (iv) $mK_X \sim -X_s$ for a unique multiple fibre X_s or any fibre and $m = 1$ if f is jacobian. Thus

$$m = -R\cdot(mK_X) = R\cdot X_s.$$

This shows that $\mathrm{ind}(f) \le m$. The opposite inequality is obvious.

Corollary 5.6.1. There is at most one genus 1 fibration on a rational surface X. If it exists, it is given by the linear system $|-mK_X|$, where $m = \mathrm{ind}(f)$ and is equal to the multiplicity of its unique multiple fibre or 1 if it does not have any.

Corollary 5.6.2. For every exceptional curve R of the first kind on X and every fibre X_s over a closed point

$$R\cdot X_s = m = \mathrm{ind}(f).$$

In particular, R is a section if $m = 1$ and conversely in this case every section is an exceptional curve of the first kind.

A rational surface X admitting an elliptic (resp. quasi-elliptic) fibration will be called a **rational elliptic** (resp. **quasi-elliptic surface**). The same name will be reserved for its unique elliptic fibration. The multiplicity of its unique multiple fibre (or 1, if it does not have multiple fibres) is called the **index** of X. By Proposition 5.5.1, the index of X is equal to the index of the unique genus 1 fibration on X. Also we know that a rational elliptic surface of index 1 has a jacobian fibration. We will call such surface also a **jacobian rational elliptic surface**.

A **Halphen pencil of index m** is an irreducible pencil of plane curves of degree 3m with 9 m-multiple base points (some of them may be infinitely near). Clearly, a Halphen pencil of index 1 is just a pencil of cubic curves whose general member is irreducible. If $p \neq 2,3$, then a general member of a Halphen pencil is a curve of (geometric) genus 1. It follows from Corollary 5.1.1 that the latter takes place if the index m is not divisible by 2 or 3.

Theorem 5.6.1. Let $f: X \to \mathbb{P}^1$ be a rational elliptic or quasi-elliptic surface of index m. Then there exists a birational morphism $\pi: X \to \mathbb{P}^2$ such that the composition of rational maps $f\circ\pi^{-1} : \mathbb{P}^2 \dashrightarrow \mathbb{P}^1$ is given by a Halphen pencil of index m. Conversely, if $\mathbb{P}^2 \dashrightarrow \mathbb{P}^1$ is a rational map given by such a pencil, then its minimal resolution of indeterminacy points is a rational elliptic or quasi-elliptic surface of index m.

Proof. Since X is rational, there is a birational map $\pi':X \to V$, where V is a minimal rational surface. Suppose that $X \ncong \mathbb{P}^2$, then $X \cong \mathbb{F}_n = \mathbb{P}(\mathcal{O}_{\mathbb{P}^1} \oplus \mathcal{O}_{\mathbb{P}^1}(-n))$, $n \neq 1$. If $n \neq 0$, then V contains a section C with $C\cdot C = -n$ whose proper inverse transform $\bar{C} = \pi'^{-1}(C)$ is a smooth rational curve on X with $\bar{C}\cdot\bar{C} \le -n$. Since

$$\bar{C}\cdot X_s = -m\bar{C}\cdot K_X = -m(-2-\bar{C}\cdot\bar{C}) \ge 0,$$

where X_s is a fibre of f, we see that $n \le 2$. Moreover, if $n = 2$ then π' is an isomorphism over C. Let $x \in V$ be a point of indeterminacy of π'^{-1}. Clearly, π' factors through the blowing-up $\varphi:\bar{V} \to V$ of this point. Blowing down the proper transform $\varphi^{-1}(L)$ of the ruling of V passing through x and then blowing down the image of $\varphi^{-1}(C)$ we get a a birational morphism $X \to \mathbb{P}^2$.

Thus, we may assume that $V = \mathbb{F}_0$. Again, π' factors through a blowing up $\varphi:\bar{V} \to V$ of a point on V. Let $\bar{V} \to \mathbb{P}^2$ be the blowing-down of the two rulings passing through this point. Then the composition $X \to \bar{V} \to \mathbb{P}^2$ is the needed π. The remaining assertions are easily proven. Since $K_{\mathbb{P}^2} \sim \pi(K_X)$, we see that

$$\pi(X_s) \sim \pi(-mK_X) \sim 3mL,$$

where L is a line on the plane. Thus, the image of fibres of f are plane curves of degree 3m. By Corollary 5.6.2 for every exceptional curve R on X, $R\cdot X_s = -m$, we see that $\pi(X_s)$ has 9 ($= \rho(X)-\rho(\mathbb{P}^2)$) m-multiple base points. So the image of the a genus 1 pencil on X is a Halphen pencil. The fact that every Halphen pencil defines a rational elliptic or quasi-elliptic surface follows from the definitions.

Corollary 5.6.3. Every Halphen pencil of index m contains a unique cubic curve with multiplicity m.

PROOF. Take the image of the unique multiple fibre of the corresponding rational elliptic surface.

Remark 5.6.1. Let $\lambda Q_3^m + \mu Q_{3m} = 0$ be a Halphen pencil of index m, where $Q_3^m = 0$ represents the image of the multiple fibre. Assume that the cubic curve $C:Q_3 = 0$ is nonsingular. Then, $m(p_1+...+p_9) - 3mp_0 = \mathrm{div}(Q_{3m}/Q_1^{3m})|C$, where $p_1,...,p_9$ are the base points of the pencil, p_0 is an inflection point on C and $Q_1 = 0$ is the equation of the tangent to C at p_0. This shows, that

$$m(p_1 \oplus \ldots \oplus p_9) = 0 ,$$

where $\otimes$ denotes the group law on C, where p_0 is taken for the zero point. Conversely, if this holds for some points on C, and $m = \mathrm{ord}(p_1 \oplus \ldots \oplus p_9)$, then there exists a Halphen pencil of index m with mC as its multiple member. In particular, we see that, if p = 2, then the curve C in the Halphen pencil of index 2 is an ordinary elliptic curve or a cuspidal cubic.

Let $\lambda Q_3 + \mu Q_3' = 0$ be a Halphen pencil of index 1 , $j: J \to \mathbb{P}^1$ be the corresponding jacobian fibration, $\pi: J \to \mathbb{P}^2$ a corresponding birational morphism. For every base point p of the pencil, $\pi^{-1}(p)$ is a (reducible) exceptional curve. It is easy to see that $\pi^{-1}(p)$ contains a unique irreducible component R_p with $R_p^2 = -1$, all other components are nodal curves. By Corollary 5.6.2, R_p is a section of j. Note that if p' is another base point and then $R_p = R_p'$ if and only if $\pi(\pi^{-1}(p')) - \pi(\pi^{-1}(p)) \geq 0$. In particular, we can fix a zero section on J_η by fixing a base point p of the pencil.

Fix a cubic curve C in the pencil and let $\bar{C}$ be its proper transform on J, a fibre of the fibration $j: J \to \mathbb{P}^1$. Let $\bar{p} = C \cap E_p$, $s = j(\bar{p})$, $\bar{C}_0$ be the irreducible component of $\bar{C}$ containing $\bar{p}$ and $\bar{q} = \bar{p} \oplus \varepsilon$, where ε is a point of order m in the group $(J_s^{\#})^0 = \bar{C}_0^{\#}$. Here we assume that $\bar{C}$ is of multiplicative type or smooth if $(p,m) = 1$, $\bar{C}$ is smooth if p|m and $p \neq m$ and C is smooth or of additive type if p = m. Let X be the surface obtained from J by blowing down the curve R_p and then by blowing up the image of the point $\bar{q}$.

Proposition 5.6.2 . In the above notation there exists a genus 1 fibration f: $X \to \mathbb{P}^1$ of index m the unique multiple fibre of which is isomorphic to $m\bar{C}$. Its local invariant at the corresponding point is given by the point $\bar{q}$ under the identification from Remark 5.4.1.

PROOF. Let C' be the image of $\bar{C}$ in X. It is easy to see that its normal bundle $L = \mathcal{O}_{C'}(C')$ is isomorphic to $\mathcal{O}_{\bar{C}}(\bar{p}-\bar{q})$ under the identification of $\bar{C}$ and C'. By the choice of the point $\bar{q}$ this sheaf is of order m in $\mathrm{Pic}(\bar{C})^0$. Thus

$$L^m \cong \mathcal{O}_{C'}(mC') \cong \mathcal{O}_{C'}.$$

Since C' is an indecomposable divisor of canonical type, $H^0(C',\mathcal{O}_{C'}) \cong k$ and the exact sequence

$$0 \to \mathcal{O}_X \to \mathcal{O}_X(mC') \to \mathcal{O}_{C'} \to 0$$

shows that $h^0(\mathcal{O}_{C'}) = 2$ and mC' moves in a genus 1 pencil. This proves the first statement. The second one is verified by using Remark 5.4.1.

Remark 5.6.2. Though it seems plausible that the jacobian fibration of the constructed fibrationf: $X \to \mathbb{P}^1$ is the original jacobian fibration j: $J \to \mathbb{P}^1$, it is not true in general.

Corollary 5.6.4. Let j: $J \to \mathbb{P}^1$ be a rational jacobian genus 1 fibration and J_s be its fibre and $m > 1$ be an integer. Assume

(a) J_s is not of additive type if $p \nmid m$;

(b) J_s is not of multiplicative type if $p | m$;

(c) J_s is not supersingular elliptic curve if $p | m$;

(d) $p = m$ if j is quasi-elliptic.

Then there exists a rational jacobian fibration f: $X \to \mathbb{P}^1$ the fibre X_s of which is multiple of multiplicity m and $\bar{X}_s \cong J_s$.

Let $j: J \to \mathbb{P}^1$ be a rational jacobian genus 1 fibration. In S5 we constructed a map

$$\varphi: J \to \mathbb{P}(V)$$

which factors through the Weiestrass model of j. In our situation

$$V \cong \mathcal{L}^2 \oplus \mathcal{O}_{\mathbb{P}^1},$$

where $\mathcal{L} = R^1 j_* \mathcal{O}_J \cong \mathcal{O}_{\mathbb{P}^1}(-1)$ (see Remark 5.5.2). Thus, $\mathbb{P}(V) \cong \mathbb{F}_2$ and φ maps the fixed zero section to the exceptional section of the minimal ruled surface $\mathbb{F}_2$.

The following result is easily checked:

Proposition 5.6.3. Let p be the base point of the Halphen pencil corresponding to j which defines the zero section. Let $p_1,\dots,p_8$ be the remaining base points and V be the Del Pezzo surface obtained blowing them up. The following diagram is commutative:

$$\begin{array}{ccc} J & \xrightarrow{\varphi} & \mathbb{F}_2 \\ \pi\downarrow & & \downarrow p \\ V & \rightarrow & \bar{\mathbb{F}}_2 \subset \mathbb{P}^3 \end{array}$$

where π is the blowing down the zero section, p blows down the exceptional section, the lower horizontal arrow is the map given by the linear system $|-2K_V|$.

Recall that the branch locus of φ is given either by Proposition 5.5.11 or by Proposition 0.3.6.

Lemma 5.6.1. Let $\pi: V \to \mathbb{P}^2$ be a birational morphism of a nonsingular surface V. Let $\rho = \rho(V)$ be its Picard number. Then the orthogonal complement of $\mathbb{Z}[K_V]$ in $\mathrm{Num}(V) \cong \mathrm{NS}(V)$ is isomorphic to the lattice $E_{\rho-1}$.

PROOF. This immediately follows from Proposition 2.5.2 and Remark 2.5.1 from Chapter 2.

Corollary 5.6.5. Let $j: J \to \mathbb{P}^1$ be a rational jacobian elliptic surface. Then $\mathrm{Num}_{fib}(J)$ is isomorphic to a primitive sublattice of the lattice $E_9 = \tilde{E}_8$.

PROOF. Since $K_J = -[J_s]$ for every closed $s \in \mathbb{P}^1$, $\mathrm{Num}_{fib}(J)$ is a primitive sublattice of $(\mathbf{Z}K_J)_{Num(J)}^{\perp}$. By corollary 5.5.1, $\rho(J) = 10$. It remains to apply the previous lemma.

Lemma 5.6.2. Suppose $\rho_f(J) = 9$. Then $Num_{fib}(J)/Rad$ is isomorphic to one of the following lattices:

$$E_8\ ,\ D_8,\ A_8,\ E_7\oplus A_1,\ A_7\oplus A_1,\ E_6\oplus A_2,\ D_6\oplus A_1^2,\ D_5\oplus A_3,$$
$$D_4\oplus A_1^4,\ A_5\oplus A_1\oplus A_2,\ D_4^2,\ A_4^2,\ A_3^2\oplus A_1^2,\ A_2^4,\ A_1^8.$$

PROOF. It follows from Corollary 5.6.2, that $Num_{fib}(J)/Rad$ is isomorphic to a sublalttice of $\tilde{E}_8/Rad \cong E_8$ (not necessary primitive). Since E_8 is unimodular, the discriminant of $Num_{fib}(J)/Rad$ is equal to the square of its index. We also know that it is isomorphic to the direct sum of the lattices A_n, D_n and E_n (Proposition 5.1.6.). It is easy to list all such lattices whose discriminant is a square. They give the above list plus one more lattice, $D_4\oplus A_2^2$. This lattice is not embeddable into E_8. This can be checked by localizing a possible embedding at the primes p = 2 and 3 and applying clasification of integral quadratic forms over p-adic numbers. On the other hand one checks directly that the other lattices are embeddable.

Theorem 5.6.2. Let $f: X \to \mathbb{P}^1$ be an elliptic fibration whose Jacobian fibration is a rational elliptic surface(for example, X is an Enriques surface). Assume that $\rho_f(F) = 9$. Then f has one of the following collections of reducible non-multiple degenerate fibres (all cases are realized):

$$\tilde{E}_8,\ \tilde{A}_8,\ \tilde{D}_8;\ \tilde{E}_7 + \tilde{A}_1\ (\tilde{A}_1^*),\ \tilde{A}_7 + \tilde{A}_1\ (\tilde{A}_1^*),\ \tilde{E}_6 + \tilde{A}_2(\tilde{A}_2^*),\ \tilde{D}_5 + \tilde{A}_3,$$
$$2\tilde{D}_4,\ 2\tilde{A}_4;\ \tilde{D}_6 + 2\tilde{A}_1(\tilde{A}_1^*),\ \tilde{A}_5 + \tilde{A}_1(\tilde{A}_1^*) + \tilde{A}_2(\tilde{A}_2^*),$$
$$2\tilde{A}_3 + 2\tilde{A}_1(\tilde{A}_1^*),\ 4\tilde{A}_2(\tilde{A}_2^*).$$

PROOF. It follows from the classification of degenerate fibres (Corollary 5.1.2) and Lemma 5.6.3 that no other combinations of reducible degenerate fibres can be realized, except maybe $\tilde{D}_4+4\tilde{A}_1^4$ and $8\tilde{A}_1$. First, note that the cases $\tilde{D}_4+4\tilde{A}_1^4$ and $8\tilde{A}_1$ cannot occur. Indeed,adding up the possible Euler-Poincare characteristics $\chi(F_s)$ of degenerate fibres, one sees that it is greater than e(J) = 12. However this contradicts Proposition 5.1.6.

To prove that the remaining cases are realized , we may assume that F = J is a jacobian rational surface. Thus, f comes from a pencil of plane cubic curves by blowing up its base points (Theorem 5.6.2).

We exhibit the pencils $\lambda Q_3 + \mu Q_3' = 0$ and let the reader verify that it defines an elliptic fibration with the given collection of reducible degenerate fibres.

$\tilde{E}_8$: $Q_3' = Q_1^3$, where $Q_1 = 0$ is the tangent to a nonsingular curve $Q_3 = 0$ at its inflection point.

For example,

$$Q_3 = x_2^2x_0 + x_1^3 + x_0^3, \quad Q_3' = x_0^3.$$

$\tilde{D}_8$: $Q_3' = Q_1^2Q_1'$, where Q_1' is the tangent to $Q_3 = 0$ at its inflection point q, $Q_1 = 0$ is the line passing through q and tangent to $Q_3 = 0$ somewhere .

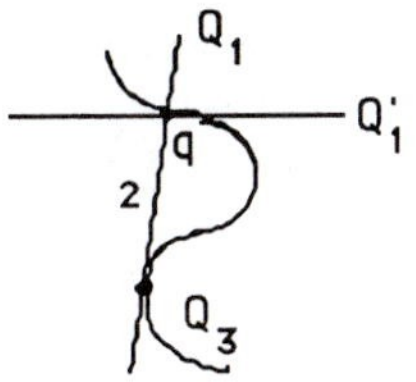

For example,

$$Q_3 = x_1x_0^2 + x_2^3 + x_2x_1^2, \quad Q_3' = x_1x_2^2.$$

$\tilde{A}_8$: $Q_3' = 0$ is an irreducible cubic with a node q, $Q_3 = 0$ is nonsingular and intersects $Q_3' = 0$ at q with multiplicity 9.

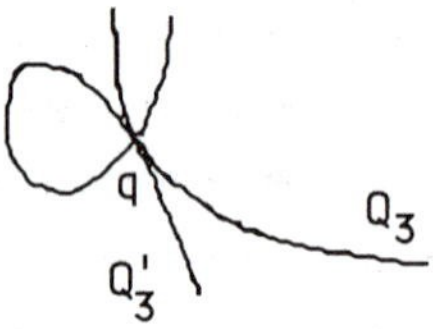

For example,

$$Q_3 = x_0^2x_1 + x_1^2x_2 + x_2^2x_0,\ Q_3' = x_1^3 - x_2^3 - x_0x_1x_2.$$

$\tilde{E}_7 + \tilde{A}_1(\tilde{A}_1{}^*)$: $Q_3' = Q_1^2Q_1'$, where $Q_1 = 0$ is a tangent line to an inflection point q of a nonsingular cubic $Q_3 = 0$, $Q_1' = 0$ is a line passing through q and tangent to $Q_3 = 0$ somewhere.

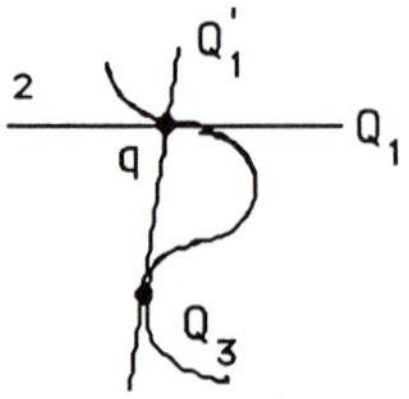

For example,

$$Q_3 = x_2^2x_0 + x_1^3 + x_0^2x_1,\ Q_3' = x_0^2x_1.$$

$\tilde{A}_7 + \tilde{A}_1(\tilde{A}_1{}^*$: $Q_3' = 0$ is an irreducible cubic with a node q, $Q_3 = 0$ is nonsingular and intersects $Q_3' = 0$ at q with multiplicity 7 and is tangent to $Q_3' = 0$ at some point.

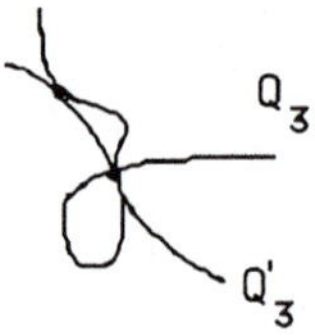

For example,

$$Q_3 = x_0^2x_1 + x_1^2x_2 + x_2^2x_0\ ,\ Q_3' = x_1^3 - x_2^3 - x_0x_1x_2.$$

$\tilde{E}_6 + \tilde{A}_2(\tilde{A}_2{}^*)$: $Q_3' = Q_1^2Q_1'$, where $Q_1 = 0$ is a tangent line to a non-inflection point q of a nonsingular cubic $Q_3 = 0$, $Q_1' = 0$ is a line passing through q and tangent to $Q_3 = 0$ at an inflection point q'.

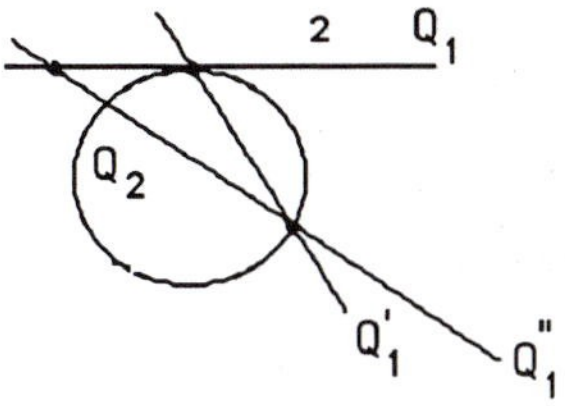

For example,

$$Q_3 = x_2^2x_0 - x_1^3 - x_1^2x_0 \ , \ Q_3' = (x_2-x_1)x_1^2.$$

$\tilde{D}_5+\tilde{A}_3$: $Q_3' = Q_1Q_1'Q_1''$ is the union of three lines intersecting at a common point p, $Q_3 = Q_2Q_1'''$, where $Q_2= 0$ is a conic tangent to the line $Q_1 = 0$ at p and intersecting the other two lines at points p' and p", Q_1''' is the line intersecting the conic at p' and p".

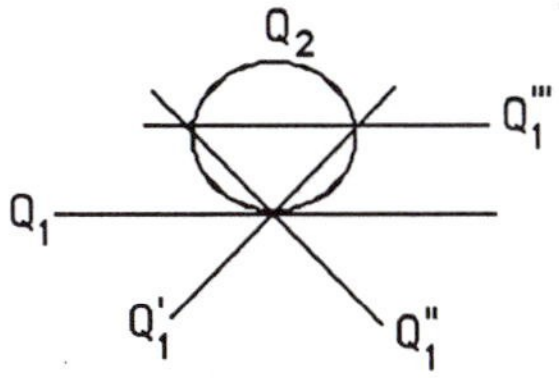

For example,

$$Q_3 = x_0x_1(x_0+x_1) \ , \ Q_3' = (x_1+x_2)(x_1^2+x_0x_2).$$

$\tilde{D}_4+\tilde{D}_4$: $Q_3 = 0$ is the union of 3 concurrent lines, $Q_3' = Q_1^2Q_1'$, where the line $Q_1' = 0$ passes through the common point of the components of $Q_3 = 0$.

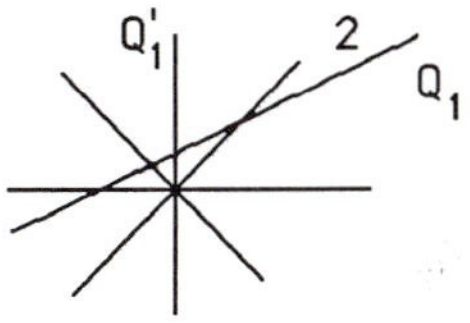

For example,

$$Q_3 = x_0x_1(x_0+x_1) \ , \ Q_3' = (x_0-x_1)x_2^2.$$

$\tilde{A}_4 + \tilde{A}_4$: $Q_3 = Q_1Q_1'Q_1'' = 0$ and $Q_3' = 0$ are the unions of three nonconcurrent lines which intersect each other as in the following picture :

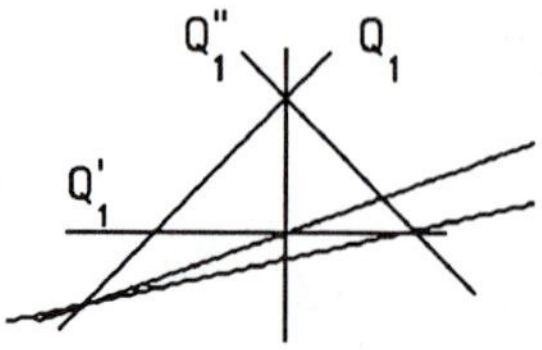

For example,

$$Q_3 = x_0x_1x_2 \ , \ Q_3' = (x_1+x_2)(x_0+x_1)(x_0+x_1+x_2).$$

$\tilde{D}_6+2\tilde{A}_1(\tilde{A}_1^*)$: $Q_3 = 0$ is a nodal cubic, $Q_3' = Q_1Q_1'Q_1'' = 0$ is the union of three lines with a common point p, $Q_1 = 0$ is the inflection tangent of $Q_3 = 0$ at p, $Q_1' = 0$ passes through the node of $Q_3 = 0$.

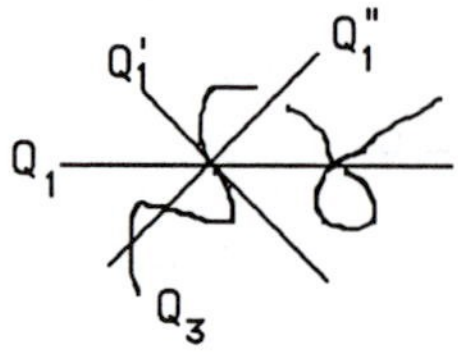

For example:

$$Q_3 = x_1^3+x_2^3+x_0x_1x_2, \ Q_3' = x_0(x_1+x_2)(-x_0+x_1+x_2).$$

$\tilde{A}_5+\tilde{A}_1(\tilde{A}_1^*)+\tilde{A}_2(\tilde{A}_2^*)$: $Q_3 = Q_1Q_1'Q_1'' = 0$ is the union of 3 lines nonconcurrent lines, two of them are inflection tangents and the third is a simple tangent to a nonsingular cubic $Q_3' = 0$.

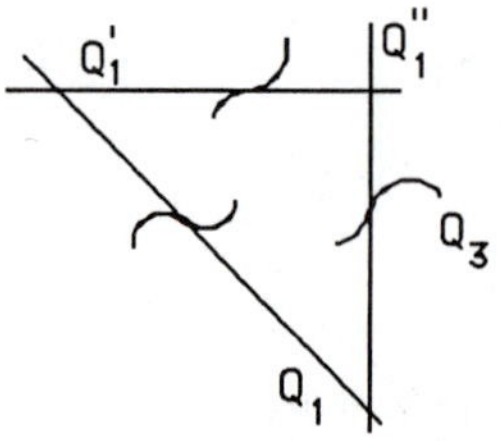

For example,

$$Q_3 = (x_0+x_1)(x_2+x_1)(ax_0+bx_1+cx_2), \quad Q_3' = x_0^3+x_1^3+x_2^3$$

for some $a,b,c \in K$.

$\tilde{A}_3+\tilde{A}_3+\tilde{A}_1(\tilde{A}_1^*)+\tilde{A}_1(\tilde{A}_1^*)$: $Q_3 = Q_1Q_1'Q_1'' = 0$ is the union of 3 nonconcurrent lines, $Q_3' = Q_2Q_1'''$, where $Q_2 = 0$ is a conic tangent to the three components of $Q_3 = 0$, $Q''' = 0$ passes through the common point of two components of $Q_3 = 0$.

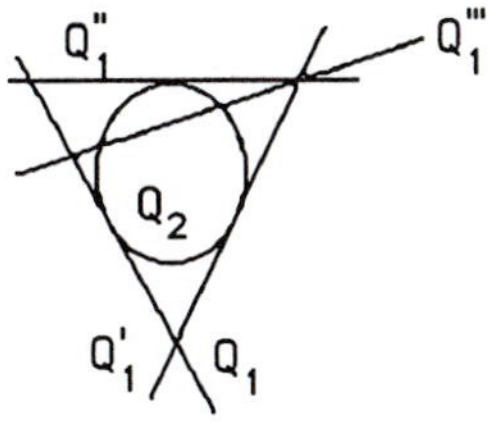

For example,

$$Q_3 = x_0x_1x_2 \ , \quad Q_3' = (x_0^2+x_1^2+x_2^2-2x_0x_1+2x_0x_2+2x_1x_2)(x_1+x_2).$$

$\tilde{A}_2(\tilde{A}_2^*)+\tilde{A}_2(\tilde{A}_2^*)+\tilde{A}_2(\tilde{A}_2^*)+\tilde{A}_2(\tilde{A}_2^*)$: Q_3 is the union of three inflection tangents to a nonsingular cubic curve $Q_3' = 0$.

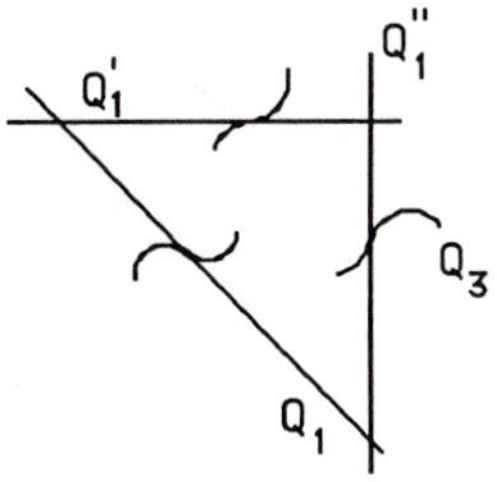

For example,

$$Q_3 = (x_0+x_1)(x_0+x_2)(x_1+x_2) \ , \quad Q_3' = x_0^3+x_1^3+x_2^3.$$

Corollary 5.6.6. Let X_s be a non-multiple fibre of an elliptic fibration whose jacobian fibration is a rational elliptic. Then $b_2(X_s) \leq 9$ and all types satisfying this condition are realized.

Corollary 5.6.7. Let $j: J \to \mathbb{P}^1$ be a jacobian rational elliptic surface. Suppose that the group of sections $J(\mathbb{P}^1)$ is finite. Then its structure is given by the following table:

Types	Order	$J(\mathbf{P}^1)$
$\tilde{E}_8$	1	$\{1\}$
$\tilde{A}_8$	3	$\mathbf{Z}/3\mathbf{Z}$
$\tilde{D}_8$	2	$\mathbf{Z}/2\mathbf{Z}$
$\tilde{E}_7 + \tilde{A}_1(\tilde{A}_1^*)$	2	$\mathbf{Z}/2\mathbf{Z}$
$\tilde{A}_7 + \tilde{A}_1(\tilde{A}_1^*)$	4	$(\mathbf{Z}/2\mathbf{Z})^2$ if $p \neq 2$
$\tilde{E}_6 + \tilde{A}_2(\tilde{A}_2^*)$	3	$\mathbf{Z}/3\mathbf{Z}$
$\tilde{D}_5 + \tilde{A}_3$	4	$\mathbf{Z}/4\mathbf{Z}$ $(p \neq 2)$
$2\tilde{D}_4$	4	$(\mathbf{Z}/2\mathbf{Z})^2$ $(p \neq 2)$
$2\tilde{A}_4$	5	$\mathbf{Z}/5\mathbf{Z}$
$\tilde{D}_6+2\tilde{A}_1(\tilde{A}_1^*)$	4	$(\mathbf{Z}/2\mathbf{Z})^2$ $(p \neq 2)$
$\tilde{A}_5+\tilde{A}_1(\tilde{A}_1^*)+\tilde{A}_2(\tilde{A}_2^*)$	6	$(\mathbf{Z}/3\mathbf{Z})\oplus(\mathbf{Z}/2\mathbf{Z})$
$2\tilde{A}_3+ 2\tilde{A}_1(\tilde{A}_1^*)$	8	$(\mathbf{Z}/4\mathbf{Z})\oplus(\mathbf{Z}/2\mathbf{Z})$ $(p \neq 2)$
$4\tilde{A}_2(\tilde{A}_2^*)$	9	$(\mathbf{Z}/3\mathbf{Z})^2$ if $p \neq 3$.

PROOF. By Corollary 5.2.1 we have $\rho_{fib}(J) = 9$. Thus, we can apply Theorem 5.6.2 to obtain all possible collections of reducible fibres of j. Together with the formula from Proposition 5.3.4 it allows us to compute the order of $J(\mathbb{P}^1)$. It is given in the second column of the table. Next, we use that $J(\mathbb{P}^1)(\neq p)$ is mapped injectively into the group $\mathfrak{a}_s/\mathfrak{a}_s^\circ$ for every degenerate fibre J_s of additive type if $p\nmid\#J(\mathbb{P}^1)$ (Lemma 5.4.2). This shows that $J(\mathbb{P}^1)\cong(\mathbf{Z}/2\mathbf{Z})^2$ in the cases $\tilde{A}_7+\tilde{A}_1(\tilde{A}_1^*)$, $2\tilde{D}_4$, $\tilde{D}_6+2\tilde{A}_1(\tilde{A}_1^*)$ and is cyclic otherwise whenever its order is 4 and $p\neq 2$. In the case of degenerations of multiplicative type, we know that $J(\mathbb{P}^1)(\neq p)$ is mapped injectively into the group $\mathfrak{a}_s$. This easily implies that $J(\mathbb{P}^1)\cong(\mathbf{Z}/2\mathbf{Z})\oplus(\mathbf{Z}/4\mathbf{Z})$ in the case $2\tilde{A}_3+2\tilde{A}_1(\tilde{A}_1^*)$, $p\neq 2$, and $J(\mathbb{P}^1)\cong(\mathbf{Z}/3\mathbf{Z})^2$ in the case $4\tilde{A}_2$ $(\tilde{A}_2^*)$ and $p\neq 3$.

Remark 5.6.3. Similarly, one can classify all possible collections of degenerate fibres with $\rho_f(F)\leq 8$. It follows from **[Dy]** or **[Cox]** that a lattice of rank ≤ 7 which is a direct sum of lattices of type A,D and E embeds into E_8 unless it is isomorphic to one of the following lattices of rank 7:

$$A_4\oplus A_1^3,\ D_4\oplus A_1\oplus A_2,\ A_3\oplus A_2^2,\ A_2^2\oplus A_1^3,\ A_2\oplus A_1^5.$$

It can be checked that all other cases are realized as the lattices $\mathrm{Num}_{fib}(F)$ except for the case A_1^7 (which is excluded by the same argument as in the case A_1^8).

Remark 5.6.4. Let C be a section of a rational jacobian elliptic surface J. Blowing down C, we obtain a minimal nonsingular model J' of a Del Pezzo surface of degree 1. Let $\pi:J\to J'$ be the corresponding map. Let J_s be a reducible fibre of J and J_s' be the union of all irreducible components of J_s which do not intersect C . Since $K_{J'}\cdot E=0$ for every component of $\pi(J_s')$, we see that $\pi(J_s')$ are blown down to a rational double point of the corresponding anti-canonical model

$$\bar{J}'=\mathrm{Proj}(\bigoplus_{m\geq 0}H^0(J',\mathcal{O}_{J'}(-mK_{J'}))\text{ of }J'.$$

Thus, we see that the problem of classification of reducible fibres in an elliptic fibration, whose Jacobian fibration is a rational elliptic surface, is equivalent to the problem of classification of singular points on the anticanonical model of a Del Pezzo surface of degree 1. For the latter we refer to **[Ur]**.

Theorem 5.6.3. Assume $p = 2$. Let $f:X \to \mathbb{P}^1$ be a quasi-elliptic fibration the jacobian fibration of which is a rational quasi-elliptic surface. Then the possible degenerate non-multiple fibres of f are the following:

$$8\tilde{A}_1^*,\ 4\tilde{A}_1^*+\tilde{D}_4,\ 2\tilde{D}_4,\ \tilde{D}_6+2\tilde{A}_1^*,\ \tilde{A}_1^*+\tilde{E}_7,\ \tilde{E}_8,\ \tilde{D}_8$$

PROOF. We know from Proposition 5.1.9 that

$$12 = e(X) = 4 + \Sigma(e(X_s)-2)$$

All possibilities are easily enumerated, since we know the possible types of degenerate fibres (Corollary 5.2.4). To show that all of them are realized we may assume that f is jacobian (Theorem 5.3.1). Then f is obtained from a pencil $\lambda Q_3+\beta Q_3' = 0$ of cubic curves. We exhibit the corresponding pencils which realize the above combinations of degenerate fibres. The easy verification is left to the reader.

$\mathbf{8\tilde{A}_1^*}$: $Q_3 = 0$ is a cuspidal cubic, $Q_3' = Q_1Q_2 = 0$, where $Q_1 = 0$ is a line passing through the cusp and tangent to a conic $Q_2 = 0$; the latter is touching the cuspidal cubic at 3 distinct points:

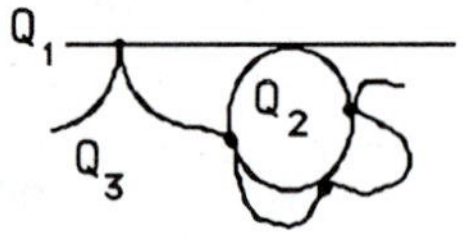

For example:

$$Q_3 = x_0x_2^2+x_1^3\ ,\ Q_3' = x_1(x_1x_0+ax_0^2+bx_1^2+cx_2^2),$$

where $a=c \neq 0$

$\mathbf{4\tilde{A}_1^*+\tilde{D}_4}$: $Q_3 = 0$ is a cuspidal cubic, $Q_3' = Q_1Q_2 = 0$, where $Q_2 = 0$ is a conic touching the cuspidal cubic at 3 distinct points p_1,p_2,p_3 none of them is an inflection point and $Q_1 = 0$ is a line tangent to the conic at p_1.

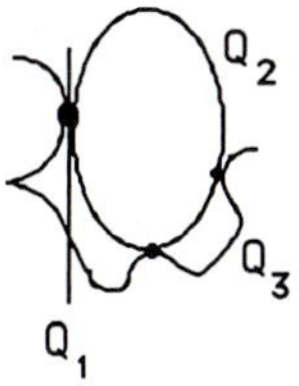

For example:

$$Q_3 = x_0x_2^2+x_1^3, \quad Q_3' = (x_1+x_0)(x_1x_0+ax_0^2+bx_1^2+cx_2^2),$$

where $a+b+c=1$, $c \neq 1$, $b \neq 0$.

$2\tilde{D}_4$: $Q_3 = 0$ is a cuspidal cubic, $Q_3' = Q_1Q_2 = 0$, where $Q_2 = 0$ is a conic touching the cuspidal cubic at a points p_1 which is not an inflection point and touching it at another point $p_2 \neq p_1$ with multiplicity 4; $Q_1 = 0$ is a line tangent to the conic at p_1

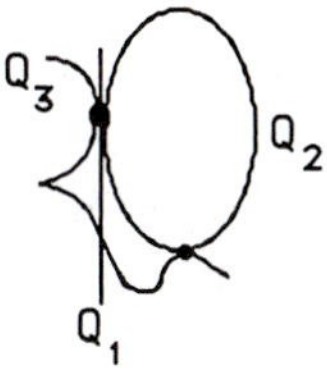

For example:

$$Q_3 = x_0x_2^2+x_1^3, \quad Q_3' = (x_1+x_0)(x_1x_0+ax_0^2+ax_1^2+x_2^2),$$

where $a \neq 1$.

$\tilde{D}_6+2\tilde{A}_1^*$: $Q_3 = 0$ is a cuspidal cubic, $Q_3' = Q_1Q_1'^2$, where $Q_1 = 0$ is the tangent line at the inflection point of the cubic and $Q_1' = 0$ is a tangent line of the cubic passing through this point.

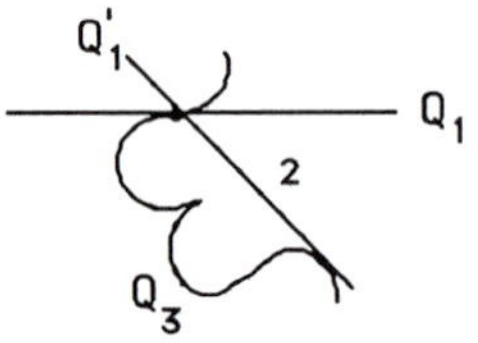

For example,

$$Q_3 = x_0x_2^2+x_1^3 \ , \ Q_3' = x_0(x_1+x_0)^2.$$

$\tilde{D}_8$: $Q_3 = 0$ is a cuspidal cubic, $Q_3' = Q_1Q_2 = 0$, where $Q_2 = 0$ is a conic touching the cuspidal cubic with multiplicity 6 at a point p_1 which is not an inflection point; $Q_1 = 0$ is a line tangent to the conic at p_1

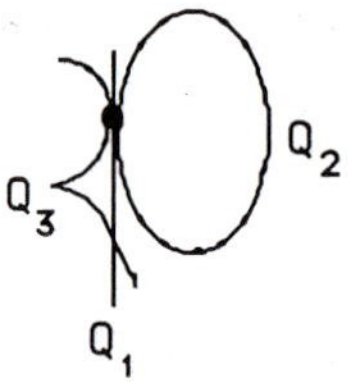

For example:

$$Q_3 = x_0x_2^2+x_1^3 \ , \ Q_3' = (x_1+x_0)(x_1x_0+x_0^2+x_1^2+x_2^2).$$

$\tilde{E}_7+\tilde{A}_1$*: $Q_3 = 0$ is a cuspidal cubic, $Q_3' = Q_1^2Q_1'$, where the line $Q_1=0$ is the tangent line at the inflection point of the cubic and the line $Q_1' = 0$ passes through this inflection point.

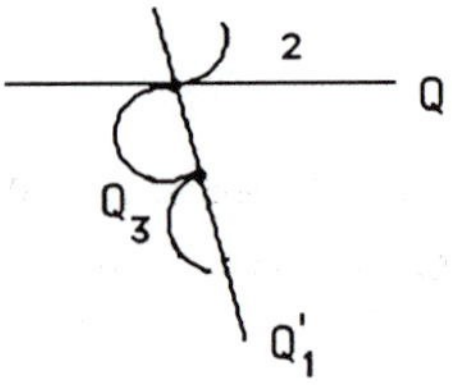

For example:

$$Q_3 = X_0X_2^2+X_1^3,\ Q_3' = X_1^2X_0.$$

$\tilde{E}_8$:$Q_3 = 0$ is a cuspidal cubic, $Q_3' = Q_1^3$, where the line $Q_1=0$ is the tangent line at the inflection point of the cubic.

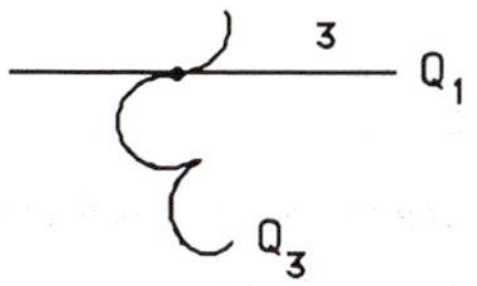

For example:

$$Q_3 = X_0X_2^2+X_1^3,\ Q_3' = X_1^3.$$

§7. Genus 1 fibrations on Enriques surfaces.

As before F denotes an Enriques surface, p = char(K).

Theorem 5.7.1. For every Enriques surface F there exists a genus 1 fibration $f:F \to \mathbb{P}^1$. Moreover there is a bijective correspondence between W_F-orbits of primitive isotropic vectors in $\mathrm{Num}(F)^+$ and elliptic or quasi-elliptic fibrations on F.

PROOF. This follows immediately from §2 of Chapter 3.

Theorem 5.7.2. Let F be an Enriques surface and $f:F \to S$ be a genus 1 fibration. Then,

(i) $S \cong \mathbb{P}^1$;

(ii) every multiple fibre is of multiplicity 2;

(iii) if $K_F \neq 0$ (i.e. F is classical), then f has two multiple fibres, both not wild;

(iv) f has one multiple fibre, which is wild if and only if F is not classical;

(v) $\mathrm{ind}(f) = 2$;

(vi) the Jacobian fibration of f is a rational genus 1 fibration.

Conversely, if $f: X \to \mathbb{P}^1$ is a genus 1 fibration satisfying (vi) and (ii) with two non-wild multiple fibres or one wild multiple fibre, then F is an Enriques surface.

PROOF. We know that $b_1(F) = 0$. This immediately implies (i). By Proposition 5.1.3,

$$\omega_F \cong f^*(\mathcal{L}^{-1} \otimes \omega_S) + \Sigma a_s \bar{F}_s,$$

where $R^1 f_* \mathcal{O}_F = \mathcal{L} \oplus T$, where T is a torsion sheaf concentrated at points $s \in S$ for which F_s is a wild multiple fibre and $\deg(\mathcal{L}^{-1} \otimes \omega_S) = \chi(\mathcal{O}_F) - 2\chi(\mathcal{O}_S) + h^0(T)$. By (i) $\deg(\mathcal{L}^{-1} \otimes \omega_S) = -1 + h^0(T)$. If $T \neq 0$, then $\mathcal{L}^{-1} \otimes \omega_S$ is effective, hence ω_F is effective, i.e. $\omega_F \cong \mathcal{O}_F$, $p = 2$ and F is not classical. This also implies that

$$\mathcal{L} \cong \mathcal{O}_F, \text{ length } T = 1 \text{ and } a_s = 0 \text{ for all } s.$$

Since $h^1(\mathcal{O}_F) = 1$, we may apply Proposition 5.1.5 to obtain that $m(s) = 2$ for a unique multiple fibre which is wild. Conversely, assume that F is not classical. Then $h^1(\mathcal{O}_F) = 1$ and the Leray spectral sequence for the morphism f and the structure sheaf $\mathcal{O}_F$ shows that

$$H^0(S, \mathcal{L} \oplus T) = H^0(S, R^1 f_* \mathcal{O}_F) \cong H^1(F, \mathcal{O}_F) \cong k.$$

This implies that either $T = 0$ and $h^0(\mathcal{L}) = 1$ or $h^0(T) = 1$ and $h^0(\mathcal{L}) = 0$. In the first case we must have $\mathcal{L} \cong \mathcal{O}_S$ and $\deg(\mathcal{L}^{-1} \otimes \omega_S) = \deg(\omega_S) = -1$ which is absurd. Thus $h^0(T) = 1$ and we finish as before.

Assume F is classical. Then $h^1(\mathcal{O}_F) = 0$ and by Proposition 5.1.5 all multiple fibres are not wild. Also we have $\deg(\mathcal{L}^{-1} \otimes \omega_S) = -1$ and $a(s) = m(s) - 1$ for all s. Since $2K_F \sim 0$, we immediately obtain that f has exactly multiple fibres of multiplicity 2. To see (v) we notice that $\mathrm{ind}(f) \geq 2$, because f has multiple fibres. On the other hand, it follows from Theorem 3.3.1 that we can always find an irreducible curve E of arithmetic genus ≤ 1 or a nodal curve E such that $E \cdot F_s = 2$ for every fibre F_s of f. This shows that $\mathrm{ind}(f) \leq 2$.

(vi) By Proposition 5.3.3,

$$\omega_J \cong f^*(\mathfrak{L}),$$

where $\deg(\mathfrak{L}) = \chi(\mathcal{O}_J) - 2$. It follows from proposition 5.3.5 that

$$e(J) = e(F) = 12.$$

This implies that $\chi(\mathcal{O}_J) = 1$, that is, $\deg(\mathfrak{L}) = -1$. Hence $\omega_J \cong \mathcal{O}_F(-F_s)$, $s\in S$. In particular, $|mK_J| = \emptyset$ for all $m \geq 1$. Since, $\chi(\mathcal{O}_J) = 1$, this also implies that $h^1(\mathcal{O}_J) = 0$. Hence, by the Castelnuovo criterion, J is rational.

To prove the converse statement, we apply Proposition 5.3.5 to obtain

$$b_2(F) = 10, \ \chi(\mathcal{O}_F) = 1.$$

The formula for the canonical class of F shows that

$$K_F = \bar{F}_s - \bar{F}_{s'},$$

where F_s and $F_{s'}$ are two non-wild multiple fibres or

$$K_F = 0$$

if f has one wild multiple fibre. In both cases $K_F \equiv 0$ and F is an Enriques surface.

Corollary 5.7.1. Let F be an Enriques surface. Then

$$\mathrm{Br}(F) \cong \mathbf{Z}/2\mathbf{Z} \text{ if F is classical,}$$
$$= 0 \text{ otherwise.}$$

PROOF. Take a genus 1 fibration $f: F \to \mathbb{P}^1$ on F and apply Proposition 5.3.4 from §3 together with the fact that $\mathrm{Br}(J) = 0$ for every rational surface J and $\mathrm{ind}(f) = 2$.

Let $F_s = 2\bar{F}_s$ be a multiple fibre of a genus 1 fibration f on F. We will call the curve $\bar{F}_s$ a **half-fibre** of f. This is an indecomposable curve of canonical type whose class in Num(F) is primitive.

Corollary 5.7.2. Let D be an indecomposable curve of canonical type on F the class [D] of which in Num(F) is primitive. Then $\dim|D| = 0$ and $|2D|$ defines a genus 1 fibration on F for which D is a half-fibre. Every such fibration is obtained in this way.

PROOF. Combine Proposition 3.1.2 with the the existence of multiple fibres following from the previous Proposition.

Proposition 5.7.1. Let $f\colon X \to \mathbb{P}^1$ be a quasi-elliptic fibration. Then X is a Zariski surface.

PROOF. This immediately follows from Proposition 5.1.7.

Proposition 5.7.2. Let F be an Enriques surface. Assume $p \neq 2$ or F is a μ_2-surface. Then F does not admit quasi-elliptic fibrations.

PROOF. We know from Theorem 1.4.1 that F is not unirational hence is not a Zariski surface.

Lemma 5.7.1. Let $p = 2$ and $f\colon F \to \mathbb{P}^1$ be a genus 1 fibration on a classical (resp. an $\boldsymbol{\alpha}_2$-surface) Then, $f^*(dt/t)$ (resp. $f^*(dt)$) is the unique up to a constant regular 1-form on F, where t is an appropriately chosen coordinate on $\mathbb{P}^1$.

PROOF. By Proposition 1.3.3

$$\dim\ H^0(F,\Omega_F^1) = 1.$$

Suppose F is classical. We may assume that its two multiple fibres lie over 0 and ∞, the simple zero and the simple pole of a rational function t on $\mathbb{P}^1$. Clearly, $f^*(dt/t)$ is regular outside F_0 and F_∞. Let x be a nonsingular point of $(F_0)_{red}$. Then, locally around $f^*(t) = \varepsilon\pi^{2n}$, where π is a local parameter at x, ε is invertible at x. Differentiating, we obtain $f^*(dt/t) = d\varepsilon/\varepsilon$, that is, $f^*(dt/t)$ is regular at x. Similarly, we show that $f^*(dt/t)$ is regular at a nonsingular point of the reduced fibre $(F_\infty)_{red}$. Thus, $f^*(dt/t)$ is regular outside a finite set of points. However, the set of non-regularity points of a rational 1-form on a regular surface is a divisor.

We deal similarly with the case of an α_2-surface.

Proposition 5.7.3. Let $\pi: \bar{F} \to F$ be the K3-cover of F. Assume $\bar{F}$ has at most isolated singularities. Then F does not admit quasi-elliptic fibrations.

PROOF. By Proposition 5.7.3, we may assume that p =2 and F is classical or an α_2-surface. By Proposition 0.1.4, the singularities of $\bar{F}$ lie over the zeroes of a certain regular 1 form α on F. Assume that F admits a quasi-elliptic fibration $f:F \to \mathbb{P}^1$. Then $\alpha = f^*(dt/t)$ or $f^*(dt)$ and the local computation show that α vanishes at every singular point of a non-multiple fibre. In particular it vanishes at the curve of cusps. This proves the assertion.

Corollary 5.7.3. Let $\pi:\bar{F} \to F$ be the K3-cover of F. Assume that p = 2 and F is classical or an α_2-surface and $\bar{F}$ is normal. Let $f: F \to \mathbb{P}^1$ be a genus 1 fibration on F. Then every non-multiple fibre of f is a reduced curve.

PROOF. Similar to the proof of the previous corollary and left to the reader.

Let $j:J \to \mathbb{P}^1$ be a jacobian rational elliptic surface, s_1 and s_2 be two points such that the corresponding fibres J_{s_1} and J_{s_2} are of multiplicative type. Let $x_i \in H^1(J_{s_i},\mu_2)$, i=1,2. By Corollary 5.2.5, there exists a unique elliptic fibration $f: X \to \mathbb{P}^1$ of an Enriques surface X whose Jacobian fibration is isomorphic to j and whose local invariants are the x_i's. Conversely, every elliptic fibration on an Enriques surface is obtained in this way for an appropriate jacobian rational elliptic surface.

Fix a zero section C_0 on J . Let $\varphi:J \to \mathbb{F}_2$ be the double cover constructed in Proposition 5.6.2 and $C = \varphi^{-1}(Z)$ for some section of $\mathbb{F}_2$ not equal to the exceptional section. We choose Z general enough to assume that C is irreducible and smooth curve on J such that $C{\bullet}X_s = 2$, for any $s \in \mathbb{P}^1$. Let $C \to \mathbb{P}^1$ be the corresponding double cover and $j' : J' \to C$ be a minimal elliptic fibration with the general fibre $J_\eta \otimes_K K(C)$. Denote by $\pi:J' \to J$ the natural projection. Clearly, $C_0' = \pi^{-1}(C_0)$ is a section of j' and $\pi^{-1}(C) \cong C \times_{\mathbb{P}^1} C$ splits into two components C_1 and C_2, each of them is a section of j'. Let G = (g) be the

Galois group of the covering $C/\mathbb{P}^1$. Obviously it acts on J' and

$$g(C_1) = C_2,\ g(C_0') = C_0'.$$

Since $(C-2C_0)\cdot J_s \sim 0$, for any $s\in\mathbb{P}^1$, $C-2C_0|J_\eta \sim 0$. This implies that

$$C_1 + C_2 - 2C_0' \mid J'_{\eta'} \sim 0,$$

or

$$C_1+g(C_1) = 0,$$

in the group of sections $J'(C)$. This shows that

$$g \to C_1,\ 1 \to 0$$

defines a cocycle $h \in H^1(G,J'(C)) \cong H^1(G,J_\eta(K(C)))$. Obviously, replacing C_1 by C_2 we get a cohomologous cocycle.

Let $\tau\in\mathrm{Aut}(J')$ be the automorphism $t_{C_1}\circ g$ of J' (see Lemma 5.2.4).

Proposition 5.7.4. Let $s_1,\ldots,s_K$ be the maximal subset of points of $\mathbb{P}^1$ satisfying the following conditions:

(i) $f|C:C \to \mathbb{P}^1$ ramifies over every s_i;

(ii) J_{s_i} is of type $\tilde{A}_{m_i}$ for some m_i, $i = 1,\ldots,K$;

(iii) C does not intersect C_0 at any point of the fibre J_{s_i};

(iv) C and C_0 intersect the same irreducible component of each fibre J_{s_i}.

Then $J'/(\tau) = X$ is a nonsingular surface and j' induces a morphism $f:X \to \mathbb{P}^1$ which is a minimal elliptic fibration the jacobian fibration of which is isomorphic to $j:J \to \mathbb{P}^1$. The local invariant of f are the elements of ${}_2H^1(J_{s_i},\mu_2) \cong {}_2\mathrm{Pic}(J_{s_i})^0$ determined by the divisor $C\cap J_{s_i}-C_0\cap J_{s_i}$. In particular, X is a classical Enriques surface (resp. a rational elliptic surface of index 2) if and only if $K = 2$ (resp. $K = 1$).

PROOF. The only assertion to check is the fact that X is smooth, the other assertions follow immediately from the constructions of §2. To check the smoothness of X, we need to analyze the set of fixed points of τ. We have to check that it does not contain isolated fixed points .If $x\in J_c'$, c is not ramified under $f|C:C \to \mathbb{P}^1$, then $\tau(x)\in J'_{g(c)}$. Hence x is not fixed.

Next, we assume that $f|C$ is ramified at c. Then each fibre J'_c is invariant under τ.The rest of the analysis depends on the structure of the fibre $J_{f(c)}$. We do it only in the case in which $J_{f(c)}$ is of type $\tilde{A}_m$. The other cases are considered similarly.

If m = 0, $J_{f(c)}$ is non-degenerate, thus J'_c is also non-degenerate. The automorphism τ leaves this fibre invariant and is either identity on it ($f(c)\notin\{s_1,\ldots,s_k\}$) or acts by a translation on a point of order 2 ($f(c)\in\{s_1,\ldots,s_k\}$). If m > 0, then J'_c is of type $\tilde{A}_{2m}$. If $f(c)\in \{s_1,\ldots,s_k\}$, then τ acts on the set of nonsingular points of J'_c by a translation on a point of order 2, hence only singular points of J'_c can be fixed. If one of these points is fixed, then all other are fixed too. Then, it is easy to see that a minimal nonsingular model X of the quotient J'/τ has a fibre of type $\tilde{A}_{2m}$ over f(c). However, this contradicts the assertion of Proposition 5.2.3. If $f(c)\notin\{s_1,\ldots,s_k\}$, then τ fixes pointwise the component of $J'_c = E_1+\ldots+E_{2m}$ which intersects C_1, say E_1. Since the set of fixed points of τ is smooth, τ must fix pointwise the components $E_3,\ldots,E_{2m-1}$ ($E_i\bullet E_{i+1} = 1$) and does not have isolated fixed points on this fibre.

Remark 5.7.1. Suppose that $k = 2$, J_{s_i} are non-degenerate and $f:C \to \mathbb{P}^1$ is non-ramified over any $s \neq s_1$ or s_2. Then J' is the double cover of J branched along the two fibres J_{s_i} and the cover $J' \to X = J'/\tau$ is non-ramified. Thus J' is the K3-cover of the Enriques surface X. Since $J' \to C \cong \mathbb{P}^1$ is a jacobian fibration, we see that the pull-back of the elliptic fibration $f:X \to \mathbb{P}^1$ to the K3-cover is a jacobian fibration.

The classification of all possible pairs of cubic curves on $\mathbb{P}^2$ such the double cover of $\mathbb{P}^2$ ramified along their union is birationally isomorphic to the K3-cover of an Enriques surface was obtained by Ulf Persson (unpublished). The following result (previously conjectured by one of the authors) is also due to him (cf. also **[Kon]**).

Theorem 5.7.3. Let $f: F \to \mathbb{P}^1$ be an elliptic fibration on an Enriques surface and $\bar{f}: \bar{F} \to \mathbb{P}^1$ it pull-back to the K3-cover $\bar{F}$ of F. Then the jacobian elliptic fibration associated to $\bar{f}$ is isomorphic to $j': J' \to \mathbb{P}^1$, where J' is birationally isomorphic to the the double cover of the $\mathbb{P}^2$ branched along the union of two cubic curves $Q_1 = 0$ and $Q_2 = 0$ such that $\lambda Q_1+\mu Q_2 = 0$ is the pencil corresponding to the jacobian elliptic surface associated to f. Moreover the two cubics correspond to the two multiple fibres of f.

Now we will study degenerate fibres of genus 1 fibrations on Enriques surfaces.

Theorem 5.7.4. Let $f:F \to \mathbb{P}^1$ be genus 1 fibration on an Enriques surface and F_s be a degenerate non-multiple fibre. Then $b_2(F_s) \le 9$. Moreover,
(i) every indecomposable divisor of canonical type with at most 9 irreducible components is realized as F_s for some elliptic fibration $f:F \to \mathbb{P}^1$ on a classical (resp. non-classical) Enriques surface F;
(ii) if f is a quasi-elliptic fibration, F_s must be of type $\tilde{A}_1^*$, $\tilde{D}_4$, $\tilde{D}_6$, $\tilde{D}_8$, $\tilde{E}_7$ or $\tilde{E}_8$ and every such type is realized for some classical Enriques surface.

PROOF. Since $\rho_f(F) \le 9$, every fibre consists of at most 9 irreducible components. We know that the jacobian fibration of f is a rational elliptic or quasi-elliptic surface. Assume f is elliptic. By Corollary 5.6.6, every indecomposable divisor of canonical type with at most 9 irreducible components is realized as J_s for some jacobian elliptic fibration on a rational surface J. We know from Remark 5.6.1 that almost all fibres of $j: J \to \mathbb{P}^1$ are ordinary elliptic curves. Let F_{s_1}, F_{s_2} be such fibres, $s_1, s_2 \ne s$. We know from Proposition 5.4.5 that the subgroup $\mathrm{Elf}_{s_i}(j)(2)^{et}$ of the group of local invariants $\mathrm{Elf}_{s_i}(j)$ is non-trivial. By Corollary 5.4.6, there exists an elliptic fibration $f: F \to \mathbb{P}^1$ from Elf(j) the local invariants of which at s_1 and s_2 are nonzero elements of $\mathrm{Elf}_{s_i}(j)(2)^{et}$ and every other local invariant is zero. Then f has exactly two multiple non-wild fibres of multiplicity 2. By Proposition 5.7.1, F is a classical Enriques surface. By Theorem 5.3.1, $F_s \cong J_s$. Similarly, if $p = 2$, we choose a non-trivial element in $\mathrm{Elf}_{s_1}(j)(2)_{rad}$ and construct an elliptic fibration $f: F \to \mathbb{P}^1$ in Elf(j) with one wild multiple fibre of multiplicity 2 at s_1. By Proposition 5.6.1, F is a non-classical Enriques surface. As before, $F_s \cong J_s$.

Assume f is quasi-elliptic. The possible types of degenerate fibres are given by Corollary 5.2.4. By Corollary 5.6.8, every such type is realized as J_s for some quasi-elliptic rational surface j: J → $\mathbb{P}^1$. Applying Corollary 5.6.4, we see that for every closed s∈$\mathbb{P}^1$ the group $Elf_s(J)$ contains non-trivial elements defining quasi-elliptic fibrations over the strict localization of $\mathbb{P}^1$ at s with a multiple non-wild fibre of multiplicity 2. Choose two such elements from $Elf_{s_1}(J)$ and $Elf_{s_2}(J)$ respectively for some $s_1,s_2 \neq s$. By Corollary 5.4.9 there exists a quasi-elliptic fibration f: F → $\mathbb{P}^1$ from Elf(j) with two non-wild multiple fibres F_{s_1} and F_{s_2} of multipliicty 2. By Proposition 5.6.1, F is a classical Enriques surface.

Theorem 5.7.5. Let f:F → $\mathbb{P}^1$ be a genus 1 fibration on a classical Enriques surface and $F_s = 2\bar{F}_s$ be a multiple fibre. Then F_s is non-wild and

(i) $\bar{F}_s$ is of type $\tilde{A}_n$, $n \leq 8$ if $p \neq 2$;

(ii) $\bar{F}_s$ is of additive type with $b_2(\bar{F}_s) \leq 9$ or a smooth ordinary elliptic curve if $p = 2$;

(iii) if f is quasi-elliptic, $\bar{F}_s$ is of type $\tilde{A}_0^*$, $\tilde{A}_0^*$, $\tilde{D}_4$, $\tilde{D}_6$, $\tilde{D}_8$, $\tilde{E}_7$ or $\tilde{E}_8$.

Every type with the above restrictions can be realized.

PROOF. Since $\rho_f(F) \leq 9$, every multiple fibre consists of at most 9 irreducible components. By Proposition 5.7.1, F_s is non-wild. Now (i) follows from Proposition 5.1.8, (ii) follows from Proposition 5.4.5 and (iii) follows from Theorem 5.3.1 and Corollary 5.2.4. It remains to prove the assertion about the realization of the possible types. Let j:J → $\mathbb{P}^1$ be a rational jacobian genus 1 fibration. The possible types of degenerate fibres are given by Corollary 5.2.4. By Corollary 5.6.8, every such type is realized as J_s for some quasi-elliptic rational surface j: J → $\mathbb{P}^1$. Applying Corollary 5.6.4, we see that for every closed s∈$\mathbb{P}^1$ the group $Elf_s(J)$ contains non-trivial elements defining genus 1 fibrations over the strict localization of $\mathbb{P}^1$ at s with a multiple non-wild fibre of multiplicity 2. Choose two such elements from $Elf_{s_1}(J)$ and $Elf_{s'}(J)$ respectively for some s'≠s. By Corollary 5.4.9 there exists a genus 1 elliptic fibration from f: F → $\mathbb{P}^1$ from Elf(j) with two non-wild multiple fibres F_s and $F_{s'}$ of multiplicity 2. By Proposition 5.6.1, F is a classical Enriques surface.

Theorem 5.7.6. Let F be a μ_2-surface (resp. an α_2-surface) and $f:F \to \mathbb{P}^1$ be a genus 1 fibration on F. Then the unique multiple fibre of f is of multiplicative type or an ordinary elliptic curve (resp. of additive type or a supersingular elliptic curve).

PROOF. Let $F_s = 2C$ for some $s \in \mathbb{P}^1$. Since

$$H^1(F,\mathcal{O}_F(-C)) \cong H^1(F,\mathcal{O}_F(C)) \cong k,$$

the exact sequence

$$0 \to \mathcal{O}_F(-C) \to \mathcal{O}_F \to \mathcal{O}_C \to 0$$

induces an isomorphism on the cohomology:

$$H^1(F,\mathcal{O}_F) \to H^1(C,\mathcal{O}_C).$$

It is easy to see that this isomorphism is compatible with the Frobenius map F. Thus F is invertible on $H^1(C,\mathcal{O}_C)$ if F is an μ_2-surface and is trivial otherwise. Now recall that $H^1(C,\mathcal{O}_C)$ is canonically identified with the Lie algebra of the Picard scheme $\mathbf{Pic}_{C/K}$. It is easy to see that F is bijective on $H^1(C,\mathcal{O}_C)$ if and only if $\mathbf{Pic}^0_{C/K} \cong \mathbf{G}_m$ or an ordinary elliptic curve.

Remark 5.7.2. One can verify the assertion in the case where F is a μ_2-surface as follows. We know from S8 of chapter 4 that the branch divisor of the double cover obtained from any superelliptic map $\varphi:F \to \mathcal{D}$ is smooth. By Remark 0.2.2, φ factors through a surface with singularities of type A_n only.

Bibliographical notes to Chapter V.

The study of elliptic pencils on algebraic surfaces was started by F.Enriques. Besides other properties of such pencils one can find in **[En 4]** the idea of the associated jacobian fibration and a formula for the canonical class. The theory was ressurected in the works of K.Kodaira **[Kod 1]** and I.Shafarevich **[AS]**. Most recently an exposition of Kodaira's approach to the theory is given in **[B-P-vdV]** and some general facts about elliptic fibrations are treated in **[G-H 1]**. In contrast to the purely analytical approach of Kodaira, the approach of Shafarevich is algebraic. In fact, it is a translation into the geometric language of the theory of principal homogeneous spaces of abelian varieties of arbitrary dimension developed independently by him **[Sh 1]** and by A.Ogg **[Ogg 1]**. An account of this approach is given in **[Do 3]**. The notion of a quasi-elliptic pencil is due to E. Bombieri and D.Mumford. It appeared in their work **[B-M 2]** in which Enriques' classification of algebraic surfaces is extended to the case of positive characteristic. A common treatment of elliptic and quasi-elliptic surfaces is given in **[Ra 6]** from which we borrowed heavily.

Let us comment in more detail on the contents of each section. §1 contains the well known facts about genus 1 fibrations. For example the formula for the canonical class from Proposition 5.1.3 is due to Kodaira **[Kod 1]** in the case $k = \mathbb{C}$. The general formula was given in **[B-M 2]** where for the first time the phenomenon of wild fibres was noticed. The classification of degenerate fibres is due to Kodaira. The observation that this classification follows immediately from the classification of symmetric root bases of affine type must be attributed to various people. One of the authors learnt it from Shafarevich long time ago. The formula for the Euler-Poincare characteristic from Proposition 5.1.6 is equivalent to the classic formula of Noether-Zeuthen in the case when $k = \mathbb{C}$ and all singularities of the fibres are at most nodes. A topological proof of the general formula (again for $k = \mathbb{C}$) can be found in **[AS]** or in **[G-H 1]**. The case char(k) > 0 was first independently discussed in **[Do 2]** and **[Del 1]**. In fact the formula is an easy consequence of the formula of A. Grothendieck for the Euler-Poincarè characteristic of a constructible sheaf in étale topology of an algebraic curvę **[Ray 2]** and in its turn is a generalization of a formula of Ogg and Shafarevich from their loc. cit. work on principal homogeneous spaces of abelian varieties.

Most of §2 is taken from **[Ray 3]**. The formula for the Picard number from Corollary 5.2.1 is attributed usually to T.Shioda (see **[Shi]**). However we believe that this formula had been known before to various people (see for example, **[Ta 2]**). One of the authors was told about this formula by I. Shafarevich in 1964. The theory of the Néron model of an abelian variety is due to **[Ne]**. More recent expositions of this theory can be found in **[Art 8]** and **[Ray 1]**. The fact that the Néron theory of elliptic curves follows simply from the theory of minimal models of two-dimensional schemes was first noticed probably by Shafarevich **[Sh 2]**. In **[Ch]** one can find a nice exposition of the theory of minimal models of two-dimensional schemes which is due to

Shafarevich **[Sh 2]** and S. Lichtenbaum **[Li]**. The classification of degenerate fibres of quasi-elliptic fibrations from Corollary 5.2.4 can be found in **[R-S 3]**.

Most of the results of §3 about the relationships between a genus 1 fibration and its Jacobian fibration are known but it is hard to give a reference for them. We tried to collect them all in full generality. Thus the relationship between the Brauer groups can be found in **[Art 6]** but its corollary that all numerical characteristics are equal for the corresponding surfaces is a new result (in positive characteristic). If $k = \mathbb{C}$ this result follows easily from the description of the topology of an elliptic surface which is based on Kodaira's notion of logarithmic transformations. The construction of elliptic fibrations by means of logarithmic transformations applied to its jacobian fibration is purely analytical and by this reason is not discussed here. We refer to **[Kod 1]** and **[B-P-vdV]** for this theory and to **[Kod 3,Mo]** for the description of the topology of elliptic surfaces. Theorem 5.3.1 is a new result. Its corollary about the equality of the types of a multiple fibre and the corresponding fibre of the Jacobian fibration in the case $p > 2$ was proven earlier by other methods in **[K-U]**. The properties of the group of the sections of a jacobian fibration presented in Proposition 5.3.4 are well-known. For example, the properties of the reduction homomorphism is due to **[Lu]** and the formula for the order of the finite group of sections can be found in **[Ta 2]**.

§4 follows an exposition of Ogg-Shafarevich's theory taken from **[Ray 2]**. The new feature is the treatment of the p-part of the theory which was not available at that time. Also, following **[La 1]**, we extend this theory to the case of quasi-elliptic fibrations. The fact that the Shafarevich group of the generic fibre is essentially the Brauer group of the surface was first noticed by A.Grothendieck **[Gro 3]**. Of course it was clear before that its corank is equal to the rank of the transendental part of the second cohomology. An explicit statement of this kind can be found in **[Sh 1,Ogg 1]**. The duality theorem for the p-part of the group of local invariants was computed in various special cases first by O.Vvedenskii **[Vv 1, Vv 2]** and later by **[Bes]** and **[Ray 3]**. It was noticed by J. Milne **[Mi 3]** that the general case can be reduced to those by applying the semi-stable reduction. The fact that the homomorphism of local invariants is surjective unless the jacobian fibration is trivial is due to Ogg and Shafarevich up to the p-part. The surjectivity of the p-part of the local invariant homomorphism in the case of a rational surface was verified first in **[La 5]**. The general case of the p-part is treated by J. Milne **[Mi 3]**. His book contains a nice exposition of the various duality theorems used in this section.

The equations of the Weierstrass models of elliptic fibrations over arbitrary bases given in

§5 were first given in **[M-S]** and later reproduced in **[Del 2]**. The case of quasi-elliptic fibration was first treated by W.Lang **[La 1]** in the case p = 3. Our exposition is very similar to his. The computation of the types of degenerate fibres of quasi-elliptic fibrations from its Weierstrass model seems to be new.

The fact that the theory of Halphen pencils from **[Hal]** is equivalent to the theory of rational elliptic surfaces was noticed in **[Do 1]**. Later on the facts about rational elliptic surfaces given in this section were rediscovered by many authors [see **[H-L,M-P]**. A construction of multiple fibres of additive type on rational elliptic surfaces is taken from **[H-L]**. A theorem classifying the types of degenerate fibres of rational elliptic surfaces with finite group of sections was also known to many people. For example, one can find this result in **[M-P]** or **[Nar]**. Its analog for quasi-elliptic fibrations is a new result. Recently a complete classification of all possible types of elliptic fibrations on rational complex elliptic surfaces was given by U.Persson. A realization of multiple fibres of additive type for rational genus 1 fibrations can be found in **[H-L]**.

The results of §7 are well-known in the case $p \neq 2$. The fact that Enriques surfaces admitting quasi-elliptic fibrations must have rational K3-cover was first proven in **[Kat 2]**. Theorem 5.7.4 (i) was proven in **[A-B]** by explicit constructions. Theorem 5.7.5 on the realization of non-wild multiple fibres of additive type of genus 1 fibrations on Enriques surfaces is new. Theorem 5.7.6 is due to W.Lang **[La 3]**.

Bibliography

[A-B] G. Angermüller, W. Barth, Elliptic fibres on Enriques surfaces, Comp. Math., 47 (1982), 317-332.

[ACGH] E. Arbarello, M. Cornalba, P. Griffiths, J. Harris, Geometry of algebraic curves, vol.I, Springer-Verlag.1984.

[Ar] E. Artin, Geometric algebra. Interscience Publ. New York. 1957. [Russian Transl.:Nauka. Moscow. 1968].

[Arn] V. Arnold, Critical points of smooth functions, Proc. Int. Congress of Math. Vancouver. 1978, vol. 1, pp. 19-39.

[Art 1] M. Artin, On Enriques surfaces, Harvard thesis. 1960.

[Art 2] M. Artin, On isolated rational singularities of surfaces, Amer. J. Math, 84 (1962), 485-496.

[Art 3] M. Artin, Algebrization of formal moduli, in "Global analysis", Princeton. Univ. Press, Princeton. 1969, pp. 21-71 [Russian transl.:Matematika, 14:4 (1970), 3-47].

[Art 4] M. Artin, Algebraic approximation of structures over complete local rings, Publ. Math. IHES, 36 (1969), 23-58 [Russian transl.:Matematika, 14:3 (1970), 3-39].

[Art 5] M. Artin, Théorème de changement de base pour un morphism propre in "Théorie des topos et cohomologie étale des schémas (SGAA 4), t.3", Lect. Notes in Math. vol. 305, Springer-Verlag, 1973, pp. 79-131.

[Art 6] M. Artin, Supersingular K3-surfaces, Ann. Scient. Éc. Norm. Sup. 4e Serie, 7 (1974), 543-570.

[Art 7] M. Artin, Coverings of the rational double points in characteristic p, in "Complex Analysis and Algebraic Geometry", Iwahami-Shoten. Tokyo. 1977, 11-22.

[Art 8] M. Artin, Néron models, in "Arithmetical Geometry", Springer-Verlag. 1986, pp. 309-325.

[AS] Algebraic surfaces (ed. by I. Shafarevich), Proc. Steklov Math. Inst., v.75, 1964. [Engl.transl.: AMS, Providence.R.I. 1967].

[A-M] M. Atiyah, I. Macdonald, Introduction to commutative algebra, Addison-Wesley, Reading. Mass. 1969 [Russian transl.: Mir. Moscow. 1972].

[Av] B. Averbukch, Kummer and Enriques surfaces of special type. Izv. Akad. Nauk SSSR, Ser. mat. 29 (1965), 1095-1118.

[Ba] W. Barth, Lectures on K3-and Enriques surfaces, in "Algebraic geometry. Sitges 1983", Lect. Notes in Math. vol. 1124, Springer-Verlag, 1985, pp. 21-57.

[B-P] W. Barth, C. Peters, Automorphisms of Enriques surfaces, Inv. Math. 73 (1983), 383-411.

[B-P-vdV] W. Barth, C. Peters, A. van de Ven, Complex algebraic surfaces, Springer-Verlag. 1984.

[Bea 1] A. Beauville, Variétes de Prym et jacobiennes intermediaires, Ann. Sci. Éc. Norm. Sup. 10 (1977), 309-331.

[Bea 2] A. Beauville, Surfaces algébriques complexes, Astérisque. vol. 54, Soc. Mat. de France. 1980 [Engl. Transl.:Complex algebraic surfaces, London Math. Soc. Lecture Notes. vol. 68, Cambridge Univ. Press. 1983].

[Ber] P. Berthelot, Le théoreme de dualité plate pour les surfaces (d'aprés J.S. Milne), in "Surfaces Algébriques", Lect. Notes in Math. vol. 868, 1981, Springer-Verlag, pp. 203-237.

[Bes] M. Bester, Local flat duality of abelian varieties, Math. Ann. 235 (1978), 149-174.

[Bl 1] P. Blass, Zariski surfaces, Thesis. Univ. of Michigan.

[Bl 2] P. Blass, Unirationality of Enriques surfaces in characteristic two, Comp. Math., 45 (1982), 393-398.

[Bog] F. Bogomolov, The theory of invariants and applications to some problems in algebraic geometry, in "Algebraic surfaces", Proc. CIME

Summer School in Cortona, Liguore, Napoli, 1981, pp. 217-245.

[B-M 1] E. Bombieri, D. Mumford, Enriques classification in char. p, II in "Complex Analysis and Algebraic Geometry", Iwanami-Shoten, Tokyo. 1977, 23-42.

[B-M 2] E. Bombieri, D. Mumford, Enriques classification in char. p, III, Invent. Math. 35 (1976), 197-232.

[Bom] E. Bombieri, Canonical models of surfaces of general type, Publ. Math. IHES. 42 (1973), 171-229.

[Bou 1] N. Bourbaki, Algébre, Livre II, Chapitre 8, Hermann. Paris. 1958 [Russian Transl.: Nauka. Moscow. 1966].

[Bou 2] N. Bourbaki, Groupes at Algébres de Lie. Chapters IV-VI. Hermann. Paris. 1966 [Engl. Transl.:, Russian. transl.: Mir. 1972].

[Br] E. Brieskorn, Rationale Singularitäten Komplexer Flächen, Inv. Math. 4 (1968), 336-358.

[Car] P. Cartier, Questions de rationalité de diviseurs en géometrie algébrique, Bull. Soc. Math. France, 86 (1958), 117-251.

[Cas] G. Castelnuovo, Sulle superficie di genere zero, Mem. delle Soc. Ital. delle Scienze, ser. III, 10 (1895).

[Cat] F. Catanese, On the rationality of certain moduli spaces of curves of genus 4, in "Algebraic geometry", Lect. notes in Math. vol. 1008, Springer-verlag. 1982, pp. 30-50.

[Ch] T.Chinburg, Minimal models for curves over Dedekind ring, in "Arithmetical Geometry", Springer-Verlag. 1986, pp. 309-325.

[Co 1] A.Coble, Theta modular groups determined by point sets, Amer.J. Math., 40 (1918), 317-340.

[Co 2] A. Coble, The ten nodes of the rational sextic and of the Cayley symmetroid, Amer. J. Math. 41 (1919), 243-265.

[Co 3] A.Coble, Algebraic geometry and theta functions, Amer. Math. Soc. Coll. Publ. vol. 10, Providence, R.I., 1929 (4d ed., 1982).

[Cos 1] F. Cossec, On the Picard group of Enriques surfaces, Math. Ann. 271 (1985), 577-600

[Cos 2] F. Cossec, Projective models of Enriques surfaces, Math. Ann.

265 (1983), 283-334.

[Cos 3] F. Cossec, Reye congruences, Trans. Amer. Math. Soc. 280 (1983),737-751.

[C-D 1] F. Cossec, I. Dolgachev, Rational curves on Enriques surfaces, Math. Ann. 272 (1985), 369-384.

[C-D 2] F. Cossec, I. Dolgachev, On automorphisms of nodal Enriques surfaces, Bull. Amer. Math. Soc. 12 (1985), 247-249.

[Cox] H.S.M. Coxeter, Finite groups generated by reflections and their finite subgroups generated by reflections, Proc. Cambridge Phil. Soc. 30 (1934), 466-482.

[Cr] R. Crew, Étale p-covers in characteristic p, Compos. Math. 52 (1984), 31-45.

[Del 1] P. Deligne, La formule de Milnor, in "Groupes de Monodromie en Géometrie Agébrique (SGA 7 II)", Lect. Notes in Math. vol. 340, Springer-Verlag, 1973, pp. 197-211.

[Del 2] P. Deligne, Courbes elliptiques: Formulaire (d'aprés J. Tate), in "Modular functions in one varibale, IV", Lect. Notes in Math. vol. 476, Springer-Verlag, 1975, pp. 53-74.

[D-I] P. Deligne, L. Illusie, Relévements modulo p^2 et decomposition du complexe de de Rham, Inv. Math. 89 (1987), 247-270.

[D-R] P. Deligne, M. Rappoport, Les schémas de modules de courbes elliptiques, in "Modular functions in one variable, II", Lect. Notes in Math. Springer-Verlag, vol. 349, 1973, pp. 143-316.

[Dem] M.Demazure, Surfaces de del Pezzo,II,III,IV,V, in "Singularités des surfaces", Lect.Notes in Math. v.777, 1980, 23-69.

[Do 1] I. Dolgachev, Rational surfaces with a pencil of elliptic curves, Izv. Akad. Nauk SSSR, Ser. Math. 30 (1966), 1073-1100.

[Do 2] I. Dolgachev, Euler characteristic of a family of algebraic varieties, Mat. Sbornik, 89 (1972), 297-312 [Engl. transl.: Mathematics of USSR-Sbornik, 18 (1972), 303-318].

[Do 3] I. Dolgachev, Algebraic surfaces with $p_g = q = 0$, in "Algebraic surfaces", Proc. CIME Summer School in Cortona, Liguore, Napoli,

1981, pp. 97-216.

[Do 4] I. Dolgachev, Integral quadratic forms: Application to algebraic geometry (after V. Nikulin), Seminaire Bourbaki 1982/83, n° 611, Astérisque, vol. 105/106. Soc. Mat. de France, Paris, pp. 251-275.

[Do 5] I. Dolgachev, Automorphisms of Enriques surfaces, Invent. Math. 76 (1984), 63-177.

[D-O] I. Dolgachev, D. Ortland, Topics in classical algebraic geometry, Astérisque, 1988 (to appear).

[DuV 1] P. Du Val, On isolated singularities which do not affect the condition of adjunction, I,II,IV, Proc. Cambridge Phil.Soc. 30 (1934), 453-491.

[DuV 2] P. Du Val, On the Kantor group of a set of points in a plane, Proc. London Math. Soc. (2) 42 (1936), 18-51.

[Dy] E. Dynkin, Semi-simple subalgebras of semi-simple Lie algebras, Mat. Sbornik, 30 (1952), 349-462 [Engl. Transl. AMS Translations, vol. 6, 1987, 111-244].

[EGA] A. Grothendieck, Elements de Géometrie Algébrique, Chapter 3, Publ. Math. IHES, vol. 17, 1963.

[Ek 1] T. Ekehdal, Foliations and inseparable morphisms, in "Algebraic Geometry", Part. 2, Proc. Symp. Pure Math. vol. 46, AMS, Providence, 1987, pp. 139-150.

[Ek 2] T. Ekehdal, Vector fields on classical Enriques surfaces (manuscript in preparation).

[En 1] F. Enriques, Introduzione alla geometria sopra le superficie algebriche, Mem Soc. Ital. delle scienze, ser. 3a, 10 (1896), 1-81 (also "Memorie Scelte di Geometria", t. 1, Zanichelli. Bologna. 1956, pp. 211-312).

[En 2] F. Enriques, Sopra le superficie algebriche di bigenere uno, Mem Soc. Ital. delle Scienze, ser. 3a, 14 (1906), 39-366 (also "Memorie Scelte di Geometria", t. 2, Zanichelli. Bologna. 1959, pp. 241-272).

[En 3] F. Enriques, Un' ossevazione relativa alle superficie di bigenere uno, Rend. Acad. Scienze Inst. Bologna, 12 (1908), 40-45 (also "Memorie Scelte di Geometria", t. 2, Zanichelli, Bologna. 1959, pp. 303-306).

[En 4] F. Enriques, Le superficie algebriche, Zanichelli, Bologna. 1949.

[Fa] G. Fano, Nuovo ricerche sulle congruenze di retta del 3 ordine, Mem. Acad. Sci. Torino, 50 (1901), 1-79.

[Fu] W. Fulton, Intersection theory, Springer-Verlag. 1984.

[Ga] P. Gabriel, Etude infinitesimal des schémas en groupes, in "Schémas en groupes (SGAA 3), t.1", Lect. Notes in Math. vol. 151, Springer-Verlag, 1970, pp. 474-560.

[GS-V] J. Gonsalez-Sprinberg, J.-L. Verdier, Construction géometrique de la correspondence de McKay, Ann. Scient. Éc. Norm. Sup. 4e Ser., 16 (19830, 409-449..

[Gri] R. Griess, Quotients of infinite reflection groups, Math. Ann, 263 (1983), 267-288.

[G-H 1] P. Griffiths, J. Harris, Principles of algebraic geometry, John Wiley and Sons, New York. 1978 [Russian transl.: Mir. Moscow. 1982].

[G-H 2] P. Griffiths, J. Harris, Residues and zero cycles on algebraic varieties, Ann. Math. 108 (1978), 461-505.

[Gro 1] A. Grothendieck, Le Théorie des classes de Chern, Bul. Soc. Math. France., 86 (1958), 137-154.

[Gro 2] A. Grothendieck, Fondéments de Géométrie Algébrique. Sem Bourbaki 1957-1962, Secr. Math. Paris. 1962.

[Gro 3] A. Grothendieck, Brauer Group I,II, III, in "Dix Exposes sur cohomologie des schémas", North-Holland, Amsterdam,1968, pp. 46-188.

[Gro 4] A. Grothendieck, Spécialization en theorie des intersections, in "Theorie des intersections et théoreme de Riemann-Roch, (SGA6)", Lect. Notes in Math. vol. 225, Springer-Verlag, 1971, pp. 560-594.

[Gro 5] A. Grothendieck, Modéles de Néron et monodromie", in "Groupes de monodromie en géométrie algébrique, (SGA 7 I)", Lect. Notes in Math. vol. 288, Springer-Verlag,· 1972, pp. 313-521.

[Hal] M.Halphen, Sur les courbes planes du sixiéme degré a neuf points doubles, Bull. Soc. Math. France, 10 (1981), 162-172.

[Ha] B.Harbourne, Blowings-up of P^2 and their blowings-down, Duke Math. J. 52 (1985), 129-148.

[H-L] B. Harbourne, W. Lang, Multiple fibres on rational elliptic surfaces (preprint).

[Har 1] R. Hartshorne, On the De Rham cohomology of algebraic varieties, Publ. Math. IHES, 45 (1976), 5-99.

[Har 2] R. Hartshorne, Algebraic Geometry. Springer-Verlag. 1977.

[Ho] E.Horikawa, On the periods of Enriques surfaces, I. Math. Ann. 234 (1978), 73-108; II, ibid. 235 (1978), 217-246.

[Il 1] L. Illusie, Complexe de De Rham-Witt et cohomologie christalline. Ann. Sci. Éc. Norm. Sup, 4e ser., 12 (1979), 501-661.

[Kac] V. Kac, Infinite-dimensional Lie Algebras. Birkhäuser. 1983.

[Kat 1] T. Katsura, Surfaces unirationnelles en caracteristique p, C.R. Acad. Sci. Paris, 288 (1979), 45-47.

[Kat 2] T. Katsura, A note on Enriques surfaces in characteristic 2. Comp. Math. 47 (1982), 207-216.

[K-U] T. Katsura, K. Ueno, Multiple singular fibres of type G_a of elliptic surfaces in characteristic p, in "Algebraic and topological theories", 1985, pp. 405-429.

[Kod 1] K. Kodaira, On compact complex analytic surfaces. I, Ann. Math. 71 (1960), 111-152; II, ibid, 77 (1963), 563-626; III, ibid, 78 (1963), 1-40.

[Kod 2] K. Kodaira, Pluricanonical systems on algebraic surfaces of general type, J. Math. Soc. Japan, 20 (1968), 180-192.

[Kod 3] K. Kodaira, On homotopy K3-surfaces, in "Essays in topology and related topics", Springer-Verlag, 1970, pp. 56-69.

[Kon] S. Kondo, Enriques surfaces with finite automorphism group, Japan J. Math. 12 (1986), 192-282.

[La 1] W. Lang, Quasi-elliptic surfaces in characteristic three, Ann. Scient. Éc. Norm. Sup. 4e Ser., 12 (1979), 473-500.

[La 2] W. Lang, Two theorems on the De Rham cohomology, Comp.

Math. 40 (1980), 417-423.

[La 3] W. Lang, On Enriques surfaces in char. p. I, Math. Ann. 265 (1983) 45-65.

[La 4] W. Lang, An analogue of the logarithmic transform in characteristic p, in "Proceedings of. 1984 Vancouver Conf. in Alg. Geometry", CMS Conf. vol. 6, 1986, pp. 337-340.

[La 5] W. Lang, On Enriques surfaces in char. p II (preprint).

[Li] S. Lichtenbaum, Curves over discrete valuation rings, Amer. J. Math. 85 (1968), 380-405.

[Lip] J. Lipman, Dualizing sheaves, differentials and residues on algebraic varieties, Astérisque, vol. 117, Soc. Math. de France. 1984.

[Lo 1] E. Looijenga, Rational surfaces with an anticanonical cycle, Ann. Math. (2) 114 (1981), 267-322.

[Lo 2] E. Looijenga, Invariant theory for generalized root systems, Invent. Math. 61 (1980), 1-32.

[Lu] E. Lutz, Sur l'equation $y^2=x^3-Ax-B$ dans le corps p-adiques, J. fur Math. 177 (1937), 238-247.

[Ma] Y. Manin, Cubic forms: algebra, geometry, arithmetic, Nauka. Moscow. 1972 [Engl. transl.:North-Holland, Amsterdam, 1974, 2nd edition 1986].

[McL] J. Maclaughlin, Some subgroups of $SL_n(\mathbb{F}_2)$, Ill. J. Math. (1969), 108-115.

[Mi 1] J. Milne, Duality in the flat cohomology of a surface, Ann. Scient. Éc. Norm. Sup. 4e ser., 9 (1976), 171-202.

[Mi 2] J. Milne, Étale cohomology, Princeton Univ. Press. 1980 [Russian Transl.: Mir. Moscow. 1983].

[Mi 3] J. Milne, Arithmetic duality theorems, Princeton Univ. Press, 1986

[M-P] H. Miranda, U. Persson, On extremal rational elliptic surfaces, Math. Zeit., 193 (1986), 537-568.

[Mo] B. Moishezon, Complex algebraic surfaces and connected sums of complex projective planes, Lect. Notes in Math. vol. 603, Springer-Verlag. 1977.

[Mu 1] D. Mumford, Lectures on curves on an algebraic surface, Ann. Math. Studies, vol. 59, Princeton Univ. Press, Princeton. 1966 [Russian transl.:Mir. Moscow. 1971].

[Mu 2] D. Mumford, Enriques' classification of surfaces in char. p, I, in "Global analysis", Princeton. Univ. Press, Princeton. 1969, pp. 325-339.

[Mu 3] D. Mumford, Abelian varieties. Oxford Univ. Press. 1970 [Russian transl.:Mir. Moscow. 1971].

[M-S] D. Mumford, K. Suomininen, Introduction to the theory of moduli, in "Algebraic geometry, Oslo 1970", Wolters-Noordhoff Publ. 1972, pp.171-222.

[Mur] J. Murre, On contravarinat functors from the category of schemes to the category of abelian groups (with applications to the Picard functor), Publ. Math. IHES, 23 (1964), 581-619.

[Na] M.Nagata, On rational surfaces, I, Mem. Coll. Sci. Univ. Kyoto, 37 (1960), 271-293 [Russ. Transl.:Matematika, 8:4 (1964), 75-94].

[Nam] Y.Namikawa, Periods of Enriques surfaces, Math. Ann. 270 (1985), 201-222.

[Nar] I. Naruki, Configurations related to maximal rational elliptic surfaces, preprint.

[Ne] A. Neron, Modéles minimaux des variétetes abéliennes, Publ. Math. Inst. IHES. vol. 21. 1964.

[Ni 1] V. Nikulin, Integral quadratic forms and some of its geometric applications, Izv. Akad. nauk SSSR, Ser. Math. 43 (1979), 103-167 [Engl. Transl.: Mathematics of USSR-Izvestia 14 (1980), 03-167].

[Ni 2] V. Nikulin, On the quotient groups of the automorphism groups of hyperbolic forms modulo subgroups generated by 2-reflections, Algebraic geometric applications, in "Current Problems of Mathematics", t. 18, VINITI, Moscow, 1981, pp. 3-114 [Engl. Transl.:J. Soviet Math. 22 (1983), 1401-1476].

[Ni 3] V. Nikulin, Surfaces of type K3 with finite automorphism group and Picard group of rank 3, Trudy. Steklov Inst. Math. 165 (1984), 119-142 [Engl. Transl. Proc. Inst. Steklov (1985), pp.131-156]

[Ni 4] V. Nikulin, On a description of the automorphism groups of an Enriques surfaces, Dokl. Akad, Nauk SSSR, 277 (1984), 1324-1327 [Engl. Transl. Soviet Math. Doklady 30 (1984), 282-285].

[Ny] N. Nygaard, The Tate conjecture for ordinary K3-surfaces over finite fields, Inv. Math. 74 (1983), 213-237.

[Od] J. Oda, The first de Rham cohomology group and Diéudonné modules, Ann. sci. Éc. Norm. sup. , 4e ser., 2 (1963), 63-135

[Oe] J. Oesterlé, Dégénerescence de la suite spectrale de Hodge vers De Rham (d'aprés Deligne et Illusie), Sem. Bourbaki. 1986/87, n° 673 , Astérisque, vol. 152-153, 1987, pp. 67-84.

[Og 1] A. Ogg, Cohomology of abelian varieties over function fields, Ann. Math. 76 (1962), 185-212.

[0g 2] A. Ogg, Elliptic curves and wild ramification, Amer. J. Math. 89 (1967), 1-21.

[Oo] F. Oort, Commutative group schemes, Lect. Notes in Math. 15, Springer-Verlag. 1966.

[Pi] H. Pinkham, Singularités rationnelles de surfaces, in "Singularités de surfaces". Lect. Notes in Math. vol. 777. Springer-Verlag. 1979, pp. 147-178.

[PS-S] I.Pyatetskii-Shapiro, I.Shafarevich, A Torelli theorem for algebraic surfaces of type K3, Izv. Akad. Nauk SSSR, ser. math., 35 (1971), 503-572 [Engl. Transl.: Math. USSR-Izvestia, 5, 547-588].

[Ram] C. Ramanujam, Remarks on Kodaira vanishing theorem, J. Indian Math. Soc., 36 (1972), 41-51.

[Ray 1] M. Raynaud, Modeles de Néron, C.R. Acad. Sci. Paris, 262 (1966), 413-414.

[Ray 2] M. Raynaud, Characteristique d'Euler-Poincaré d'un faisceau et cohomologie des varietés abeliennes. In "Dix Exposés sur cohomologie des schémas", North-Holland. Amsterdam, 1968, pp. 12-30.

[Ray 3] M. Raynaud, Spécialization du foncteur de Picard, Publ. Math. IHES, 38 (1970), 27-76.

[Ray 4] M. Raynaud, Contre-exemple au "vanishing theorem" en characteristic p>0, in "C.P.Ramanujam". Publ. Tata Insitute. 1978, pp. 273-278.

[Ray 5] M. Raynaud, Fibrés vectorielles instables (d'aprés Bogomolov), in "Surfaces Algébriques", Lect. Notes in Math. vol. 868, 1981, Springer-Verlag, pp. 293-314.

[Ray 6] M. Raynaud, unpublished notes

[Rec] S. Recillas, Jacobians of curves with g_1^4's are the Prym's of trigonal curves. Bol. de la Soc. Mat. Mexicana, 19 (1974),

[Rei] I. Reider, Vector bundles of rank 2 and linear systems on algebraic surfaces, Ann. Math.,127 (1988), 309-316.

[R-S 1] A. Rudakov, I. Shafarevich, Inseparable morphisms, Izv. Akad. Nauk SSSR, Ser. Math., 40 (1976), 1269-1307 [Engl. Transl.: Mathematics of USSR-Izvestija, 10 (1976), 1205-1243].

[R-S 2] A. Rudakov, I. Shafarevich, Supersingular surfaces of type K3 in characteristic 2, Izv. Akad. Nauk SSSR, Ser. Math., 42 (1978), 848-869 [Engl. Transl.: Mathematics of USSR-Izvestija, 13 (1976), 147-165].

[R-S 3] A. Rudakov, I. Shafarevich, Surfaces of type K3 over fields of finite characteristic, in "Current Problems of Mathematics", t. 18, VINITI, Moscow, 1981, pp. 3-114 [Engl. Transl.: J. Soviet Math. 22 (1983), 1477-1580].

[Ru] P. Russell, Forms of the affine line and its additive group, Pacif. J. Math. 32 (1970), 527-539.

[SD] B.Saint-Donat, Projective models of K3-surfaces, Amer. J. Math.,96 (1974), 602-639.

[Se] C. Segre, Surfaces du 4e ordre a conique double, Math. Ann. 24 (1884), 313-344.

[Ser 1] J. P. Serre, Groupes algébrique et corps de classes, Hermann. Paris.

1959 [Russ. Transl.: Mir. Moscow. 1968].

[Ser 2] J. P. Serre, Groupes proalgébriques, Publ. Math. IHES, 7 (1960).

[Ser 3] J. P. Serre, Cohomologie Galoissiennes, Lect. Notes in Math. vol. 5, Springer-Verlag, 1964 [Russian Transl. : Mir. Moscow. 1968].

[Ser 4] J. P. Serre, Cours de Arithmétique, Pres. Univ. de France, Paris. 1970 [Engl. Transl.:Springer-Verlag. 1973].

[Sh 1] I. Shafarevich, Principal homogeneous spaces over function fields, Proc. Steklov Inst. Math. 64 (1961), 316-346[Engl. Transl.:AMS Translations, 37 (1964), 85-114].

[Sh 2] I. Shafarevich, Lectures on minimal models and birational transformations of two-dimensional schemes, Tata Lect. Notes Math. Bombay. 1966.

[Sha] J.Shah, Projective degenerations of Enriques surfaces, Math. Anal. 256 (1981), 475-495.

[Shi] T. Shioda, On elliptic modular surfaces, J. Math. Soc. Japan, 24 (1972), 20-59.

[Sie] C. Siegel, Einheiten quadratischer Formen, Math. Sem. Hansischen Univ. 13 (1940), 209-239 (also in "Gesammelten Abhandlungen, B. II", Springer-Verlag, 1966, pp. 138-169).

[SS] Seminaire sur les singularités des surfaces. Lect. Notes in Math. vol. 777. Springer-Verlag. 1979.

[Sta] E. Stagnaro, Constructing Enriques surfaces from quintics in P^3_K, in "Algebraic Geometry-Open Problems", Lect. Notes in Math. vol. 997, Springer-Verlag, 1983, pp. 400-403.

[Ste] R. Steinberg, Finite reflection groups, Trans. Amer. Math. Soiic. 91 (1959), 493-504.

[Ta 1] J. Tate, Genus change in purely inseparable extensions of function fields, Proc. Amer. Math. Soc. 3 (1952), 400-406.

[Ta 2] J. Tate, On the conjecture of Birch and Swinnerton-Dyer and a geometric analog, Sem. Bourbaki 1965/66, n° 306 (also in "Dix exposés sur cohomologie des schémas", North-Holland. Amsterdam, 1968, pp. 189-214.

[Ta 3] J. Tate, Algorithm for determining the type of a singular fiber in an elliptic pencil, in "Modular functions in one variable, IV", Lect. Notes in Math. vol.476, Springer-Verlag, 1975, pp 33-52.

[Ti] G. Timms, The nodal cubic surfaces and the surfaces from which are derived by projections, Proc. Roy. Soc, Ser.A , 119 (1928), 213-248.

[Tyu] G.Tyurina, On a type of contractible curves, Doklady Akad. Nauk SSSR,, 173 (1967), 529-531 [Engl. Transl.:Soviet Math. Doklady 8 (1967), 441-443].

[Ur] T. Urabe, On singularities on degenerate Del Pezzo surfaces of degree 1,2, in "Singularities", Part 2, Proc. Symp. Pure Math.,v.40, Part I, 1983, pp.587-591.

[Ve 1] A. Verra, On Enriques surfaces as a fourfold cover of $\mathbb{P}^2$, Math. Anal. 266 (1983), 241-250.

[Ve 2] A. Verra, The étale double covering of an Enriques' surface, Rend. Sem. Mat. Univ. Polyt. Torino, 41 (1983), 131-166.

[Vi 1] E. Vinberg, Discrete groups generated by reflections, Izv. Akad. Nauk SSSR, Ser. Math. 51 (1971) [Engl. Transl.: Math. USSR-Izvestija, 5 (1971), 1083-1119].

[Vi 2] E. Vinberg, Some discrete groups in Lobachevskii spces, in "Discrete subgroups of Lie groups", Oxford Univ. Press. 1973, pp. 323-348.

[Vi 3] E. Vinberg, The two most algebraic K3-surfaces, Math Ann. 265 (1983), 1-21.

[Vv 1] O.Vvedenskii, Duality in elliptic curves over a local field, II, Izv. Akad. Nauk SSSR, Ser. Math. , 30 (1966), 891-922.

[Vv 2] O.Vvedenskii, On the galois cohomology of elliptic curves defined over a local field, Math. Sbornik, 83 (1970), 474-484 [Engl. Transl.: Math. USSR-Izvestia, 12 (1970), 477-488].

Index

absolute invariant 329

admissible triple 11

splittable 11

Artin-Schreier sequence 18

Bertini involution 151, 156,

branch divisor 13,14

Cartan matrix 106

constant variety 295.

Coxeter graph 107

Coxeter group 107

cubic surface 57

Cayley 60

symmetric 58

of type $\mathcal{C}_1, \mathcal{C}_1', \mathcal{C}_1''$ 58

of type $\mathcal{C}_2, \mathcal{C}_2', \mathcal{C}_2''$ 60

of type $\mathcal{C}_3, \mathcal{C}_3', \mathcal{C}_3''$ 60

curve of cusps 290

cycle map 79

de Rham

Betti number 88

cohomology 88

-Hodge spectral sequence 88

degeneracy set 286

degenerate fibre286

additive type 301

multiple 287

multiplicative type 301

non-multiple 287

types of 288

wild 287

Del Pezzo surface 35

anticanonical 35

symmetric 39

4-nodal quartic ($\mathcal{D}_1$)43

degenerate 4-nodal quartic ($\mathcal{D}_1'$) 43

of type $\mathcal{D}_2$, $\mathcal{D}_3$ 50

of type $\mathcal{D}_2', \mathcal{D}_3'$ 54

discriminant

of an elliptic curve 328

of an elliptic fibration 338

of a quasi-elliptic fibration 340

divisor

indecomposable 166

m-connected 232

nef 281

of canonical type 166,170

double cover 9
  inseparable 9
  principal 20
  separable 9
double cubic surface 272
double $\mathbb{F}_1$ 279
double plane 269
  non-special type 269
  special type 269
double quadric 271
double quartic 277
double $S_{0,2}$ 279
Dynkin diagram 106
  affine type 114
  finite type 114
  rank 114
Dynkin graph 27
elliptic curve 329
elliptic fibration 285
elliptic surfaces
  properly
  rational
Enriques lattice 117
Enriques sextic 273
  classical 274
  degenerate 274
  non-classical 275
  non-degenerate 276
Enriques surface 73
  classical 74
  $\tilde{A}_1+\tilde{E}_7$-special 186
  α_2-surface 77
  $\tilde{D}_8$-special 186
  $\tilde{E}_8$-special 182
  extra special 186
  μ_2-surface 77
  nodal 178
  ordinary 77
  unnodal 178
  supersingular 77
exceptional vector124
  strongly 150
exceptional curve 24
exceptional r-sequence 124
Fano model 279
Fano variety 35
fundamental chamber 107
fundamental cycle 25, 227
fundamental polyhedron 113
  vertex of 114
  improper vertex 114
  proper vertex 114
fundamental weights 127
Geiser involution 161
genus 1 fibration 285
  cohomologically flat 287
  constant 314
  elliptic 285
  index of 306
  isotrivial 302
  jacobian 302
  minimal 285
  quasi-elliptic 285

trivial 302
genus 1 pencil 172
Gram matrix 104
half-fibre 172,365
Halphen pencil 347
of index m 347
Hodge numbers 88
hyperbolic space 113
integral quadratic form 104
isotropic r-sequence 122,178
c-degenerate 180
canonical
c-degenerate 180
canonical 180
degenerate 180
maximal 179
non-degenerate
canonical 178
isotropic r-set 179
isotropic vector 105
Jacobian fibration 300
Jacobian variety 70
K3-cover 84
K3-surface 84
Kantor involution 156
Kodaira dimension 72
Kummer exact sequence 17
lattice 103
(non)degenerate 103
discriminant group of 104
discriminant of 104
even 104
isomorphic 103
isometry 103
hyperbolic 103
embedding 103
(in)definite 103
finite index 103
orthogonal
complement 104
orthogonal group of103
orthogonal sum 104
primitive element of
radical 105
semi-definite 103
signature 103
sublattice 103
sum 104
unimodular 104
A_n, E_n, D_n 105
$\tilde{E}_6$, $\tilde{E}_7$,$\tilde{E}_8$, 105
$Q_{p,q,r}$ 105
$U_{[n]}$, U 105
linear system
hyperelliptic 228, 229
non-special 243
Prym canonical 62
special 243
superelliptic 228
Lobachevsky space113
local invariant 313
minimal surface 72
multiple fibre 287
elimination of 291

multiplicity of 287
wild 287
Néron model 300
Néron sheaf 3298
néronian group scheme 297
Néron-Severi group 69
Nikulin lattice 116
nodal chamber 175
nodal curve 29
nodal curve 30,182
node 30
non-degeneracy invariant 182
non-degenerate fibre 285
non-multiple fibre 287
nondegenerate subvariety 33
non-normal octic 278
ordinary double point 30
Picard
lattice 69
number 77
relative functor 65
scheme 66
variety 69
primitive sublattice 103
principal
double G-cover 21
local G-cover 31
maximal local G-cover 31
Prym canonical
linear system 62
map. 62
model 62
purity theorem 20
quadrangle 45
degenerate 47
vertices of45
A_2-vertex 47
simple vertices 47
quadruple plane 270
quasi-elliptic curve 329
ramification divisor 13, 14
rational double point 25
of type A_n, D_n, E_6, E_7, E_8 27
rational elliptic surface 347
quasi-elliptic surface 347
index of 347
rational scroll 34
reducibility lemma 176
Reye lattice 128
Reye root basis 132
Riemann-Roch formula 95
2-reflection group 115
root 106
irreducible 106
positive 109
root basis 106
affine type 110
canonical 107
crystallographic 113
finite type 110
hyperbolic 112
irreducible 106
crystallographic
extension of 130

semi-stable reduction 292
Serre duality 95
Shafarevich (-Tate) group 315
simple reflection 106
singularity 24
genus of 24
Gorenstein 12
minimal resolution 24
multiplicity of 25
normal 24
rational 24
resolution of 24
simple 29
standard hyperbolic plane 105
strictly local scheme 285
Tits cone 107
total degree 70
U-marking 181
degenerate 181
non-degenerate 181
U-pair 181
degenerate 181
non-degenerate 181
$U_{[r]}$-marking 181
non-degenerate 181
isomorphic 181
unirational 88
vector field 20
of additive type 21
multiplicity of 22
of multiplicative type 21
quotient by 21
isolated singularities 22
zero divisor of 22
p-closed 21
Weierstrass model 335
Weyl group
l-congruence subgroup 140
of a root basis 106
of a lattice 115
of a surface 175
length function on 109
Wirtinger sextic 65
Zariski variety 87

Glossary of Notations

Standard notations:

$\mathfrak{L}$	invertible sheaf
$\|D\|$, $\|\mathfrak{L}\|$	complete linear system associated to a divisor D or an invertible sheaf $\mathfrak{L}$
K_X	canonical class of X
ω_X	canonical sheaf of X
Ω_X^i	sheaves of differentials
$\mathcal{O}_X$	structure sheaf of X
$\mathcal{O}_X(D)$	invertible sheaf associated to a divisor D
$\sim$	linear equivalence of divisors
f^* (f_*)	inverse (direct) image
R^if_*	higher direct images
$H^i(X,\mathfrak{F})$	cohomology of X with coefficients in a sheaf $\mathfrak{F}$
$h^i(\mathfrak{F})$ $(h^i(D))$	$\dim H^i(X,\mathfrak{F})$ $(\dim H^i(X,\mathcal{O}_X(D))$
$\Gamma(\mathfrak{F})$	group of sections of a sheaf $\mathfrak{F}$
$\chi(X,\mathfrak{F})$	Euler characteristic
$\mathrm{Pic}(X)$	Picard group of classes of invertible sheaves on X
$D\cdot D'$, D^2	intersection index of divisors on a surface
$\mathrm{char}(\kappa)$	characteristic of a field
p	positive characteristic of a field
$\mathbb{P}^n$	n-dimensional projective space
A^*	invertible elements of a ring A
$\mathbb{Z}$	ring of integers
$\mathbb{R}$	field of real numbers
$\mathbb{C}$	field of complex numbers
$\mathbb{Z}_l$ $(\mathbb{Q}_l)$	ring (field) of l-adic numbers
$\mathbb{F}_p$	finite field of p elements
μ_n	group scheme of n-th roots of unity
$\mathbb{G}_m$	multiplicative group scheme
$\mathbb{G}_a$	additive group scheme

Other notations:

(l,V,s)	11
$\mathbb{V}(\mathfrak{I})$	12
$b_i(Z), e(Z)$,	15
$H^i_{et}(Z,\mathbb{Q}_l)$	15
$c_d(Z)$	15
$\mathbf{Z}/2,\ \mu_2,\ \alpha_2$,	16,39
G_Z	16
$\mathbf{F}$	17
$\mathfrak{p}$	18
$H^i_Z(X,G)$	20
Θ_X	20
X^D, A^D, π_D	21
X^p	21
$mult_x(D)$	22
μ^D	23
$k[\varepsilon]$	23
$p_a(Z)$	25
A_n, D_n, E_6, E_7, E_8	26,27,105
$D_n^{(r)}, E_n^{(r)}$	28
$\mathbb{F}_n,\ S_{n,k}$	34
$X(\Sigma)$	35
$\mathfrak{a}_X$	40
$\mathcal{D}_1, \mathcal{D}_1'$	43
$\mathcal{D}_2, \mathcal{D}_3$	50
$\mathcal{D}_2', \mathcal{D}_3'$	53
Aut(X), Aut(X)°	56
φ_ε	62
$\mathcal{P}ic_{X/S}$	65
Br(X)	66
$Pic_{X/S}$	66
G°, G^τ	67
Lie(G)	67
Num(X), NS(X)	69
J(X)	70
$deg(\mathfrak{I})$	70
$\varkappa(X)$	72
$\equiv$	77
c_1	77
$\rho,\ \rho(X)$	77
T_1	79
t_p	79
$b_i^{DR},\ H^i_{DR}$	88
$h^{p,q}$	88
$W\mathcal{O}_F,\ W(k)$	94
$M_{\mathbb{R}}$	103
(t_+, t_-, t_0)	103
O(M)	103
$M_1+M_2,\ M_1\perp M_2$	104
$M^\perp$	104
$D(M),\ i_M$	104
discr(M)	104
$M(n),\ M_n$	105
$\langle n\rangle,\ M^n$	105
$T_{p,q,r}$	105
$U,\ U_{[n]}$	105
$Q_{p,q,r}$	105
$\tilde{E}_6,\ \tilde{E}_7,\ \tilde{E}_8$	105,169,289
B	106
$\alpha,\ s_\alpha$	106
$\Gamma(B)$	106

R_B, R	106
$W_B(M)$, W	106
$K(B)$	107
$C(B)$, $C(B)^\circ$	107
1	109
R_B^+, R_B^-	109,110
$\tilde{D}_n$, $\tilde{A}_n$	111,168,288
V^+, V^-	112
$P(B)$	113
$H(V)$	113
$W(M)$	115
$A(P(M))$	116
W_n	117
E_n	117
$\mathbf{E}$	117
f	118
κ_n	119
$H^{1,n}$	119
f_i	122
Δ	23,327, 338,339
$\mathfrak{E}_n$, $\mathfrak{E}_n'$	124
ω_i	126,132
$f_{i,j}$	127
$\mathbf{R}$	128
R_n	128
$\mathfrak{r}$	129
κ_n'	130
$'H^{1,n}$	130
$V(2)$	131
φ_M	134
E_α	138
φ_α	138
$G(1)$	140
$W(2)$	141
W_0	141
$O^\pm(V,q)$, $O^\pm(2n,\mathbb{F}_2)$	143
$Sp(V)$	143
$Sp(2n,\mathbb{F}_2)$	143
$A(i,j,\kappa)$	151
σ_0	151
K	155
B	156
$\bar{W}(2)$	161
$\tilde{A}_0^*$, $\tilde{A}_0^{**}$	167,288
$\tilde{A}_1^*$, $\tilde{A}_2^*$	168,288
C_X, C_X^+	175
$\mathfrak{N}(X)$	175
W_X	175
$Num(X)^+$	175
V_X	175
Φ	178
$d(F)$	182
φ_C	226
$\mathfrak{N}_C$	226
Z_E	227
η	285
X_η, $X_{\bar{\eta}}$	286
X_s	286
$m(s)$	287
$\bar{X}_s$	288
$\mathfrak{P}$	292
Γ_s	292
$\mathfrak{E}$	296

ρ_f	296
$\mathfrak{a}$	296
$\mathfrak{A}$	298
$\mathbf{A}$	298
$J^{\#}$	300
ind(f)	306
Elf(j)	313
φ_s	313
Ш$(\mathfrak{A})$	315
${}_nA$, $A(\neq 1)$, $A(1)$	315
$\mathfrak{g}(G)$	318
$\pi_i(\mathfrak{g}(G))$	318
j	329
V, V'	331
E	334
W	335
φ, $\bar{\varphi}$	345

Progress in Mathematics

Edited by:

J. Oesterlé
Departement des Mathematiques
Université de Paris VI
4, Place Jussieu
75230 Paris Cedex 05
France

A. Weinstein
Department of Mathematics
University of California
Berkeley, CA 94720
U.S.A.

Progress in Mathematics is a series of books intended for professional mathematicians and scientists, encompassing all areas of pure mathematics. This distinguished series, which began in 1979, includes authored monographs and edited collections of papers on important research developments as well as expositions of particular subject areas.

All books in the series are "camera-ready", that is they are photographically reproduced and printed directly from a final-edited manuscript that has been prepared by the author. Manuscripts should be no less than 100 and preferably no more than 500 pages.

Proposals should be sent directly to the editors or to: Birkhauser Boston, 675 Massachusetts Avenue, Suite 601, Cambridge, MA 02139, U.S.A.

1 Gross. Quadratic Forms in Infinite-Dimensional Vector Spaces
2 Pham. Singularités des Systèmes Différentiels de Gauss-Manin
3 Okonek/Schneider/Spindler. Vector Bundles on Complex Projective Spaces
4 Aupetit. Complex Approximation. Proceedings, Quebec, Canada, July 3–8, 1978
5 Helgason. The Radon Transform
6 Lion/Vergne. The Weil Representation. Maslov Index and Theta Series
7 Hirschowitz. Vector Bundles and Differential Equations Proceedings, Nice, France, June 12–17, 1979
8 Guckenheimer/Moser/Newhouse. Dynamical Systems, C.I.M.E. Lectures, Bressanone, Italy, June 1978
9 Springer. Linear Algebraic Groups
10 Katok. Ergodic Theory and Dynamical Systems I
11 Balslev. 18th Scandinavian Congress of Mathematicians, Aarhus, Denmark, 1980
12 Bertin. Séminaire de Théorie des Nombres, Paris 1979–80
13 Helgason. Topics in Harmonic Analysis on Homogeneous Spaces
14 Hano/Marimoto/Murakami/Okamoto/Ozeki. Manifolds and Lie Groups: Papers in Honor of Yozo Matsushima
15 Vogan. Representations of Real Reductive Lie Groups
16 Griffiths/Morgan. Rational Homotopy Theory and Differential Forms
17 Vovsi. Triangular Products of Group Representations and Their Applications
18 Fresnel/van der Put. Géometrie Analytique Rigide et Applications
19 Oda. Periods of Hilbert Modular Surfaces
20 Stevens. Arithmetic on Modular Curves
21 Katok. Ergodic Theory and Dynamical Systems II
22 Bertin. Séminaire de Théorie des Nombres. Paris 1980–81
23 Weil. Adeles and Algebraic Groups

24 LE BARZ/HERVIER. Enumerative Geometry and Classical Algebraic Geometry
25 GRIFFITHS. Exterior Differential Systems and the Calculus of Variations
26 KOBLITZ. Number Theory Related to Fermat's Last Theorem
27 BROCKETT/MILLMAN/SUSSMAN. Differential Geometric Control Theory
28 MUMFORD. Tata Lectures on Theta I
29 FRIEDMAN/MORRISON. Birational Geometry of Degenerations
30 YANO/KON. CR Submanifolds of Kaehlerian and Sasakian Manifolds
31 BERTRAND/WALDSCHMIDT. Approximations Diophantiennes et Nombres Transcendants
32 BOOKS/GRAY/REINHART. Differential Geometry
33 ZUILY. Uniqueness and Non-Uniqueness in the Cauchy Problem
34 KASHIWARA. Systems of Microdifferential Equations
35 ARTIN/TATE. Arithmetic and Geometry: Papers Dedicated to I.R. Shafarevich on the Occasion of His Sixtieth Birthday, Vol. 1
36 ARTIN/TATE. Arithmetic and Geometry: Papers Dedicated to I.R. Shafarevich on the Occasion of His Sixtieth Birthday. Vol. II
37 DE MONVEL. Mathématique et Physique
38 BERTIN. Séminaire de Théorie des Nombres, Paris 1981–82
39 UENO. Classification of Algebraic and Analytic Manifolds
40 TROMBI. Representation Theory of Reductive Groups
41 STANELY. Combinatories and Commutative Algebra
42 JOUANOLOU. Théorèmes de Bertini et Applications
43 MUMFORD. Tata Lectures on Theta II
44 KAC. Infinite Dimensional Lie Algebras
45 BISMUT. Large Deviations and the Malliavin Calculus
46 SATAKE/MORITA. Automorphic Forms of Several Variables Taniguchi Symposium, Katata, 1983
47 TATE. Les Conjectures de Stark sur les Fonctions L d'Artin en s = 0
48 FRÖHLICH. Classgroups and Hermitian Modules
49 SCHLICHTKRULL. Hyperfunctions and Harmonic Analysis on Symmetric Spaces
50 BOREL, ET AL. Intersection Cohomology
51 BERTIN/GOLDSTEIN. Séminaire de Théoire des Nombres. Paris 1982–83
52 GASQUI/GOLDSCHMIDT. Déformations Infinitesimales des Structures Conformes Plates
53 LAURENT. Théorie de la Deuxième Microlocalisation dans le Domaine Complèxe
54 VERDIER/LE POTIER. Module des Fibres Stables sur les Courbes Algébriques Notes de l'Ecole Normale Supérieure, Printemps, 1983
55 EICHLER/ZAGIER. The Theory of Jacobi Forms
56 SHIFFMAN/SOMMESE. Vanishing Theorems on Complex Manifolds
57 RIESEL. Prime Numbers and Computer Methods for Factorization
58 HELFFER/NOURRIGAT. Hypoellipticité Maximale pour des Opérateurs Polynomes de Champs de Vecteurs
59 GOLDSTEIN. Séminarie de Théorie des Nombres, Paris 1983–84
60 PROCESI. Geometry Today: Giornate Di Geometria, Roma. 1984
61 BALLMANN/GROMOV/SCHROEDER. Manifolds of Nonpositive Curvature
62 GUILLOU/MARIN. A la Recherche de la Topologie Perdue
63 GOLDSTEIN. Séminaire de Théorie des Nombres, Paris 1984–85
64 MYUNG. Malcev-Admissible Algebras
65 GRUBB. Functional Calculus of Pseudo-Differential Boundary Problems

66 CASSOU-NOGUÈS/TAYLOR. Elliptic Functions and Rings and Integers

67 HOWE. Discrete Groups in Geometry and Analysis: Papers in Honor of G.D., Mostow on His Sixtieth Birthday

68 ROBERT. Antour de L'Approximation Semi-Classique

69 FARAUT/HARZALLAH. Deux Cours d'Analyse

70 ADOLPHSON/CONREY/GHOSH/YAGER. Number Theory and Diophantine Problems: Proceedings of a Conference at Oklahoma State University

71 GOLDSTEIN. Séminaire de Théories des Nombres, Paris 1985–1986

72 VAISMAN. Symplectic Geometry and Secondary Characteristics Classes

73 MOLINO. Riemannian Foliations

74 HENKIN/LEITERER. Andreotti–Grauert Theory by Integral Formulas

75 GOLSTEIN. Séminaire de Théories des Nombres, Paris 1986–87

76 COSSEC/DOLGACHEV. Enriques Surfaces I